Umesh Kumar Lilhore, Vishal Dutt, T. Ananth Kumar, Martin Margala,
Kaamran Raahemifar (Eds.)

Math Optimization for Artificial Intelligence

Mathematical Methods in the Digital Age

Edited by
Abhishek Kumar, Satya Prakash Yadav,
Ashutosh Kumar Dubey, Andrei Chtcheprov,
Sayavur Bakhtiyarov

Volume 2

Math Optimization for Artificial Intelligence

Heuristic and Metaheuristic Methods for Robotics and Machine Learning

Edited by
Umesh Kumar Lilhore, Vishal Dutt, T. Ananth Kumar,
Martin Margala, Kaamran Raahemifar

DE GRUYTER

Editors

Umesh Kumar Lilhore
Professor, Department of Computer
Science and Engineering
Galgotias University
Greater Noida, UP,
India

Vishal Dutt
University Institute of Engineering
Chandigarh University
Mohali, Punjab, India
vishaldutt53@gmail.com

T. Ananth Kumar
Indo French Educational Trust (IFET)
College of Engineering
IFET Road
605108 Nagar, Gangarampalaiyam, Tamil Nadu
India
tananthkumar@ieee.org

Martin Margala
University of Louisiana at Lafayette
School of Computing
104 E University Avenue
Lafayette, LA 70504
Louisiana, USA
martin.margala@louisiana.edu

Kaamran Raahemifar
Pennsylvania State University
College of Information Sciences and Technology
Westgate Building
University Park, PA 16802
Pennsylvania, USA
kvr5517@psu.edu

ISBN 978-3-11-143605-0
e-ISBN (PDF) 978-3-11-143618-0
e-ISBN (EPUB) 978-3-11-143624-1
ISSN 2943-2731

Library of Congress Control Number: 2024949494

Bibliographic information published by the Deutsche Nationalbibliothek
The Deutsche Nationalbibliothek lists this publication in the Deutsche Nationalbibliografie; detailed bibliographic data are available on the Internet at http://dnb.dnb.de.

Typesetting: Integra Software Services Pvt. Ltd.

www.degruyter.com
Questions about General Product Safety Regulation:
productsafety@degruyterbrill.com

Foreword

Living in a time of rapid technological change, artificial intelligence, and robotics are right at the forefront. These are areas in which not only will the way business gets done in industries change but the nature of everyday interaction with the world will also dramatically change. Optimization is at the core of many of the advances that occur and is, in that sense, a very important tool driving both the efficiency and intelligence of AI systems. Whereas the basic ones are foundational, they cannot usually handle nonlinear challenges, common in current AI and robotics applications. This is a point that heuristic and metaheuristic methods have come into play, offering their dynamic and adaptive ways of solution.

This book assembles a collection of state-of-the-art research that goes deep inside these new optimization techniques. This book also fills an important gap within the existing literature, elaborating on practical and theoretical aspects of heuristic and metaheuristic methods and showing their relevance and potential across many AI and robotics domains. The increasing complexity of machine learning models and the expanding demand for smart, autonomous robotic systems urgently require new ways to solve optimization problems, which this book tries to provide.

It covers, within its chapters, a wide variety of topics from the very basics of mathematical optimization to real-life AI and robotics applications. It shall inspire various necessary methods, including evolutionary algorithms, swarm intelligence, and nature-inspired optimization techniques, which are designed to exactly deal with complex problems with precision and efficiency. Furthermore, this book highlights how these methods can be used in many areas, such as tuning machine learning models, autonomous systems, and real-time decision-making. Each chapter is written by renowned researchers and professionals with experience in artificial intelligence, robotics, and optimization, giving this book significant weight in content. Their combined knowledge ensures that this book will be important not only at the academic level but also as a practical guide in the implementation of optimization techniques for real-world applications. We are confident that their contributions will attract both new and seasoned researchers to continue the push toward harnessing the full potential of heuristic and metaheuristic approaches to overcome most of today's complex AI problems.

In this future, optimization will also stand as one of the powerful driving forces behind further development of smart and autonomous systems. This book represents an important step in the use of some of the advanced techniques of optimization for efforts to be pursued in research and innovation by those working in the fields of artificial intelligence and robotics. We would like to encourage readers with these considerations in the pages that follow, to consider how these powerful methods can be applied in their research and projects. We are, therefore, very grateful to the contributing authors and to the collaborative work of all persons and agencies that have made this project possible amidst many competing demands on their time. We are proud that we can offer this book as a contribution to important works in AI and robotics.

https://doi.org/10.1515/9783111436180-202

Contents

List of Authors

Dr. A. Mohammed Shapique
Department of Science and Humanities (Mathematics)
IFET College of Engineering
Villupuram
Tamil Nadu, India
shapique@gmail.com
Chapter 1

Dr. A. Vaithiyanathan
Department of Science and Humanities (Mathematics)
IFET College of Engineering
Villupuram
Tami Nadu, India
vaithi05@gmail.com
Chapter 1

Nancy Jasmine Goldena
Department of Computer Applications and Research Centre
Sarah Tucker College (Autonomous)
Affiliated to Manonmaniam Sundaranar University
Abishekapatti, Tirunelveli 627 012
Tamil Nadu, India
nancy_lordwin@rediffmail.com
Chapter 2, 4, 7

Thangapriya
Department of Computer Science
St. John's College, affiliated to Manonmaniam Sundaranar University
Abishekapatti, Tirunelveli 627 012
Tamil Nadu, India
priyasaravananrex@gmail.com
Chapter 2, 7

Dankan Gowda
Department of Electronics and Communication Engineering
BMS Institute of Technology and Management
Bangalore
Karnataka, India
dankan.v@bmsit.in
Chapter 3

Gedas Kučinskas
BPN LT, Tumeno 4
Vilnius 01109
Lithuania
Gedas.kucinskas@gmail.com
Chapter 3

Pullela SVVSR Kumar
Department of Computer Science and Engineering
Aditya College of Engineering and Technology
Surampalem 533437
Andhra Pradesh, India
pullelark@yahoo.com
Chapter 3

Rahul Vadisetty
U.S. Bank
Dallas,
Texas, USA
rahulvy91@gmail.com
Chapter 3

Aparna Atul Junnarkar
Department of Information Technology
Vishwakarma Institute Of Information Technology (VIIT)
Pune 411058
Maharashtra, India
aparna.junnarkar@gmail.com
Chapter 3

Rashia Subashree R.
Department of Computer Application and Research Centre
Sarah Tucker College (Autonomous)
Affiliated to Manonmaniam Sundaranar University
Tirunelveli
Tamil Nadu, India
rasiyakrishnan@gmail.com
Chapter 4

https://doi.org/10.1515/9783111436180-204

Leena Arya
Department of Computer Science and Engineering
Koneru Lakshmaiah Education Foundation
Vaddeswaram, Guntur 522502
Andhra Pradesh, India
leenaarya18@gmail.com
Chapter 5

Yogesh Kumar Sharma
Department of Computer Science and Engineering
Koneru Lakshmaiah Education Foundation
Vaddeswaram, Guntur 522502
Andhra Pradesh, India
dr.sharmayogeshkumar@gmail.com
Chapter 5

Ravi Rastogi
Department of Computer Science and Engineering
Koneru Lakshmaiah Education Foundation
Vaddeswaram, Guntur 522502
Andhra Pradesh, India
ravikumarrastogi@gmail.com
Chapter 5

Jasmine Sugil
Department of Computer Applications and Research Centre
Sarah Tucker College (Autonomous)
Affiliated to Manonmaniam Sundaranar University
Abishekapatti, Tirunelveli 627012
Tamil Nadu, India
findsugil@gmail.com
Chapter 6

K. Merriliance
Department of Computer Applications and Research Centre
Sarah Tucker College (Autonomous)
Affiliated to Manonmaniam Sundaranar University
Abishekapatti, Tirunelveli 627007
Tamil Nadu, India,
merriliance@gmail.com
Chapter 6

Yogesh Kumar Sharma
Koneru Lakshmaiah Education Foundation
Green Fields, Vaddeswaram, Guntur 522302
Andhra Pradesh, India
dr.sharmayogeshkumar@gmail.com
Chapter 8, 9

Uppuluri Lakshmi Soundharya
Koneru Lakshmaiah Education Foundation
Green Fields, Vaddeswaram, Guntur 522302
Andhra Pradesh, India
mail2usoundharya@gmail.com
Chapter 8, 9

Ravi Rastogi
Koneru Lakshmaiah Education Foundation
Green Fields, Vaddeswaram, Guntur 522302
Andhra Pradesh, India
ravikumarrastogi@gmail.com
Chapter 8, 9

Anantha Raman G. R.
Koneru Lakshmaiah Education Foundation
Green Fields, Vaddeswaram, Guntur 522302
Andhra Pradesh, India
granantha.raman@gmail.com
Chapter 8, 9

Umesh Kumar Lilhore
Department of CSE
School of Computing
Galgotias University
Greater Noida
Uttar Pradesh, India
umeshlilhore@gmail.cim
Chapter 10, 16

Sarita Simaiya
Department of CSE
School of Computing
Galgotias University
Greater Noida
Uttar Pradesh, India
saritasimaiya@gmail.com
Chapter 10

Surjeet Dalal
Department of Computer Science and Engineering
Amity University Gurugram
Haryana, India
sdalal@ggn.amity.edu
Chapter 10

Yash Mahajan
Department of Computer Science and Engineering
Chandigarh University
Mohali
Punjab, India
gyash724@gmail.com
Chapter 11

Muskan Sharma
Department of Computer Science and Engineering
Chandigarh University
Mohali
Punjab, India
muskaanvishambera@gmail.com
Chapter 11

Alba Meça
Faculty of Engineering
Universita degli Studi di Padova
Padova, Italy
ameca18@epoka.edu.al
Chapter 11

Narayan Vyas
Department of Computer Science and Application
Vivekananda Global University
Jaipur
Rajasthan, India
narayanvyas87@gmail.com
Chapter 11

Rakesh Sharma
Department of Computer Science and Application
Vivekananda Global University
Jaipur
Rajasthan, India
srak.911@gmail.com
Chapter 11

P. Manju bala
Department of Artificial Intelligence and Data Science
IFET College of Engineering
Villupuram
Tamil Nadu, India
pkmanju26@gmail.com
Chapter 12

S. Usharani
Department of Artificial Intelligence and Machine Learning
IFET College of Engineering
Villupuram
Tamil Nadu, India
ushasanchu@gmail.com
Chapter 12

A. Balachandar
Department of Computer Science and Engineering
IFET College of Engineering
Villupuram
Tamil Nadu, India
balaifet20@gmail.com
Chapter 12

R. Rajmohan
Department of Computer Science and Engineering
School of Computing Technologies
SRM institute of Technology
Chennai
Tamil Nadu, India
rjmohan89@gmail.com
Chapter 12

R. Manjula Devi
Department of ECE
National Engineering College
Kovilpatti
Tamil Nadu, India
rmdece@nec.edu.in
Chapter 13

Shenbagavalli
Department of ECE
National Engineering College
Kovilpatti
Tamil Nadu, India
A-shenbagavalli@nec.edu.in
Chapter 13

S. Dheenathayalan
Department of Computer Science and Engineering
National Engineering College
Kovilpatti
Tamil Nadu, India
ddhayalan@nec.edu.in
Chapter 13

Prabakaran D.
IFET College of Engineering
Villupuram
Tamil Nadu, India
dprabakaranmtech@gmail.com
Chapter 14

Senthil Kumaran R.
SRM Institute of Science and Technology
Kattankulathur, Chennai
Tamil Nadu, India
sen19841@gmail.com
Chapter 14

Shyamala R.
University College of Engineering
Tindivanam
Tamil Nadu, India
drshyamalaucet@gmail.com
Chapter 14

S. Saranya
IFET College of Engineering
Gangarampalayam
Villupuram 605108
Tamil Nadu, India
sahaasaranya@gmail.com
Chapter 15

Neha
School of Computing Science and Engineering
Galgotias University
Greater Noida
Uttar Pradesh, India
neha.scse@galgotiasuniversity.edu.in
Chapter 16

Shailja
School of Computing Science and Engineering
Galgotias University
Greater Noida
Uttar Pradesh, India
shailja.scse@galgotiasuniversity.edu.in
Chapter 16

Roselin Kiruba
Department of Computer Science and Engineering
Vel Tech Rangarajan Dr. Sagunthala R&D Institute of Science and Technology
Tamil Nadu, India
kirubaroselin@gmail.com
Chapter 17

L. Sharmila
Department of Computer Science and Engineering
Vel Tech Rangarajan Dr. Sagunthala R&D Institute of Science and Technology
Tamil Nadu, India
shar.hariharan@gmail.com
Chapter 17

M. Misba
Department of Computer Science and Engineering
Vel Tech Rangarajan Dr. Sagunthala R&D Institute of Science and Technology
Tamil Nadu, India
misbarajan2007@gmail.com
Chapter 17

J. Kavitha
Department of Computer Science and Engineering
Vel Tech Rangarajan Dr. Sagunthala R&D Institute of Science and Technology
Tamil Nadu, India
vsskavitha@gmail.com
Chapter 17

J. Jude Moses Anto Devakanth
Department of Computer Applications
Madanapalle Institute of Technology and Science
Andhra Pradesh, India
judedevaj@gmail.com
Chapter 17

S. S. Uma
Arunachala College of Engineering for Women
Nagercoil
Tamil Nadu, India
dhatshirn@gmail.com
Chapter 18

K. Suresh Kumar
Sri Krishna College of Technology
Coimbatore
Tamil Nadu, India
sureshkumarkskphd@gmail.com
Chapter 18

P. Suresh
Sri Krishna College of Technology
Coimbatore
Tamil Nadu, India
psuresh82@gmail.com
Chapter 18

P. Dhivya
Sri Krishna College of Technology
Coimbatore
Tamil Nadu, India
dhivyainfo@gmail.com
Chapter 18

Dankan Gowda V.
Department of Electronics and Communication Engineering
BMS Institute of Technology and Management
Bangalore
Karnataka, India
dankan.v@bmsit.in
Chapter 19

Anand Polamarasetti
Senior Systems Architect (Cloud/DevOps)
Booz Allen Hamilton Herndon
Virginia, United States
exploretechnologi@gmail.com
Chapter 19

Pullela S. V. V. S. R. Kumar
Department of Computer Science and Engineering
Aditya College of Engineering and Technology
Surampalem 533437
Andhra Pradesh, India
pullelark@yahoo.com
Chapter 19

Aparna Atul Junnarkar
Department of Information Technology
Vishwakarma Institute Of Information Technology (VIIT)
Pune 411058
Maharashtra, India
aparna.junnarkar@gmail.com
Chapter 19

Bhaskar Roy
Department of CSE (AIML)
Asansol Engineering College
Asansol 713305
West Bengal, India
bhaskar.ec11@gmail.com
Chapter 19

A. Mohammed Shapique and A. Vaithiyanathan

1 The Role of Mathematical Optimization in Advanced AI Applications

Abstract: This chapter explores the significance of mathematical optimization across diverse fields, focusing on its critical role in artificial intelligence (AI) applications. Optimization techniques are essential for finding the best solutions while considering constraints, and they are widely applied in AI to improve decision-making and performance. We discuss key optimization methods, including linear programming, gradient descent, and evolutionary algorithms, demonstrating their applications in machine learning, robotics, and natural language processing. Practical examples illustrate how these techniques optimize model training, clustering, and policy development in reinforcement learning. Additionally, the chapter examines the emerging role of quantum computing in solving complex optimization problems and highlights ethical considerations in AI development.

Keywords: Mathematical optimization, AI applications, linear programming, nonlinear programming, optimization techniques

1.1 Introduction to Mathematical Optimization

Mathematical optimization, often referred to simply as optimization, is a powerful and versatile field of mathematics with applications spanning a wide range of disciplines, including engineering, economics, computer science, and operations research. At its core, optimization is concerned with finding the best solution to a problem from a set of possible choices, where "best" is defined according to specific criteria or objectives. This process involves the systematic exploration of feasible solutions to identify the optimal one, often under constraints that must be satisfied. Optimization problems arise naturally in various real-world scenarios where resources are limited, and decisions need to be made efficiently. For instance, a manufacturing company may seek to minimize production costs while meeting the demand and capacity constraints, or a transportation network may aim to minimize the travel time or cost between various locations. In each case, the mathematical optimization provides a rigorous framework for modelling the problem, formulating objectives and constraints, and finding optimal solutions. The scope of optimization is broad, encompassing different types of problems and solution techniques. Optimization problems can be classified based on the nature of variables, objective functions, and constraints. Depending on these characteristics, various optimization algorithms and methodologies are employed to efficiently solve the problem and obtain optimal solutions [1].

https://doi.org/10.1515/9783111436180-001

Optimization also plays a pivotal role in artificial intelligence (AI), serving as a fundamental mechanism for enhancing decision-making processes and refining the performance of AI systems [2–4]. At its core, optimization within AI involves the pursuit of the optimal solution from a range of feasible options, typically situated within a complex problem landscape characterized by numerous objectives, constraints, and uncertainties. Optimization techniques are indispensable across various AI applications, from machine learning and robotics to natural language processing and computer vision. By leveraging optimization algorithms, AI systems can efficiently navigate vast search spaces, enabling them to discover optimal solutions to intricate problems promptly. AI and optimization techniques are widely used across various fields of science and technology [5]. Applications include predicting and optimizing the length and number of microshoots and biomass in plant cell cultures [6], image processing [7], detecting intrusion attacks in network systems [8], stability analysis of various systems [9], and renewable energy resources [10].

One of the primary applications of optimization in AI lies within machine learning, where models are trained to discern patterns and make predictions from data. Optimization algorithms serve as the backbone of training procedures, iteratively adjusting model parameters to minimize a defined loss function and enhance predictive accuracy. Optimization techniques extend beyond the realm of supervised learning, finding utility in unsupervised and reinforcement learning paradigms as well. In unsupervised learning, optimization algorithms facilitate the discovery of meaningful representations and structures within data, enabling tasks such as clustering and dimensionality reduction. Reinforcement learning, on the other hand, relies on optimization strategies to identify optimal policies that maximize cumulative rewards in sequential decision-making.

The interplay between optimization and AI is greatly enriched by contributions from diverse fields such as operations research, statistics, and applied mathematics. This interdisciplinary approach fosters the development of more sophisticated and resilient optimization methods, thereby propelling the advancement of AI. These collaborative efforts lead to the creation of innovative techniques that enable AI systems to tackle increasingly complex problems effectively. In the realm of AI, the computational efficiency of optimization algorithms holds paramount importance, particularly for applications requiring rapid data processing. Efficient algorithms ensure that AI systems can operate in real time and respond swiftly to dynamic environments, enhancing their effectiveness in various domains ranging from autonomous vehicles to real-time decision-making systems.

Optimization in AI raises critical ethical questions as AI integrates into society. This involves addressing potential biases in decision-making processes and ensuring that algorithms are transparent and equitable. Optimization is integral to the creation of adaptive and self-improving AI systems. Through continuous learning and optimization, these systems can adapt to changing environments and improve their performance over time autonomously. By harnessing optimization techniques, AI systems can evolve dynamically, enhancing their capabilities and resilience in response to new challenges and opportunities. The emergence of quantum computing holds the potential to revolutionize optimization in AI. With its superior computational capabilities, quantum computing offers the prospect of solving optimization problems that are currently intractable for classical computers. This groundbreaking technology opens new horizons for AI development, enabling the exploration of complex problem spaces and the discovery of innovative solutions with unprecedented speed and efficiency.

1.2 Fundamentals of Optimization Techniques

Fundamentals of optimization encompass several key components, including problem definition, types, objective functions, and constraints. Optimization is the process of selecting the best option from a feasible set [11]. It involves maximizing or minimizing a target function while adhering to specified limitations. The aim is to achieve the optimal outcome, such as profit maximization, cost minimization, or performance enhancement, within given resource constraints.

Objective Function: This represents the quantity to be optimized, which could be either maximized or minimized. Mathematically, it is typically denoted as $f(x)$, where x represents the decision variables.

Decision Variables: These are the controllable factors within an optimization problem that influence the objective function's value. They are often denoted as $x_1, x_2, x_3, \ldots, x_n$.

Constraints: These are conditions that must be satisfied for a solution to be considered feasible. Constraints can be either equality constraints $(Ax = B)$ or inequality constraints $(Ax \leq B)$ or $(Ax \geq B)$, where A is a matrix of coefficients, x is the vector of decision variables, and b is a vector of constants.

Feasible Region: This region comprises all potential decision variable values that adhere to problem constraints. It is defined by the overlap of these constraints.

Optimization Problem: An optimization problem involves finding the values of decision variables that optimize the objective function while satisfying the given constraints. Mathematically, it can be formulated as

Minimize or maximize $f(x)$

Subject to optimize $z = f(x_1, x_2, \ldots, x_n)$

Subject to $g_i(x_1, x_2, \ldots, x_n) \leq, =, \geq b_i$ $(i = 1, 2, \ldots, m)$ (called structural constraints) and $x_1, x_2, \ldots, x_n \geq 0$. (called the non-negative restrictions or constraints).

Example: Consider a simple linear optimization problem where we aim to maximize the objective function.

Maximize $Z = 3x_1 + 2x_2$

Subject to constraints $2x_1 + x_2 \leq 2 \;\; 3x_1 + 4x_2 \geq 12$ and $x_1, x_2 \geq 0$.

1.3 Optimization Techniques

To achieve this "best solution," optimization problems are categorized as follows in Table 1.1.

Table 1.1: Comparing different types of optimization problems [12].

Feature	Linear optimization	Nonlinear optimization	Discrete optimization	Continuous optimization
Objective function	Linear	Nonlinear	Linear/nonlinear	Continuous function
Constraints	Linear	Nonlinear	Linear/nonlinear	Continuous functions
Decision variables	Continuous	Continuous	Discrete	Continuous values
Solution techniques	Linear programming	Iterative numerical methods	Specialized algorithms	Calculus-based methods
Examples	Resource allocation	Machine learning model fitting	Travelling salesman problem	Minimizing distance travelled

1.4 Importance of Mathematical Optimization Techniques in Various AI Applications

Mathematical optimization is a critical component in the field of AI, providing a structured approach to find the best possible solution among many feasible options. This process is essential in decision-making, performance enhancement, resource management, and solving complex problems across various AI applications [13–15].

- **Enhanced Decision-Making**: Optimization techniques empower AI systems to make informed decisions, optimizing outcomes in critical areas such as resource allocation, scheduling, and logistics. For instance, in route optimization, these techniques can significantly reduce travel times and costs, leading to more efficient operations.

- **Improved Performance:** In machine learning, optimization algorithms like gradient descent are used to fine-tune model parameters, effectively reducing error rates and improving the model's ability to generalize from training data to new, unseen datasets.
- **Resource Allocation and Management:** Optimization plays a vital role in managing limited resources in sectors such as logistics and supply chain management. By optimizing inventory and distribution, companies can achieve cost savings while maintaining high service levels.
- **Complex Problem-Solving:** AI tackles multidimensional problems by formulating them as mathematical optimization challenges. For example, in finance, portfolio optimization models help in maximizing returns while minimizing risks.
- **Robustness and Adaptability:** Optimization methods significantly contribute to enhancing the robustness and adaptability of AI systems. These techniques, such as robust optimization and online optimization, empower AI systems to effectively address uncertainties and seamlessly adapt to evolving environments, thereby ensuring consistent performance across varying conditions. Robust optimization is tailored to optimize AI performance amidst uncertainty. For instance, within supply chain management, this technique enables the prediction and mitigation of disruptions, ensuring uninterrupted operations even in the face of unforeseen challenges. In parallel, online optimization equips AI systems with the capability to make real-time decisions, dynamically adjusting to new information as it emerges. Particularly advantageous in dynamic settings like stock trading, this approach allows AI systems to swiftly respond to market fluctuations, thereby maximizing effectiveness in volatile environments.
- **Efficient Resource Utilization:** Optimization techniques help in optimizing the utilization of computational resources, such as memory, processing power, and energy consumption. This is crucial for AI systems deployed in resource-constrained environments, where efficient resource utilization is essential for scalability and cost-effectiveness.
- **Innovation and Discovery:** Optimization facilitates innovation and discovery by enabling the exploration of vast solution spaces. In AI research, optimization algorithms are used to train and optimize neural network architectures, discover novel patterns in data, and optimize hyperparameters for model performance.

The basic workflow of the classification and optimization techniques is shown in Figure 1.1.

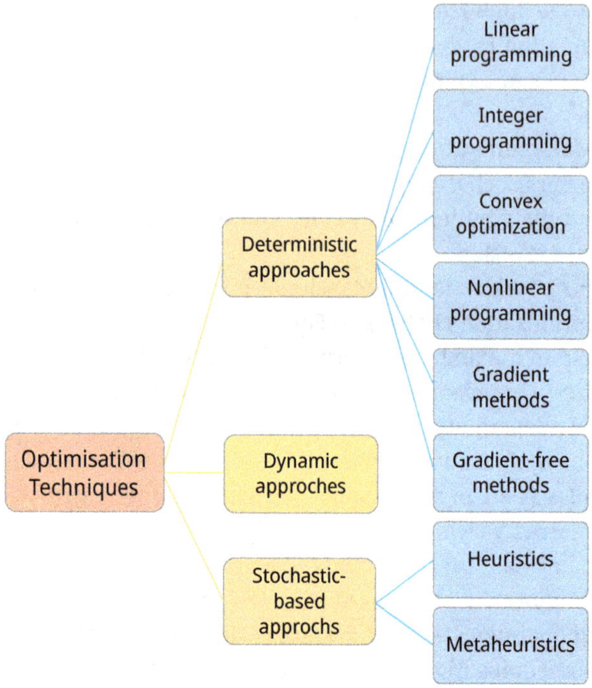

Figure 1.1: Classification of optimization techniques [28].

1.5 Linear Optimization

In linear optimization, both the objective function and constraints are linear functions of the decision variables. These problems are typically represented by linear equations or inequalities and are solved using techniques such as linear programming (LP).

Simplex Algorithm: This method iterates through vertices of the feasible region to find the optimum.

The simplex method is a systematic procedure used in LP to determine the optimal solution subject to linear constraints. Initially, the method requires transforming inequalities into equations by introducing slack variables. For instance, to maximize profit with constraints on machine time and labour hours, equations might look like $2x + 3y + s_1 = 30$ and $x + 2y + s_2 = 20$, where s_1 and s_2 are the slack variables ensuring that these are equations, not inequalities.

The simplex method initiates with a tableau encapsulating the objective function and constraint coefficients. Iterative improvements are made by selecting variables to enter and exit the basis. The entering variable is determined by the most negative coefficient in the objective row, indicating potential objective enhancement. The exiting variable is chosen based on the minimum non-negative ratio of the right-hand side to the pivot column, ensuring feasibility. This process continues until no negative coefficients remain in the objective row, signifying the optimal solution.

Integrating this method with AI can substantially enhance decision-making in production planning. AI can dynamically update production models in response to real-time data on market demand and supply chain status, and employ predictive analytics to anticipate future trends. This allows companies to adapt their production plans pre-emptively, optimizing resource utilization and maximizing profitability. However, managing the complexity of such systems and ensuring sufficient data quality and quantity are significant challenges. The ability of AI to process extensive data can transform static planning methods like the simplex method into dynamic, adaptive strategies, making operations more flexible and responsive to external changes. This combination of traditional optimization methods and modern AI capabilities represents a powerful tool for modern businesses. The complete procedure is visualized in Figure 1.2.

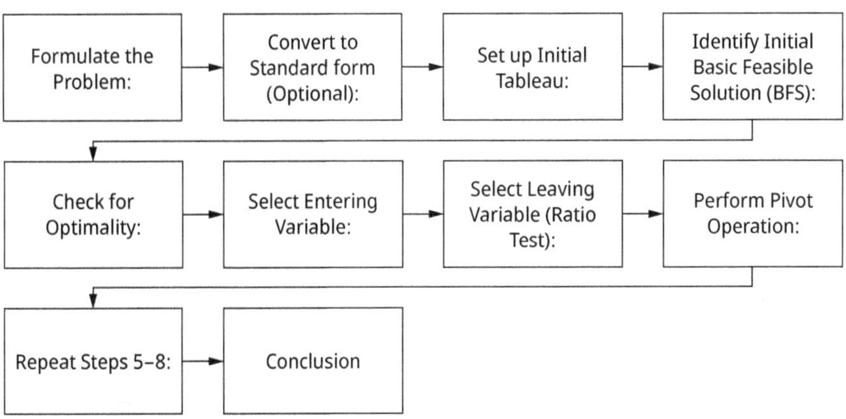

Figure 1.2: Optimization procedure for linear programming [29].

The following example explores how linear optimization tackles a resource allocation problem in an AI-powered marketing campaign.

Example 1:
A company is launching a marketing campaign to promote a new product across different media channels (e.g. TV, social media, and billboards). They have a limited budget and want to allocate their resources effectively to maximize brand awareness.

Channel	Cost per unit (c_i)	Reach per unit (r_i)
TV	Rs 100	100,000
Social media	Rs 10	20,000
Billboards	Rs 50	50,000

Relevance to AI:
Linear optimization plays a crucial role in various AI applications that involve resource allocation and decision-making under constraints. Here, AI models can analyse historical data, customer demographics, and media channel effectiveness to predict campaign reach and engagement for different allocation strategies.

Mathematical Formulation:
Media Channels: These represent the media channels (TV, social media, and billboards) as variables $x_i, i = 1, 2, 3, \ldots, n$.

Cost per Channel: Define the cost per unit of advertising on each channel as constants c_i, $i = 1, 2, 3, \ldots, n$.

Reach Per Unit: Define the estimated reach per unit of advertising on each channel as constants $r_i, i = 1, 2, 3, \ldots, n$.

Target Reach: Set a target minimum reach for the campaign, denoted by R.

Objective Function: We want to maximize the total reach of the campaign (maximizing the number of people exposed) within the budget constraint. This can be expressed as follows:

$$\text{Maximize: } \sum(r_i x_i) \text{ for all channels } i.$$

Constraints:
The total budget spent cannot exceed a predefined amount B: $\sum_{i=1}^{n} c_i x_i \leq B$.
We cannot spend negative amounts on any channel: $x_i \geq 0$ for all channels i.

Solution:
There are three channels (TV, social media, and billboards) with costs and reach per unit as follows:
The target reach (R) is 5 million people.
Budget (B) is 1,000,000. Using an LP solver, we can find the optimal solution:
Allocate 70,000 to TV advertising (x_1 = 70,000).
Allocate 930,000 to social media advertising (x_2 = 930,000).
Allocate 0 to billboards (x_3 = \$0).
This solution achieves a total reach of slightly above 5 million people (100,000 × 0.7 + 20,000 × 46.5 + 50,000 × 0) and stays within the budget constraint (70,000 + 930,000 = 1,000,000).

Benefits of Using AI:
AI models can analyse vast amounts of data to predict the reach and engagement of different media channels for various audience segments. This data-driven approach can lead to more targeted and effective marketing campaigns compared to traditional, rule-based allocation methods. By continuously learning from campaign performance data, AI can refine allocation strategies over time, maximizing the campaign impact within budget limitations.

Conclusion:
This example showcases how linear optimization, combined with AI-powered data analysis, helps businesses make data-driven decisions for optimal resource allocation in marketing campaigns and various other AI applications.

Example 2: Problem:

An intersection in a city experiences heavy traffic congestion during rush hour. The goal is to determine the optimal traffic light timings for each direction that minimizes the total waiting time for vehicles.

Imagine a four-way intersection with the following data:

a) Traffic light cycle time (C): 120 s
b) Red light duration: 10 s per cycle
c) Average traffic flow (vehicles/s):
d) North: 2 vehicles/s
e) South: 3 vehicles/s
f) East: 1 vehicle/s
g) West: 4 vehicles/s

Relevance to AI

Linear optimization plays a valuable role in various AI applications for traffic management. Here, AI can analyse traffic data and predict future patterns to dynamically adjust traffic light timings, leading to smoother traffic flow and reduced congestion.

Mathematical Formulation:

- **Traffic Directions:** These define directions (north, south, east, and west) as variables (*x_i*).
- **Green Light Duration:** This represents the green light duration for each direction (in seconds) by variables (x_i).
- **Traffic Flow:** This captures the average traffic flow (vehicles per second) for each direction (a_i).
- **Objective Function:** This minimizes the total waiting time, which is proportional to the product of green light duration and traffic flow for each direction, summed across all directions. Minimize: $\sum(x_i * a_i)$.
- **Constraints:**
  - **Cycle Time Constraint:** The total duration of all green lights plus the red light duration (fixed) must equal the traffic light cycle time (C).
  - $\sum x_i +$ red_light_duration $= C$.
- **Non-negativity Constraints:** Green light durations cannot be negative:
  - $x_i \geq 0$ for all directions i.
- **Solution Process:**
  - Formulate the objective function and constraints as an LP.
- **Possible Solution (Using a Solver):**
  - Green light north: 30 s
  - Green light south: 40 s
  - Green light east: 15 s
  - Green light west: 35 s

Benefits of Using Linear Optimization

- **Reduced Traffic Congestion:** AI can optimize traffic light timings to minimize waiting times and improve traffic flow.
- **Increased Efficiency:** Smoother traffic flow leads to reduced travel times and improved overall traffic efficiency.
- **Scalability:** Linear optimization can be applied to intersections of various sizes and complexities.

Challenges and Considerations

- **Dynamic Traffic Flow:** Traffic patterns can change throughout the day. Real-time traffic data integration is crucial for continuous optimization.
- **Infrastructure Limitations:** The effectiveness of traffic light optimization depends on the existing infrastructure and intersection design.
- **Safety Considerations:** Optimal traffic flow needs to be balanced with pedestrian safety and emergency vehicle priority:
  - Linear optimization empowers AI-based traffic management systems to optimize traffic light timings and improve traffic flow in smart cities. By minimizing waiting times and congestion, AI can contribute to a more efficient and sustainable transportation system.

Example 3: An AI assistant helps a user create a daily meal plan that meets their specific nutritional requirements (calories, vitamins, and minerals) while staying within budget constraints. Imagine a user needs a daily meal plan with the following constraints: daily calorie requirement is 2,000 calories, minimum vitamin C intake is 75 mg, budget is $20. We have data on various food items F_j like their calorie content (C_{1j}), vitamin C content (C_{2j}), and cost (P_j).

Relevance to AI:
Linear optimization helps AI personalize meal plans for individuals. Here, the AI can consider various food options, their nutritional content, and costs to create a plan that meets the user's specific needs and preferences. This is a linear optimization problem because all relationships between variables (amount of food) and constraints (nutrient intake and cost) are linear. We can use LP solvers to find the optimal combination of foods that minimizes the total cost while satisfying the nutritional and budget constraints.

Solution Process:
Formulate the objective function and constraints based on the food data and user requirements. Use the LP solver to find the optimal values for the amount of each food item (F_j) to be included in the meal plan.

Possible Solution (Using a Solver):
Include specific quantities of:
a) Chicken breast (high protein and low calories)
b) Broccoli (rich in vitamin C)
c) Brown rice (a good source of complex carbs)
d) Other healthy options

1.6 Integer Programming Optimization

Integer programming (IP) is a mathematical optimization approach where decision variables are constrained to integer values, offering a robust framework for tackling discrete and combinatorial optimization problems. In its standard form, an IP problem aims to maximize an objective function subject to linear constraints, where the decision variables are restricted to integers [16].

In the realm of AI, IP plays a pivotal role in optimization tasks that involve discrete decisions, combinatorial structures, and complex constraints. For instance, in combinatorial optimization problems like the travelling salesman problem (TSP) or knapsack problem, IP provides rigorous methods to find optimal or near-optimal solutions. Furthermore, in machine learning applications, IP formulations facilitate feature selection by identifying optimal subsets of features through binary decision variables that represent the inclusion or exclusion of each feature, thereby enabling the construction of more efficient and interpretable predictive models.

Moreover, IP is instrumental in resource allocation scenarios where limited resources need to be allocated among competing demands, ensuring that operational constraints such as capacity limitations and demand requirements are adhered to. Additionally, in network design and routing optimization, IP enables the creation of efficient network structures by minimizing costs or maximizing throughput while considering factors like distance, capacity, and connectivity. Recent advancements have also incorporated mixed IP (MIP) techniques in neural network architectures, allowing for discrete decisions in network design, such as node selection and pruning, leading to more interpretable and efficient networks. Lastly, the strength of IP lies in its ability to provide globally optimal solutions under constraints by systematically exploring the solution space, thereby avoiding the pitfalls of settling for suboptimal local solutions. In conclusion, IP serves as a foundational and versatile tool in AI optimization, addressing a wide array of challenges through its rigorous mathematical formulations and algorithms.

MIP extends the concept of IP by allowing some decision variables to take continuous values, while others are restricted to integer values.

1.6.1 Machine Learning for Initialization

Machine learning models can be trained to provide good initial solutions for the MIP problem. This can significantly reduce the search space and improve the convergence speed of traditional MIP solvers.

Heuristic Methods: AI-based heuristic algorithms can be used to find near-optimal solutions for MIP problems, especially when exact methods are computationally prohibitive.

Reinforcement Learning: Reinforcement learning can be employed to learn optimal decision-making policies for solving MIP problems iteratively. This can be particularly useful for dynamic or stochastic optimization problems where the optimal solution changes over time.

Constraint Programming: AI techniques like constraint satisfaction and logic programming can complement MIP by efficiently handling complex constraints and reducing the problem size through constraint propagation and relaxation.

Deep Learning Approximations: Deep learning models, such as neural networks, can be trained to approximate the objective function of the MIP problem. This can help in accelerating the solution process, especially for problems with complex or nonlinear objective functions.

Hybrid Approaches: Combining the strengths of AI techniques with MIP solvers, hybrid approaches can be developed to tackle specific classes of optimization problems more effectively. For example, a hybrid algorithm might use machine learning for feature selection and dimensionality reduction before applying MIP to find the optimal solution.

Scenario: You manage a delivery drone company with a small fleet of drones. Your AI assistant is tasked with optimizing delivery routes to minimize total flight time while ensuring all deliveries are completed within a specific time frame.

Problem Formulation:
We have **N** packages to be delivered to different locations. Each package delivery has a weight (W_i) and a specific delivery deadline (d_i). Each drone has a maximum weight capacity (**W**) and a fixed flight speed (**v**). The travel time (t_{ij}) between locations i and j is known.

Decision Variables:
x_{ij} (binary): The value is 1 if drone visits location j after location i, 0 otherwise.

Objective Function (Minimize Total Flight Time):
Minimize: $\sum(t_{ij} * x_{ij})$ for all i, j ($i \neq j$).

Constraints:
Flow Conservation: Each location (except the starting and ending points) must be visited exactly once by a drone. Mathematically:

$\sum(x_{ij} = 1)$ for all j (excluding starting and ending points) and $\sum(x_{ki} = 1)$ for all i (excluding starting and ending points).

Capacity Constraint: The total weight of packages on a drone cannot exceed its capacity (W).

$\sum(w_i * x_{ki}) \leq W$ for all k (representing each drone).

Delivery Deadline: Each package must be delivered before its deadline. Since flight times are known (t_{ij}) and the drone has a fixed speed (v), we can calculate the arrival time at each location. This can be expressed as a constraint, ensuring the arrival time is less than or equal to the deadline.

IP Aspect:
The decision variables (t_{ij}) represent binary choices – whether a drone goes from location i to j (1) or not (0).

Solution Approach:

This problem can be solved using various IP solvers available in software packages such as Gurobi, CPLEX, or Python libraries like PuLP. These solvers use branch-and-bound algorithms or other techniques to find the optimal solution (minimum total flight time) that satisfies all constraints. The algorithm for the IP optimization is shown in Figure 1.3.

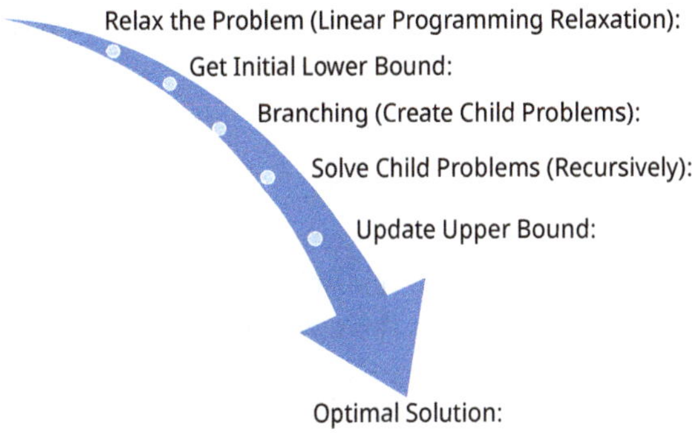

Figure 1.3: Problem formulation [30].

Example with Numerical Values:

Data:

$N = 4$ packages ($P1$, $P2$, $P3$, $P4$), drone capacity (W) = 10 kg, and drone speed (v) = 20 m/s.

Travel times (t_{ij}) and package weights (W_i) are given in the following table:

Location	Package weight (w_i)	*A*	*B*	*C*	*D*
*P*1 (start)	–	–	t_{AB}	t_{AC}	t_{AD}
*P*2	w_B	t_{BA}	–	t_{BC}	t_{BD}
*P*3	w_C	t_{CA}	t_{CB}	–	t_{CD}
*P*4 (end)	w_D	t_{DA}	t_{DB}	t_{DC}	–

Solving with a Solver:

You would input the objective function, constraints, and data into an IP solver. The solver would then find the optimal assignment of packages to drones and the corresponding delivery route, minimizing the total flight time while satisfying all constraints (capacity and deadlines).

1.7 Convex Optimization

Convex optimization is a specialized field focusing on problems where both the objective function and the constraints exhibit a convex shape. Convexity, denoting the property where any line segment connecting two points within the set remains entirely within the set, plays a vital role in simplifying the optimization process. Convex optimization focuses on problems where both the desired outcome (objective function) and the limitations (constraints) have a specific mathematical shape called convexity. This special shape guarantees that any solution that is locally the best will also be the overall best (global minimum), making the search for the optimal answer much easier and reliable. This property has widespread applications across various fields, including machine learning (e.g. support vector machines), signal processing, control theory, and finance (e.g. portfolio optimization). In convex optimization, specialized solution methods, such as gradient descent, interior point methods, and subgradient methods, leverage the convex structure of the problem to efficiently find the global optimum [17, 18]. An illustrative example, such as least squares regression or LP, can help demonstrate how convexity manifests in practical optimization problems, further elucidating its importance and applicability.

Scenario: You are developing an AI system for classifying images into different categories (e.g. cat, dog, and bird). The system uses softmax regression, a popular technique for multi-class classification, which relies on convex optimization.

Problem Formulation:
We have a dataset of N training images, each represented by a feature vector x_i. Each image belongs to one of K possible classes (e.g. $K = 3$ for cat, dog, and bird). We want to learn a weight matrix W (dimensions: feature vector size × number of classes) and a bias vector b (dimensions: $1 \times$ number of classes) that map an image's feature vector to probabilities belonging to each class.

Objective Function (Minimize Cross-Entropy Loss):
We want the model to predict the correct class for each training image. A common way to achieve this is by minimizing the cross-entropy loss between the predicted probabilities and the actual class labels. Mathematically, the objective function becomes:

$$\text{Minimize: } \sum(y_i * \log(\sigma(W * x_i + b)) \quad \forall i = 1, 2, 3, \ldots, N.$$

where σ is the sigmoid function (ensures output probabilities between 0 and 1) and y_i is a one-hot encoded vector representing the actual class label for image I denotes element-wise multiplication.

Solution Approach:
Several algorithms can solve this convex optimization problem, including gradient descent, stochastic gradient descent, and Adam (adaptive moment estimation). These al-

gorithms iteratively update the weight matrix (W) and bias vector (b) to minimize the cross-entropy loss across the training data.

Example with Numerical Values:
Data:
$N = 100$ training images, feature vector size = 100 (representing image features), and $K = 3$ classes (cat, dog, and bird).

Training Data:
Each training image would have a corresponding feature vector (100 numbers) and a one-hot encoded class label vector (e.g. (1, 0, 0) for cat).

Solving with an Optimization Library:
We would implement the softmax regression model with the cross-entropy loss function in a programming language like Python. We would use an optimization library like SciPy or TensorFlow to define the objective function and leverage built-in optimization algorithms. The library would iteratively update the weight matrix (W) and bias vector (b) based on the training data, minimizing the overall cross-entropy loss.

Conclusion:
After training, the softmax regression model with the learned weights (W) and bias (b) can be used to predict class probabilities for new unseen images. The image with the highest predicted probability would be classified as the corresponding class.

1.8 Nonlinear Technique

1.8.1 AI's Role in Enhancing Nonlinear Optimization

AI, particularly through advancements in machine learning and deep learning, is fundamentally transforming the approach to nonlinear optimization, addressing its complexities with remarkable efficiency. One significant contribution of AI is in predictive analytics, where models trained on historical data predict outcomes and trends that are crucial for the optimization process [19]. For example, in supply chain management, AI can anticipate demand fluctuations and accordingly adjust inventory levels in a nonlinear fashion. Additionally, AI excels in enhancing traditional optimization techniques with heuristic and metaheuristic algorithms such as genetic algorithms (GAs), simulated annealing, and swarm optimization. These are particularly effective for navigating large, complex search spaces more efficiently, which is essential for identifying global optima in nonlinear scenarios.

AI also plays a vital role in real-time decision-making, processing vast amounts of data instantaneously to make critical optimization decisions in dynamic environments like financial trading or emergency services, where conditions and requirements change rapidly. Moreover, deep learning, with its robust neural networks, is particularly suited for managing nonlinear data, effectively approximating complex nonlinear functions far better than many conventional models. This capability allows for the development of sophisticated optimization algorithms that can adapt to and operate within highly intricate problem spaces. Collectively, these AI-driven strategies are enhancing the scope and accuracy of nonlinear optimization, leading to smarter, faster decision-making across various fields.

Scenario:
You are developing an AI system to control a robot arm for tasks like object manipulation or assembly. The robot arm needs to move its end effector (gripper) along a specific path while avoiding obstacles and reaching the target location efficiently.

Problem Formulation:
We define the desired path for the robot arm's end effector using a set of waypoints (x, y, and z coordinates) in 3D space. Each joint of the robot arm has limitations on its range of motion (joint angles). We want to find the optimal joint angles (θ) for each waypoint that minimizes a specific cost function.

Cost Function (Example):
A common cost function considers two objectives:

Tracking Error: Minimize the distance between the actual end-effector position and the desired waypoint position.

Joint Effort: Minimize the overall energy consumption of the robot arm motors to achieve the desired motion. This cost function can be a nonlinear function of the joint angles (θ) due to the complex kinematics (relationship between joint angles and end-effector position) of the robot arm.

Constraints:
Joint Angle Limits: Each joint angle (θ_i) must be within its defined minimum and maximum values $\left(\theta_{i_{\min}}, \theta_{i_{\max}}\right)$.

Obstacle Avoidance (Optional):
If obstacles are present in the workspace, additional constraints can be incorporated to ensure that the robot arm avoids collisions during its movement. These constraints might involve maintaining a minimum distance between the robot arm and the obstacles for safety reasons.

Nonlinearity:
The relationship between joint angles (θ) and the end-effector position is typically nonlinear due to the trigonometric functions involved in robot arm kinematics. Additionally, the cost function that combines the tracking error and joint effort might also be nonlinear.

Solution Approach:
Several optimization algorithms are suitable for nonlinear problems. Here are a few examples:

Gradient Descent with Momentum: This iterative method uses the gradient of the cost function to update the joint angles in the direction that minimizes the cost. Momentum helps the algorithm converge faster by considering the direction of previous updates.

Levenberg–Marquardt algorithm: This method combines gradient descent with a trust-region approach, balancing exploration and exploitation during the search for the optimal solution.

The detailed algorithm is shown in Figure 1.4.

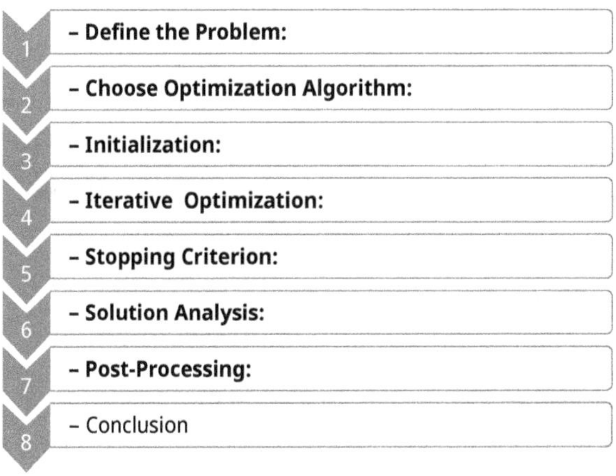

Figure 1.4: Algorithm for nonlinear technique [31].

Data:
A robot arm with three joints $(\theta_1, \theta_2, \theta_3)$, the desired path with five waypoints (*x*, *y*, *z* coordinates) for the end effector, and joint angle limits $\left(\theta_{i_{\min}}, \theta_{i_{\max}}\right)$ for each joint.

Cost Function:
A weighted combination of tracking error and joint effort (specific formulas would depend on robot arm properties).

Solving with an Optimization Library:
We would implement the robot arm kinematic model (relating joint angles to end-effector position) and the cost function in a programming language like Python. We would choose a nonlinear optimization library like SciPy or TensorFlow that provides implementations of suitable algorithms (e.g. gradient descent with momentum and Levenberg–Marquardt). The library would iteratively update the joint angles (θ)

based on the cost function gradient, aiming to minimize the overall cost while satisfying the joint angle limits and any obstacle avoidance constraints (if applicable).

Outcome:
The optimization process would provide the optimal joint angles (θ) for each waypoint along the desired path. This information can be used to control the robot arm motors and achieve the task efficiently while avoiding obstacles (if considered).

Example: 2 Portfolio optimization with AI

Problem Statement: Optimize a portfolio of two assets to maximize the expected return while minimizing the risk, where risk is measured as the standard deviation of the portfolio's return. The expected returns and covariance matrix of the assets are given below: Expected return of asset A $E(A) = 8\%$, expected return of asset B is $E(B) = 12\%$, covariance between assets A and B is $\sigma_{AB} = 0.02$, variance of asset A is $\sigma_A^2 = 0.04$, and variance of asset B is $\sigma_B^2 = 0.09$.

Solution:
Step 1: Mathematical formulation
The expected return of the portfolio is $E(R_P) = W_A \times E(R_A) + W_B \times E(R_B)$, where W_A and W_B are the weights of asset A and asset B in the portfolio, respectively. The variance of the portfolio σ_P^2 is given by

$$\sigma_P^2 = W_A^2 \times \sigma_A^2 + W_B^2 \times \sigma_B^2 + 2W_A \times W_B \times \sigma_{AB}$$

Step 2: Optimization with AI
We can use AI optimization algorithms to solve this nonlinear optimization problem. The objective function to maximize is $E(R\_P)$ (expected return), subject to the constraint that the sum of the weights is $W_A + W_B = 1$.

Step 3: Solution
Using an AI optimization algorithm, we find the optimal weights for asset A (W_A) and asset B (W_B) that maximize the expected return while considering the risk (standard deviation).

Let us say that our AI algorithm determines the optimal weights as $W_A = 0.4$ and $W_B = 0.6$.

Step 4: Results
Expected return of the portfolio is $E(R_p) = 15.2\%$, variance of the portfolio is $\sigma_P^2 = 0.0436$, and standard deviation of the portfolio is $\sigma_P = 0.209\%$.

In this optimized portfolio, we achieve an expected return of 15.2% with a standard deviation (risk) of 0.209%. This example showcases how AI can be leveraged to solve nonlinear optimization problems effectively, finding optimal solutions that balance risk and return in complex scenarios like portfolio management.

1.9 Gradient Method Optimization

Scenario:
You are developing an AI system to remove noise from digital images. The system uses a denoising filter based on linear regression, and the parameters of this filter are optimized using gradient descent.

Problem Formulation:
We have a noisy image represented as a matrix I_{noisy}, where each pixel intensity is corrupted by noise. We want to find a denoised image I_{denoised} by applying a linear filter. The filter can be a simple mask (e.g. averaging neighbouring pixels) or a more complex convolutional filter with learnable parameters (weights and biases).

Linear Regression Model:
We can formulate the denoising process as a linear regression problem, where the noisy pixel intensity $\sum\sum\left(\left(I_{\text{denoised}_{ij}}\right)-\left(w * I_{\text{neighbor}_{ij}}+b\right)\right)^2$ at position (i, j) is modelled as a linear combination of its neighbouring pixels in the original clean image $\left(I_{\text{clean}_{ij}}\right)$.

Objective Function (Mean Squared Error (MSE)):
We want the denoised image (predicted by the filter) to be as close as possible to the original clean image (ground truth). A common way to measure this is by minimizing the MSE between the noisy image and the filtered (denoised) image:

$$\text{Minimize: } \sum\sum\left(\left(I_{\text{noisy}_{ij}}\right)-\left(w * I_{\text{neighbor}_{ij}}+b\right)\right)^2 \text{ for all pixels } (i, j)$$

where w is the weight of the filter (e.g. averaging weight for a simple filter), b is the bias term of the filter (can be used to adjust brightness), and $I_{\text{neighbor}\ ij}$ represents the intensity of neighbouring pixels used by the filter.

Gradient Descent Approach:
We can use gradient descent to optimize the filter parameters (w and b) that minimize the MSE. The gradient descent algorithm iteratively updates these parameters based on the negative gradient of the objective function (MSE).

Steps:
a) Start with initial values for w and b.
b) Calculate the MSE for the current filter parameters.
c) Compute the gradients of the MSE with respect to w and b.
d) Update w and b by subtracting a small learning rate (α) multiplied by the respective gradients.
e) Repeat steps 2–4 until convergence (MSE stops decreasing significantly).

Solution with Example:
Data:
A small grayscale image with noise added to each pixel intensity.

Filter: A simple averaging filter with a 3 × 3 mask (weights and bias to be learned).

Implementation: You can implement the filter and the gradient descent algorithm in Python using libraries like NumPy and Scikit-learn.

The code would involve iteratively updating the filter weights and bias based on the calculated gradients of the MSE with respect to these parameters.

After convergence, the filter parameters (w and b) would represent the optimal weights and bias for the denoising filter. Applying this filter to the noisy image would result in a denoised image with reduced noise and a clearer visual appearance.

1.10 Gradient-Free Method Optimization

Scenario:
In developing an AI system to optimize traffic light schedules in a city with multiple intersections, the goal is to minimize traffic congestion and improve overall traffic flow. Gradient descent methods might not be suitable due to the complex and potentially discontinuous nature of the traffic flow function.

Problem Formulation:
We have a network of intersections with traffic lights. Each intersection has multiple traffic light phases (e.g. green light for specific directions). The objective is to find the optimal scheduling of light phases (durations) at each intersection for a specific time period.

Objective Function (Minimize Traffic Congestion): We want to minimize the average travel time for vehicles within the network during the considered time period. However, directly calculating the gradient of this objective function with respect to individual light phase durations might be challenging due to complex traffic dynamics.

Gradient-Free Optimization Approach: Several gradient-free optimization algorithms can be used in this scenario.

Simulated Annealing: This algorithm mimics the process of annealing in metals. It explores different scheduling options, accepting worse solutions with a certain probability to escape local optima and potentially find better solutions overall.

Particle Swarm Optimization (PSO): This method simulates a swarm of particles (potential solutions) that move through the search space, influenced by the best solutions found by individual particles and the swarm as a whole. Over time, the particles converge towards promising regions of the search space.

GA: This approach draws inspiration from biological evolution. It iteratively creates new generations of scheduling options by combining existing solutions (crossover) and introducing random variations (mutation). The new generation is then evaluated based on the objective function, and better-performing solutions are selected for further iterations.

Example with Numerical Values:
Consider a city network with five intersections. Each intersection has four traffic light phases with varying durations. Traffic data (e.g. vehicle counts and average speeds) for each road segment is available.

Solution Approach:
You would implement the chosen gradient-free optimization algorithm (e.g. simulated annealing) in a programming language like Python. Each potential solution (represented by a chromosome in a GA or a particle's position in PSO) would encode the scheduling information for all traffic lights (durations of each phase). The objective function would be evaluated for each solution by simulating traffic flow within the network using traffic flow models and the provided traffic data. The simulation would estimate the average travel time. The chosen optimization algorithm would then iteratively explore different scheduling options, aiming to minimize the average travel time across the network.

Outcome:
The optimization process would identify the best traffic light schedule (durations for each phase at each intersection) for the given time period. This schedule would be expected to minimize traffic congestion and improve overall traffic flow within the network.

1.11 Dynamic Optimization

Dynamic optimization represents a powerful combination for solving complex, time-varying problems across various domains. Dynamic optimization deals with problems where decisions are made over time, taking into account the dynamic nature of the system, evolving constraints, and changing objectives. AI techniques, especially machine learning, reinforcement learning, and neural networks, can enhance dynamic optimization by providing adaptive, learning-based solutions that can handle uncertainty and nonlinearity effectively [20].

1.11.1 Challenges in Dynamic Optimization

Dynamic optimization problems are inherently complex due to the following factors:

Temporal Dependencies: Decisions made at one time affect future outcomes, requiring a sequential decision-making approach.

AI and Dynamic Optimization: Future states and outcomes are often uncertain, introducing complexities in modelling and decision-making.

Complexity: Dynamic systems with evolving constraints and objectives can be challenging to model and solve using traditional optimization techniques.

1.11.2 AI's Role in Dynamic Optimization

AI offers several advantages when applied to dynamic optimization problems. First, adaptive learning through machine learning and reinforcement learning algorithms allows the optimization model to adapt and learn from new data and experiences, continuously improving and adjusting its strategies over time. Second, predictive modelling capabilities enable AI models to forecast future states, outcomes, or uncertainties based on historical data, providing better foresight and facilitating proactive planning in dynamic environments. Third, AI algorithms excel in real-time decision-making, processing information instantaneously to make immediate decisions and adjustments, thereby optimizing system performance dynamically. Lastly, the use of neural networks and deep learning models enables the approximation of complex, nonlinear functions and high-dimensional data structures, leading to more accurate and efficient solutions in dynamic optimization scenarios.

1.11.3 Applications of AI in Dynamic Optimization

Here are some examples illustrating the integration of AI and dynamic optimization across various domains:

Finance: AI-driven dynamic portfolio optimization can adjust asset allocations in real time based on market conditions, risk profiles, and investment goals, maximizing returns while minimizing risks dynamically.

Manufacturing: AI-enhanced dynamic scheduling and control systems can optimize production processes in real time, considering machine availability, resource constraints, and demand fluctuations to maximize throughput and efficiency.

Energy Management: AI can optimize energy generation, storage, and distribution in smart grids dynamically, adapting to variable renewable energy sources, fluctuating demand, and equipment failures to ensure stability and efficiency.

Healthcare: AI-driven dynamic treatment planning can adapt treatment regimens for patients based on real-time monitoring data, changing health conditions, and treatment responses, optimizing healthcare outcomes and resource utilization.

Example:
A robot needs to navigate from a point A to a point B in an environment with obstacles that can move or appear/disappear unexpectedly. The goal is to find the fastest or most energy-efficient path while adapting to these dynamic changes.

Relevant to AI
Dynamic optimization plays a crucial role in various AI applications that involve planning and decision-making in environments that change over time. Here, AI models can leverage sensor data and past experiences to adapt the robot's path in real time, ensuring safe and efficient navigation.

Challenges of Static Path Planning:
Traditional path planning algorithms might pre-compute a path based on a static map. However, in dynamic environments, obstacles might move, or new obstacles might appear, rendering the pre-computed path unusable.

Dynamic Formulation:
Robot State: The robot's position and orientation are represented by state variables.

Environment: The environment state includes the positions of obstacles and any other relevant factors that might change (e.g. slippery surfaces).

Actions: The robot can take different actions (move forward, turn left, and turn right) to navigate the environment.

Objective Function:
The objective function depends on the specific task. It could be:
a) minimizing travel time to reach the goal,
b) minimizing energy consumption during navigation, and
c) maximizing safety by avoiding collisions with obstacles.

Constraints:
a) The robot's movement capabilities (speed and turning radius)
b) Physical limitations of the environment (walls and boundaries)

Solution with Dynamic Programming:
Here is a simplified explanation of how dynamic programming can be used for dynamic optimization: Discretize the environment into a grid. Define a value function that represents the "cost" of reaching each grid cell from the starting point, considering the objective function. The robot iteratively updates the value function based on its current state, possible actions, and the perceived state of the environment (from sensors). At each step, the robot chooses the action that leads to the cell with the minimum cost (based on the updated value function).

Numerical Example (Simplified):
Imagine a robot navigating a grid-based world with obstacles. The robot receives sensor data indicating a previously clear path is now blocked by a moving obstacle. The dynamic programming algorithm updates the value function for neighbouring cells, reflecting the newly blocked path. Based on the updated value function, the robot chooses a new path that bypasses the obstacle and continues towards the goal.

Benefits of Using AI:
AI models can process sensor data in real time to perceive changes in the environment and adapt the robot's path accordingly. This dynamic approach allows robots to navigate complex and unpredictable environments safely and efficiently. By learning from past experiences and adapting to new situations, AI can improve the robot's navigation performance over time. Dynamic optimization, combined with AI's real-time perception capabilities, empowers robots to navigate dynamic environments effectively. This is just one example, and dynamic optimization finds applications in various AI tasks, where planning and decision-making need to adapt to changing conditions.

1.12 Stochastic-Based Optimization Techniques

Stochastic optimization offers a transformative approach to addressing complex and uncertain real-world problems. Stochastic optimization, by nature, deals with optimization challenges where some inputs or parameters are uncertain and can be represented as random variables. This uncertainty introduces complexities that traditional deterministic optimization methods find challenging to manage. Factors such as market fluctuations, demand variability, or incomplete information can significantly impact the outcomes of optimization models.

The integration of AI techniques, particularly machine learning and deep learning, into stochastic optimization provides several advantages that enhance the robustness, efficiency, and effectiveness of solutions. Machine learning models can be trained on historical data to predict uncertain variables, such as future demand patterns or market prices. These predictive analytics enable the formulation of more informed and resilient optimization models that can adapt to changing conditions.

Furthermore, AI algorithms' adaptive learning capabilities allow optimization models to continuously evolve and improve over time as new data becomes available. This adaptability is crucial for addressing uncertainties and dynamically adjusting to unforeseen changes, providing organizations with more agile and responsive decision-making tools. Additionally, AI's ability to analyse and assess risks associated with uncertain variables offers valuable insights into the potential impacts of different decision alternatives, aiding in informed decision-making. Deep learning models, with their capacity to capture intricate nonlinear relationships and process high-dimensional data, are particularly well-suited for modelling the complex structures inherent in stochastic optimization problems. This capability enables more accurate, efficient, and scalable solutions across various domains, from supply chain management and finance to healthcare and energy systems.

1.12.1 AI's Role in Enhancing Stochastic Optimization

Let us consider a numerical example to illustrate how AI and stochastic optimization can be applied to a supply chain management problem involving inventory control. A retailer needs to determine optimal inventory levels for two products (A and B) to minimize costs while considering uncertain demand. The retailer has historical sales data for both products, which can be used to train machine learning models for demand forecasting. The expected demand for product A is $E(D_A) = 100$ units and for product B is $E(D_B) = 150$ units, the variance of demand for product A is $\sigma_A^2 = 50$ and for product B is $\sigma_B^2 = 75$, the holding cost per unit per day is $H = 0.5$, and the stockout cost per unit is $P = 10$.

Objective:
Minimize the TEC, which includes holding costs and potential stockout costs.

1.12.2 AI-Enhanced Stochastic Optimization Approach

Demand Forecasting: Train machine learning models to forecast demand for each product based on historical sales data. These models can predict the mean and variance of demand distributions for both products.

Stochastic Optimization Model: Formulate a stochastic optimization model to minimize the total expected cost considering uncertain demand. The model should determine the optimal inventory levels (Q_A and Q_B) for both products.

Minimize $Z = h \times \left(\frac{Q_A + Q_B}{2}\right) + P \times (\max(0, E[D_A] - Q_A) + \max(0, E[D_B] - Q_B))$

Subject to $\mathrm{Var}(D_A) = \sigma_A^2,\ \mathrm{Var}(D_B) = \sigma_B^2$

Solution
Let us assume that after running the AI-driven stochastic optimization algorithm, the optimal inventory levels are determined as follows: Optimal inventory level for prod-

uct A is $Q_A = 110$ units and for product B is $Q_B = 160$ units. The total expected cost calculation is $Z = 67.5$. In this optimized scenario, the retailer achieves a total expected cost of $67.5 by maintaining inventory levels of 110 units for product A and 160 units for product B. By leveraging AI-driven stochastic optimization, the retailer can make more informed decisions that consider the uncertainties in demand, leading to cost savings and improved inventory management efficiency.

1.13 Metaheuristic Optimization

Metaheuristic optimization techniques synergize to tackle complex, combinatorial, and high-dimensional optimization problems across various domains. Metaheuristic algorithms, such as GAs and PSO, are heuristic search methods that provide robust and efficient solutions for problems where exact algorithms are impractical due to computational complexity or problem size. AI techniques, including machine learning, reinforcement learning, and neural networks, enhance metaheuristic optimization by providing adaptive, learning-based approaches that can handle uncertainties, nonlinearities, and dynamic environments effectively.

1.13.1 Advantages of Combining AI and Metaheuristic Optimization

AI and metaheuristic optimization form a powerful alliance in tackling complex optimization challenges across diverse domains. One of the key strengths of this collaboration lies in adaptive learning. AI algorithms can evolve and learn from new data, enhancing the performance of metaheuristic optimization techniques by fine-tuning their search strategies based on specific insights and feedback from the problem at hand. This adaptability enables the algorithms to improve over time, continuously refining their approaches for better results. Additionally, AI plays a pivotal role in balancing the exploration and exploitation phases of metaheuristic algorithms. By guiding these algorithms, AI ensures a harmonious mix of searching diverse solution spaces and refining promising solutions. This balance leads to more effective and efficient optimization processes, enabling the algorithms to navigate complex landscapes with agility and precision. Furthermore, the integration of machine learning models offers problem-specific customization capabilities. These models can learn the underlying patterns and structures of a particular optimization problem, tailoring metaheuristic algorithms to address the unique challenges presented by specific instances. This customization results in optimized solutions that are finely tuned to meet the distinct requirements and constraints of individual problems.

1.13.2 Applications of AI in Metaheuristic Optimization

Here are some examples illustrating the integration of AI and metaheuristic optimization in different domains:

- **Supply Chain Optimization**: AI-enhanced metaheuristic algorithms can optimize inventory management, production scheduling, and logistics by efficiently exploring large solution spaces and adapting to dynamic supply chain dynamics, uncertainties, and constraints.
- **Finance**: In portfolio optimization, AI-guided metaheuristic algorithms can identify optimal asset allocations by balancing risk and return in complex and volatile financial markets, considering various investment constraints and objectives.
- **Engineering Design**: AI-driven metaheuristic optimization can optimize engineering designs by exploring innovative and efficient design configurations, considering multiple objectives, constraints, and uncertainties to achieve superior performance and reliability.
- **Healthcare**: AI-enhanced metaheuristic algorithms can optimize healthcare resource allocation, treatment planning, and scheduling by considering patient-specific needs, dynamic healthcare environments, and evolving constraints to improve patient outcomes and operational efficiency.

1.14 Discrete Optimization

Discrete optimization involves finding the best solution from a finite set of possibilities. Unlike continuous optimization where variables can take any value within a range, discrete optimization deals with variables that can only assume specific, distinct values. Examples of discrete optimization problems include:

- **Combinatorial Optimization Problems**: These problems involve finding the best arrangement, combination, or selection of discrete items or entities. Examples include TSP, knapsack problem, job scheduling, and graph colouring.
- **IP Problems**: These are a mathematical optimization technique used to find the best possible solution to a problem where some or all of the decision variables must be whole numbers. Essentially, these involve maximizing or minimizing a linear function subject to linear constraints while ensuring that the variables take integer values. These are particularly useful for solving real-world problems such as facility location, production planning, and resource allocation.
- **Binary Optimization Problems**: These problems deal with variables that can only take binary values (0 or 1). Examples include Boolean satisfiability (SAT) problems, Boolean logic synthesis, and binary decision diagrams.
- **Network Optimization Problems:** These problems involve optimizing the flow of resources or information through a network.

- **Constraint Satisfaction Problems:** These involve finding a solution that satisfies a set of constraints that specify relations among variables. Examples include Sudoku, map colouring, and scheduling problems.

1.15 Continuous Optimization

Continuous optimization is a mathematical process aimed at finding the best possible value (maximum or minimum) of a function where the input variables can be any real number within a specified range. Unlike discrete optimization, which deals with whole numbers or specific options, continuous optimization explores an infinite set of possibilities. These problems are prevalent in fields like engineering, economics, and machine learning, where finding optimal solutions within continuous ranges is crucial. Mathematically, these problems can be formulated as $y = f(x)$ subject to $x \in \mathbb{R}^n$.

Continuous optimization is the basis of machine learning. Models are refined by adjusting their parameters to minimize a chosen error metric, known as the loss function. For instance, in linear regression, the goal is to find the model parameters that produce predictions closest to the actual values. This is achieved by minimizing the MSE, which quantifies the average squared difference between predicted ($\hat{y}$) and observed (y) outcomes.

Minimize MSE as follows:

$$(y, \hat{y}) = \frac{1}{n}\sum_{i=1}^{n}(y_i - \hat{y}_i)^2$$

Gradient descent plays a key role in continuous optimization, providing an efficient method to iteratively refine model parameters. By calculating the gradient (slope) of the loss function, it guides the optimization process towards regions of lower error. This technique, coupled with its various adaptations, has become indispensable in training complex machine learning models. The gradient descent iteratively updates the parameter vector $\bar{x}$ in the opposite direction of the gradient of the objective function $y_i = f(x_i)$, until convergence $x_{i+1} = x_i - \alpha \nabla y_i$, where α represents the learning rate and ∇y_i denotes the gradient of the objective function $y_i = f(x_i)$ with respect to x evaluated at the current point x_i.

1.15.1 Evolutionary Algorithms

Evolutionary algorithms (EAs) are a powerful subset of AI inspired by the principles of natural selection. By simulating biological processes like selection, mutation, and recombination, EAs excel at tackling complex optimization and search problems. These algorithms maintain a population of potential solutions, which evolve over time through iterative refinement. This adaptive nature empowers EAs to discover optimal or near-optimal solutions in challenging and dynamic environments. GAs and PSO are prominent examples of EAs that have found widespread application in diverse fields, from engineering design to machine learning.

1.15.2 Optimizing Parameters in Complex Models

One of the primary applications of EAs is the optimization of parameters in complex models, including but not limited to, neural networks, simulation models, and various machine learning algorithms. These models often involve a multitude of parameters that need fine-tuning to achieve optimal performance, a task that can be daunting and computationally intensive through traditional means.

1.15.3 Genetic Algorithms (GAs)

GAs are particularly suited for this task due to their ability to explore a vast search space efficiently. They operate by encoding a potential solution to an optimization problem as a string or chromosome, thus forming a population of solutions. Through iterative processes mimicking natural selection, including crossover (mating) and mutation, GAs evolve these solutions towards better fitness values.

Example: A neural network's performance in recognizing patterns or making predictions can be significantly influenced by its weights and architecture. By employing GAs to optimize these weights, one can achieve improvements in accuracy and efficiency. Researchers have successfully used GAs to fine-tune hyperparameters and architecture designs of neural networks for tasks ranging from image recognition to natural language processing, often resulting in models that outperform those optimized through conventional methods.

1.16 Particle Swarm Optimization (PSO)

PSO is a computational technique inspired by the social behaviour of animals. It involves a group of potential solutions, called particles, which move through a problem space. Each particle adjusts its position based on its own best-found solution and the best solution found by the entire group. This collective learning process helps the particles converge towards optimal solutions. Essentially, PSO mimics the way flocks of birds or schools of fish cooperate to find the best feeding or nesting grounds:

- **Example:** In optimizing the layout of a wind farm, PSO can be used to adjust the positions of turbines to maximize total energy production while accounting for

land constraints and aerodynamic effects. By modelling each turbine's position as a particle, PSO iteratively adjusts the swarm of turbines to find an optimal arrangement that achieves the highest efficiency. Recently, Shapique et al.'s [21] PSO algorithm was applied to optimize the total expected cost in problems related to queuing systems.
- **Feature Selection:** This is another crucial area where EAs excel, particularly in machine learning and data analysis. Selecting the right set of features can significantly impact the performance of predictive models by reducing over fitting, improving accuracy, and decreasing training times [22, 23].

1.17 Bayesian Optimization

Bayesian optimization represents a sophisticated approach to optimizing objective functions that are expensive to evaluate. It stands out in the field of mathematical optimization by leveraging Bayesian inference to efficiently navigate the search for the optimum, especially in scenarios where the evaluations of the objective function consume significant resources, time, or both. This optimization technique is particularly well-suited to applications with limited budgets for function evaluations, such as hyperparameter tuning in machine learning models, where each evaluation can be computationally intensive [24].

The core of Bayesian optimization lies in its use of a probabilistic model to predict the performance of the objective function at untested points. By incorporating prior belief about the objective function and updating this belief based on the outcomes of function evaluations, Bayesian optimization effectively learns about the function over time [25]. This process involves two main components:
- **Surrogate Model:** Bayesian optimization employs a surrogate probabilistic model to approximate the objective function. Gaussian processes are commonly used as the surrogate model due to their flexibility and the rich information they provide about the uncertainty in predictions. The surrogate model provides a belief about the function's behaviour across the search space, which is updated as more evaluations are conducted.
- **Acquisition Function:** This function is used to determine the next point to evaluate by balancing the exploration of uncharted areas of the search space against the exploitation of known promising areas. The acquisition function calculates the expected utility of sampling each point, given the current surrogate model. The common acquisition functions include expected improvement, probability of improvement, and upper confidence bound.

1.17.1 Applications of Bayesian Optimization

- One of the most prominent applications of Bayesian optimization is hyperparameter tuning in machine learning. Hyperparameters are the configuration settings used to optimize the learning process of a model. The optimal set of hyperparameters can significantly enhance the model's performance, but identifying this set is notoriously challenging due to the vastness of the hyperparameter space and the expensive nature of evaluating model performance.
- Traditional grid or random search methods for hyperparameter tuning can be exceedingly inefficient, especially when the search space is large. Bayesian optimization, in contrast, builds a model of the objective function (e.g. validation set accuracy) and intelligently selects the most promising hyperparameters to evaluate next. This results in finding better solutions with fewer evaluations. The use of a probabilistic model allows for the incorporation of prior knowledge about the hyperparameters' likely performance. This can be particularly useful when some intuition or results from previous experiments can guide the search. Bayesian optimization is naturally suited to handle noise in the objective function evaluations, a common scenario in machine learning experiments due to varying data splits, initialization, and inherent stochasticity in learning algorithms [26, 27].
- Designing the architecture of neural networks involves selecting hyperparameters. Bayesian optimization has been used to efficiently navigate this complex search space, significantly reducing the computational resources required to find high-performing architectures. Beyond individual models, Bayesian optimization can optimize entire machine learning pipelines, including feature selection, model selection, and hyperparameter tuning of all components within the pipeline.

1.18 Conclusion

The synergy between mathematical optimization and AI has transformed problem-solving across diverse fields. Optimization offers a structured approach that empowers AI systems to tackle complex challenges effectively, serving as a cornerstone in areas like machine learning, robotics, and natural language processing. Its adaptability shines through its ability to handle various problem types, from linear to nonlinear and discrete to continuous, with techniques like gradient descent and the simplex algorithm showcasing its versatility in AI applications. As AI becomes increasingly integrated into our lives, ethical considerations gain prominence. Balancing innovation with ethical values is crucial for responsible AI development, ensuring transparency, fairness, and accountability in decision-making processes. Looking ahead, the advent of quantum computing heralds a new frontier in optimization, promising unprecedented computational power to address previously insurmountable challenges and spur innovative advancements in AI. In essence, mathematical optimization is foundational to AI, enabling systems to navigate complex problems, make data-driven decisions, and achieve desired outcomes across various applications. This partnership underscores a transformative era in problem-solving, pushing boundaries and shaping a future where intelligent systems contribute to addressing societal challenges both efficiently and ethically.

References

[1] Snyman, J. A., & Wilke, D. N. (2005). Practical mathematical optimization. Springer Science+ Business Media, Incorporated.

[2] Chong, E. K., & Żak, S. H. (2013 Jan 14). An introduction to optimization. John Wiley & Sons.

[3] Ertel, W. (2018 Jan 18). Introduction to artificial intelligence. Springer.

[4] Everitt, B. Introduction to optimization methods and their application in statistics.

[5] Diwekar, U. M. (2020 Oct 29). Introduction to applied optimization. Springer Nature.

[6] Hesami, M., & Jones, A. M. (2020 Nov). Application of artificial intelligence models and optimization algorithms in plant cell and tissue culture. *Applied Microbiology and Biotechnology, 104*(22), 9449–9485.

[7] Boopathi, S., Pandey, B. K., & Pandey, D. (2023). Advances in artificial intelligence for image processing: Techniques, applications, and optimization. In Handbook of research on thrust technologies' effect on image processing (pp. 73–95). IGI Global.

[8] Le, D. N., Pandey, A. K., Tadepalli, S., Rathore, P. S., & Chatterjee, J. M. (2019). Network modeling, simulation and analysis in MATLAB: Theory and practices. John Wiley & Sons.

[9] Hu, D., Sun, H., Mehrabi, P., Ali, Y. A., & Al-Razgan, M. (2023 Apr). Application of artificial intelligence technique in optimization and prediction of the stability of the walls against wind loads in building design. *Mechanics of Advanced Materials and Structures, 24*, 1–8.

[10] Sayed, E. T., Olabi, A. G., Elsaid, K., Al Radi, M., Semeraro, C., Doranehgard, M. H., Eltayeb, M. E., & Abdelkareem, M. A. (2023). Application of artificial intelligence techniques for modeling, optimizing, and controlling desalination systems powered by renewable energy resources. *Journal of Cleaner Production, 413*, 137486.

[11] Nayak, S. (2020 Aug 25). Fundamentals of optimization techniques with algorithms. Academic Press.

[12] Yang, X. S. (2018 Sep 19). Optimization techniques and applications with examples. John Wiley & Sons.

[13] Zahraee, S. M., Assadi, M. K., & Saidur, R. (2016 Dec 1). Application of artificial intelligence methods for hybrid energy system optimization. *Renewable and Sustainable Energy Reviews, 66*, 617–630.

[14] Wang, F., Sohail, A., Wong, W. K., Ul Ain Azim, Q., Farwa, S., & Sajad, M. (2023 Jul 1). Artificial intelligence and stochastic optimization algorithms for the chaotic datasets. *Fractals, 31*(6).

[15] Yu, C., Zhu, X., Li, Z., Ma, Y., Yang, M., & Zhang, H. (2023 May 15). Optimization of elliptical pin-fin microchannel heat sink based on artificial neural network. *International Journal of Heat and Mass Transfer, 205*, 123928.

[16] Wolsey, L. A. (2020 Sep 10). Integer programming. John Wiley & Sons.

[17] Bertsekas, D. (2015 Feb 1). Convex optimization algorithms. Athena Scientific.

[18] Luo, Z. Q., & Yu, W. (2006 Jul 31). An introduction to convex optimization for communications and signal processing. *IEEE Journal on Selected Areas in Communications, 24*(8), 1426–1438.

[19] Dhiman, H., Vyas, N., Pagrotra, A., & Dutt, V. (2023). Real-time traffic analysis and driver safety enhancement via IoT and computer vision with high-resolution project lamps for intelligent road illumination [Internet]. In 2023 1st International Conference on Circuits, Power and Intelligent Systems (CCPIS). IEEE. Available from: http://dx.doi.org/10.1109/CCPIS59145.2023.10291872.

[20] Hinderer, K., Rieder, U., & Stieglitz, M. (2016). Dynamic optimization. Springer.

[21] Mohammed Shapique, A., Sudhesh, R., & Dharmaraja, S. (2024 Sep). Transient analysis of a modified differentiated vacation queueing system for energy-saving in WiMAX. *Methodology and Computing in Applied Probability*, *26*(3), 23.

[22] Lilhore, U. K., Dalal, S., Faujdar, N., Simaiya, S., Dahiya, M., Tomar, S., & Hashmi, A. (2024 Jan). Unveiling the prevalence and risk factors of early stage postpartum depression: A hybrid deep learning approach. *Multimedia Tools and Applications*, *25*, 1–35.

[23] Simaiya, S., Lilhore, U. K., Sandhu, J. K., Snehi, J., Garg, A., & Manhar, A. (2023 Jan 1). LoRa-based IoT architecture using ant colony optimization for intelligent traffic system. In Machine Learning, Image Processing, Network Security and Data Sciences: Select Proceedings of 3rd International Conference on MIND 2021 (pp. 751–761). Springer Nature Singapore, Singapore.

[24] Kukkar, A., Lilhore, U. K., Frnda, J., Sandhu, J. K., Das, R. P., Goyal, N., Kumar, A., Muduli, K., & Rezac, F. (2023 Jan 1). ProRE: An ACO-based programmer recommendation model to precisely manage software bugs. *Journal of King Saud University-Computer and Information Sciences*, *35*(1), 483–498.

[25] Dalal, S., Lilhore, U. K., Faujdar, N., Simaiya, S., Agrawal, A., Rani, U., & Mohan, A. (2024 Jul). Enhancing thyroid disease prediction with improved XGBoost model and bias management techniques. *Multimedia Tools and Applications*, *1*, 1–32.

[26] Trivedi, N. K., Anand, A., Lilhore, U. K., & Guleria, K. (2022 Jul). Deep learning applications on edge computing. In Machine Learning for Edge Computing (pp. 143–168 Vol. 29). CRC Press.

[27] Simaiya, S., Lilhore, U. K., Prasad, D., & Verma, D. K. (2021 Jan 28). MRI brain tumour detection & image segmentation by hybrid hierarchical K-means clustering with FCM based machine learning model. *Annals of the Romanian Society for Cell Biology*, 88–94.

[28] Mashwani, W. K., Haider, R., & Brahim Belhaouari, S. (2021). A multiswarm intelligence algorithm for expensive bound constrained optimization problems. *Complexity*, *2021*(1), 5521951.

[29] Bertsimas, D., & Tsitsiklis, J. N. (1997 Jan). Introduction to linear optimization. Athena Scientific, Belmont, MA.

[30] Meerschaert, M. M. (2013). Chapter 3 – Computational methods for optimization. *Mathematical Modeling*, 57–112.

[31] Bazaraa, M. S., Sherali, H. D., & Shetty, C. M. (2006 May 5). Nonlinear programming: Theory and algorithms. John Wiley & Sons.

Nancy Jasmine Goldena and Thangapriya

2 An Overview of Mathematical Optimization in Artificial Intelligence

Abstract: Mathematical optimization (MO) in "reality" is an area of applied mathematics that has significance in a wide range of fields, including the areas of operations research, soft computation, surveillance, and image processing, particularly in the field of artificial intelligence (AI) as well as other technological areas. Optimization's components, the connection between AI and mathematical intelligence, and optimization levels are all extensively explained. Basic optimization concepts such as learning rate, vector norms, derivatives, maxima, and minima are presented to help understand how AI systems perform. These concepts serve as a building block for learning and using optimization algorithms in AI. This chapter examines the connection between optimization and AI, focusing on primary concepts and techniques. Many optimization algorithms and types are explained with examples. MO is increasingly being utilized to address a wide range of real-world problems. One of the most important uses of AI technology is intelligent optimization. To combine, create, and provide an in-depth understanding of MO in AI, the chapter provides an up-to-date, thorough, and creative examination of MO algorithms.

Keywords: Artificial intelligence, constrained optimization, convex optimization, maximizing, minimizing, optimization methodologies, technological areas

2.1 Introduction

The words "optimal" and "optimization" have the same root, referring to "finest." However, the term "finest" varies. Optimization is the iterative process of training a model to produce both maximum and minimum function evaluations. In optimization problems, maximizing and minimizing are the two important terms as shown in Figure 2.1.

It is one of the most crucial processes in artificial intelligence (AI) for achieving improved outcomes. The goal of optimization is to identify the minimal number of inputs required to produce an effective result [1]. These inputs and outputs vary in different domains, but the fundamental concept remains the same for an enormous variety of problems across various studies. The objectives of mathematical optimization (MO) in AI are as follows:

Acknowledgements: We are expressing our gratitude to the GOD ALMIGHTY for providing guidance and inspiration throughout the process, allowing us to complete the work in a timely manner. We would like to express the blessings and love received, which contributed to the successful completion of the chapter.

https://doi.org/10.1515/9783111436180-002

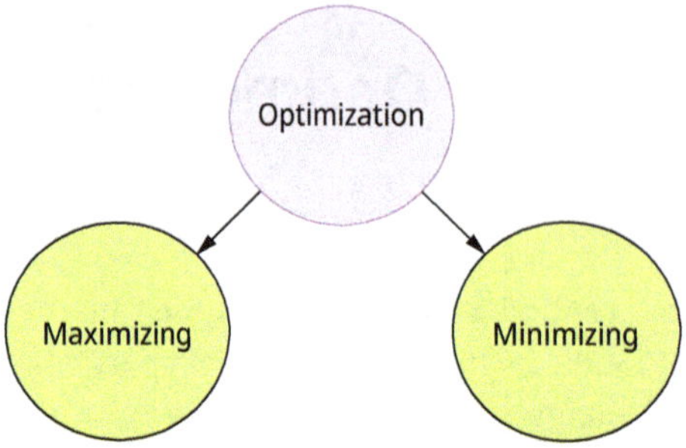

Figure 2.1: Two optimization methodologies.

- To understand the types of various optimization issues
- To recognize the category to which a given problem belongs
- To develop a logical formulation for a practical issue
- To understand the first-order procedures, including their features, feasibility, and fundamental validation
- To understand discrete optimization, the concept of submodularity, and its relationship to convex optimization problems
- To understand specific second-order methods
- To understand the nonconvex optimization methods and their applicability in AI
- To understand the applicable areas of widely used optimization algorithms

The term "mathematical programming," which is basically similar with optimization, originated in the 1940s. The process of selecting the optimal solution based on specific criteria from a set of potential solutions to a given problem is known as MO. The exploration of the mathematical framework of optimization issues, the establishment of techniques for resolving them, an examination of the mathematical aspects, and the computational execution of MO methods are all included in the area of mathematical programming [3]. Faster computers have substantially increased the complexity of the optimization issues. Two categories of optimization constraints are minimizing and maximizing. The computing science would not have advanced as far as if MO is not present. It is interesting to note that nearly all AI algorithms are considered as solutions to optimization problems. The majority of AI research involves analysing the outcomes of each iteration of an algorithm after adjusting the hyperparameters until the optimal results, characterized by high accuracy and a low error rate, are achieved [2].

2.2 Components of MO

The three components in an MO problem are given in Table 2.1.

Table 2.1: Components of MO.

Name	Description	Notation
Objective function	The function that needs to be either minimized or maximized required for resolving the optimization problems within the constraints.	$f(x)$
Variables	These variables are also known as input variables and an optimizer for modification.	$x1$, $x2$, $x3$, and so on
Constraints	Constraints are formulas for managing each variable's range.	$g_n(x)$

2.2.1 Understanding Optimization Basics

Understanding MO needs an in-depth knowledge of optimization problems. The goal of an optimization problem is to maximize or minimize a given function. MO, which plays a crucial role in computer science, is one of the most significant areas in the field of intelligent optimization. The term "intelligent" refers to responding intellectually. To get desired results, this kind of optimization involves intelligently modifying a problem's inputs. The intelligent optimization algorithms are evolutionary and metaheuristic approaches [4]. These algorithms are mostly based on counting methods. These techniques are capable of resolving highly complex issues. The fundamental notions of optimization are described below.

2.2.1.1 Maxima and Minima

A function's maxima are the largest value, and minima is the smallest value within a specified range. Furthermore, it can be divided into local and global sections as shown in Figure 2.2.

- ***Global Maxima and Minima:*** These are the highest and lowest values over the whole function's domain.
- ***Local Maxima and Minima:*** These are the function's maximum and minimum values inside a specified range.

There is only one global minima and maxima; however, there could be multiple local minima and maxima.

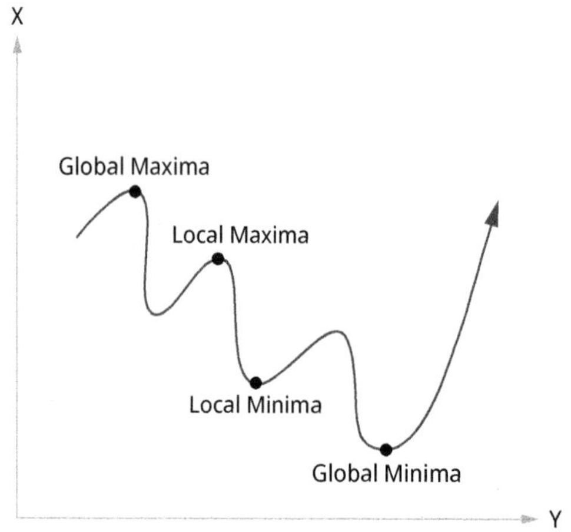

Figure 2.2: Maxima and minima representation of a function.

2.2.1.2 Learning Rate

A hyperparameter, also known as a tuning parameter, or learning rate generates the number of steps for each iteration as the function hits minima. The term r represents a tuning parameter. To choose the right r-value, the cross-validation technique is used. To carry out this, plot a graph with several learning rates, measure each value's training loss, and select the value with the least amount of loss. Mostly, parameter tuning is performed in machine learning, deep learning, and AI [5].

2.2.1.3 Norms of Vectors

Norms of vectors are initially required for characterizing optimal position in the first place, determining derivatives for functions involving multiple variables, and detailing methods and their convergence to optimal points. A norm is a function constructed on n-dimensional vectors (x) that assesses their magnitude and meets the following conditions:

- For each vector x, $||x|| > = 0$
- For any vector x and scalars, a and b, $||ax|| = |a|||x||$
- For any two vectors x and y, $||x|| + ||y|| > = ||x + y||$
- For the zero vector x, $||x|| = 0$ only

2.2.1.4 Derivatives

A linear operator that represents a function's local behaviour is called a derivative. Whenever assessed at a position x, it provides a quantifiable approximation of the function value at nearby points. A derivative is a constant that is calculated at each point in a function of a single variable. The generalization of the second derivative is the Hessian matrix, which is defined as a symmetric matrix of second-order partial derivatives evaluated at a given point.

2.3 Types of Optimization Problems

The fundamental components of MO are already discussed in Table 2.1. Types of optimization problems are discussed in Figure 2.3.

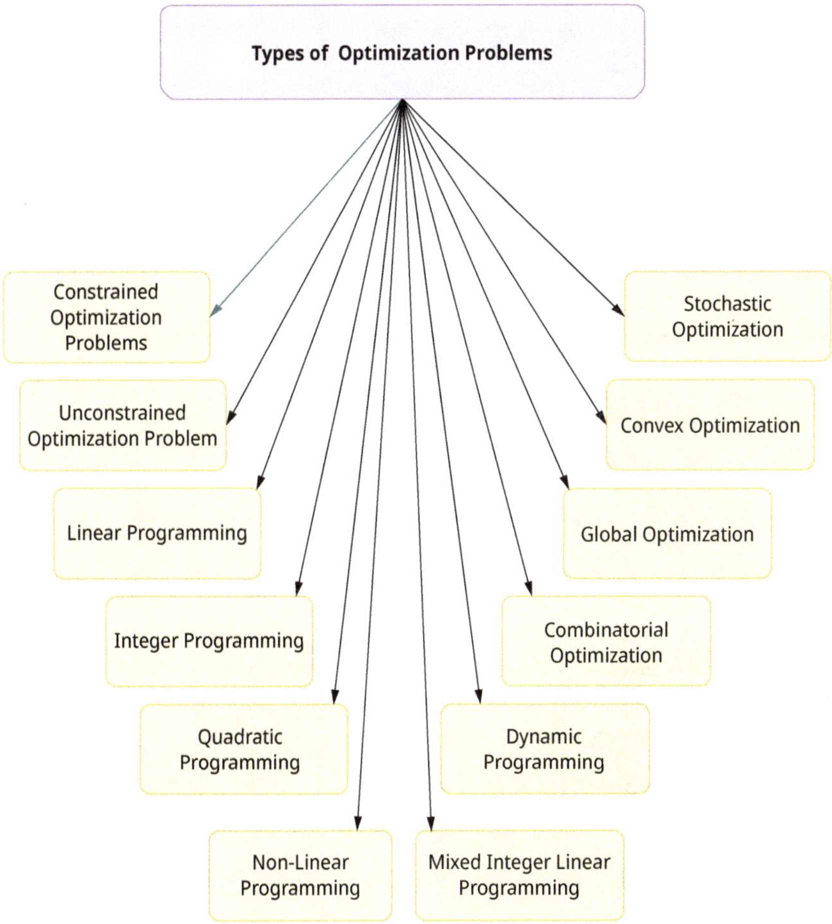

Figure 2.3: Types of optimization problems in AI.

2.3.1 Constrained Optimization Problems in AI

Constrained optimization problems involve finding the optimal value of a function while satisfying certain constraints. These constraints can either be linear or nonlinear in nature and can involve a combination of integer, binary, and mixed variables. Constrained optimization problems are used to optimize resource allocation, production schedules, and design processes while ensuring that certain constraints or limitations are met. For example, in engineering, constrained optimization is used to design efficient structures while adhering to material and weight limitations. Constrained optimization problems play a crucial role in various areas of AI [6]. Here are some common applications of constrained optimization problems in AI.

2.3.2 Machine Learning

- **Regularization**: Techniques such as LASSO and ridge regularization in linear regression and logistic regression involve optimizing an objective function subject to constraints on the model parameters to prevent overfitting:

$$\begin{gathered}\text{L1 Regularization}\\ \text{Cost} = \sum_{i=0}^{N}\left(y_i - \sum_{j=0}^{M} x_{ij}W_j\right)^2 + \lambda\sum_{j=0}^{M}|W_j|\\ \text{L2 Regularization}\\ \text{Cost} = \underbrace{\sum_{i=0}^{N}\left(y_i - \sum_{j=0}^{M} x_{ij}W_j\right)^2}_{\text{Loss function}} + \underbrace{\lambda\sum_{j=0}^{M} W_j^2}_{\text{Regularization Term}}\end{gathered} \tag{2.1}$$

- **Support Vector Machines (SVMs)**: SVMs involve solving a constrained optimization problem to identify the optimal hyperplane that separates different classes while maximizing the margin between them.

  Neural Network Training: Training neural networks often involve optimizing a loss function subject to constraints on the network's parameters, such as weight bounds or sparsity constraints.

2.3.3 Reinforcement Learning

Policy Optimization: In reinforcement learning (RL), agents learn to maximize long-term rewards by solving a constrained optimization problem, where the constraints may include resource limits or safety constraints.

Constrained Markov Decision Processes (CMDPs): CMDPs extend standard Markov decision processes (MDPs) to include constraints on the agent's behaviour, leading to constrained optimization problems in the context of RL.

2.3.4 Optimal Control

Robotics: Constrained optimization problems arise in robot motion planning and control, where constraints on robot dynamics, actuator limits, and environmental obstacles must be considered.

Autonomous Vehicles: Optimal control techniques are used in autonomous vehicles to plan safe and efficient trajectories while adhering to traffic regulations and vehicle dynamic constraints.

2.3.5 Natural Language Processing (NLP)

Text Generation: Generating text sequences using techniques like sequence-to-sequence models involves optimizing an objective function subject to constraints on sequence length or vocabulary usage.

Machine Translation: Training translation models involves solving a constrained optimization problem to maximize the likelihood of translating source sentences to target sentences while ensuring fluency and grammatical correctness.

2.3.6 Computer Vision

Object Detection and Recognition: Optimizing object detection and recognition models involves balancing accuracy with computational efficiency, leading to constrained optimization problems where computational resources are limited.

In all these applications, constrained optimization techniques are used to find optimal solutions while satisfying various constraints, leading to more robust and reliable AI systems.

2.4 Unconstrained Optimization Problems in AI

Unconstrained optimization problems involve optimizing the objective function without explicit constraints, often in mathematical modelling and theoretical analysis. They are common in calculus, where the goal is to find the extrema of functions. Techniques like gradient descent and Newton's method are used to solve these problems, iteratively refining the solution until the derivative is zero. Unconstrained optimization problems are also prevalent in various areas of AI, where the aim is to find the optimal solution without any specific constraints [9]. Here are some common applications of unconstrained optimization problems in AI.

2.4.1 Deep Learning

2.4.1.1 Neural Network Training

Training deep neural networks typically involves optimizing an objective function, such as cross-entropy loss or mean squared error, to minimize prediction errors without explicit constraints on the model parameters. Methods like stochastic gradient descent (SGD) and its variants are commonly employed for this optimization. The following equation is used for SGD:

$$\begin{aligned}&\text{Loop}\{\\&\qquad x := x - \alpha(4x - 2)\\&\}\quad \text{until convergence at minimum}\end{aligned} \tag{2.2}$$

2.4.1.2 Natural Language Processing (NLP)

Word Embeddings: Learning distributed representations of words (word embeddings) using techniques like Word2Vec or GloVe involves optimizing an objective function to capture semantic relationships between words without any constraints.

2.4.1.3 Computer Vision

Image Classification: Training CNNs for image categorization involves optimizing an objective function to minimize classification errors without constraints on the network parameters.

Image Generation: Generating realistic images using generative adversarial networks (GANs) involves optimizing an objective function to minimize the discrepancy between generated and real images without explicit constraints. The following formula is used for the GAN model:

$$\min_G \max_D (E_x{\sim}P_{\text{data}}[\log(D(x))] + E_x{\sim}P_{\text{model}}[\log(1-D(G(x)))]) \tag{2.3}$$

Figure 2.4 shows a simple GAN model, especially used to create new data that looks like the training data.

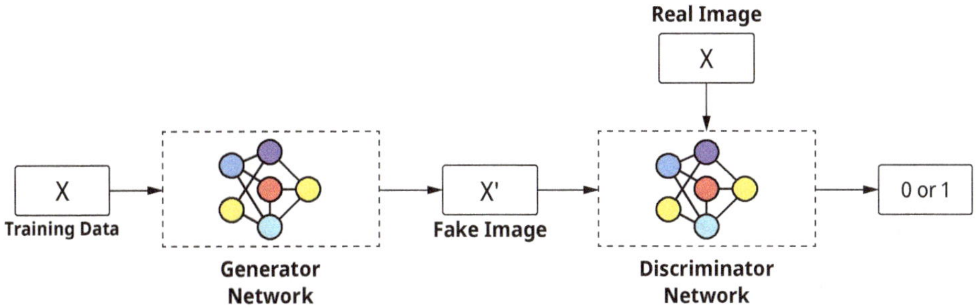

Figure 2.4: Simple GAN model.

2.4.2 Reinforcement Learning

Value Function Approximation: Solving RL problems often involves optimizing an objective function to approximate the value function or policy function without explicit constraints. Techniques such as temporal difference learning and Q-learning are commonly used for this purpose.

2.4.3 Optimization algorithms

Hyperparameter Optimization: Tuning hyperparameters of machine learning models involves optimizing an objective function (e.g., validation error) to find the best combination of hyperparameters without constraints on their values.

In all these applications, unconstrained optimization techniques play a crucial role in training AI models, optimizing performance, and improving the effectiveness of AI systems.

2.5 Linear Programming in AI

Linear programming is a robust optimization technique used to solve problems where both the objective function and the constraints are linear. Linear programming problems can be addressed using various algorithms, including the simplex method, interior-point methods, and specialized software packages such as CPLEX, Gurobi, and MATLAB's Optimization Toolbox. Linear programming is widely used in AI for various optimization tasks. Here are some common applications of linear programming in AI.

2.5.1 Resource Allocation

Supply Chain Management: Optimize the allocation of resources such as raw materials, inventory, and transportation to minimize costs or maximize profits.

Project Scheduling: Schedule tasks and allocate resources to complete projects within constraints such as deadlines, resource availability, and dependencies.

2.5.2 Network Optimization

Telecommunications: Optimize the routing of data packets or telephone calls in communication networks to minimize latency, maximize throughput, or optimize resource utilization.

Transportation: Optimize the flow of goods and vehicles through transportation networks to minimize costs, reduce congestion, or improve delivery times.

2.5.3 Production Planning

Manufacturing: Optimize production processes, including production scheduling, workforce allocation, and inventory management, to maximize efficiency and minimize costs.

Energy Systems: Optimize the generation, transmission, and distribution of energy resources to meet the demand while minimizing costs and environmental impact.

2.5.4 Machine learning

Linear Regression: Solving linear regression problems involves minimizing the least squares error between the observed and predicted values, which can be formulated as a linear programming problem.

2.5.5 Game Theory

Optimal Strategies: Finding optimal strategies in zero-sum games, such as two-player games like chess or poker, can be formulated as linear programming problems.

Linear programming provides a powerful framework for modelling and solving optimization problems with linear objectives and constraints. It is widely used in AI for decision-making, resource allocation, and optimization tasks across various domains.

2.6 Integer Programming in AI

Integer programming is like linear programming but requires some or all of the decision variables to be whole numbers. This constraint introduces discrete decision variables into the optimization problem, leading to more complex and often combinatorial optimization challenges. Solving integer programming problems is usually harder than linear programming because the decision variables are whole numbers. Integer programming is a valuable tool in AI for solving optimization problems where some or all of the decision variables are whole numbers [11]. Here are some common applications of integer programming in AI.

2.6.1 Scheduling and Planning

Job Scheduling: Determine the optimal schedule for completing a set of tasks with precedence constraints and resource limitations. Integer programming can be used to model and solve various scheduling problems.

Project Management: Allocate resources, schedule tasks, and manage dependencies to optimize project timelines and resource utilization.

2.6.2 Network Design and Routing

Facility Location: Select optimal locations for facilities (e.g., warehouses and factories) to minimize transportation costs or maximize coverage. Integer programming can be used to model facility location problems.

Network Flow: Optimize the flow of goods, information, or resources through networks subject to capacity constraints, demand, and supply considerations.

2.6.3 Portfolio Optimization

Portfolio Selection: Allocate investments across different assets to maximize returns while managing risk. Integer programming can be used to model portfolio optimization problems with integer constraints on asset allocation.

2.6.4 Machine Learning

Feature Selection: Select a subset of features or variables to improve the performance of machine learning models. Integer programming techniques can be used to perform feature selection with constraints on the number of selected features.

2.6.5 Game Theory

Nash Equilibrium: Finding Nash equilibria in games with discrete action spaces, such as strategic board games or economic models, can be formulated as integer programming problems.

Integer programming provides a flexible and powerful framework for solving optimization problems with discrete decision variables. It allows AI systems to make optimal from resource allocation and scheduling to machine learning and game theory.

2.7 Quadratic Programming in AI

Quadratic programming is an optimization problem where the objective function is quadratic and the constraints are linear. This means both the objective function and the constraints involve quadratic expressions of the decision variables. Techniques for solving quadratic programming problems include interior-point methods, active set methods, sequential quadratic programming, and specialized algorithms tailored to specific problem structures. Quadratic programming plays a significant role in various applications of AI [12]. Here are some common applications of quadratic programming in AI.

2.7.1 Support Vector Machines (SVMs)

The optimization problem in SVM involves solving a quadratic programming problem to find the hyperplane that separates classes with the maximum margin or to minimize classification errors with soft-margin constraints. Figure 2.5 shows the SVM hyperplane.

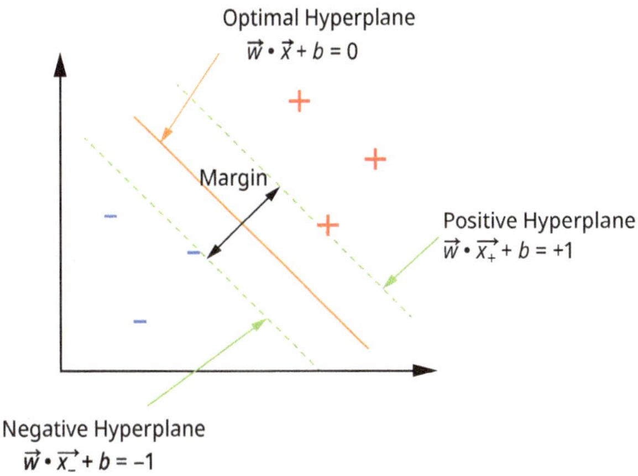

Figure 2.5: SVM hyperplane.

2.7.2 Portfolio Optimization

In finance, investors often seek to optimize their investment portfolios to achieve a desirable risk-return trade-off. Quadratic programming can be used to model portfolio optimization problems where the objective is to minimize portfolio risk (variance) subject to constraints on expected returns and portfolio weights.

2.7.3 Control Systems

In control theory, quadratic programming is used to design optimal control policies for dynamical systems subject to quadratic performance criteria.

2.7.4 Image Processing

Quadratic programming can help with image processing tasks like removing noise, segmenting images, and aligning images. For example, quadratic programming formulations are used in certain variational image processing models to minimize energy functions that consist of data fidelity terms and regularization terms.

2.7.5 Computer Vision

Quadratic programming is used in computer vision applications such as camera calibration, stereo matching, and object tracking. For instance, quadratic programming formulations are used in camera calibration algorithms to minimize reprojection errors between observed image points and their corresponding 3D world points.

Quadratic programming provides a flexible framework for solving optimization problems with quadratic objectives and constraints, making it applicable to a wide range of AI applications. It allows AI systems to make optimal decisions and solve complex optimization tasks efficiently.

2.8 Nonlinear Programming in AI

Nonlinear programming is an optimization technique used for problems where the objective function or constraints are nonlinear. Unlike linear programming, which only deals with straight-line relationships, nonlinear programming handles more complex relationships like quadratic, exponential, logarithmic, or trigonometric functions. Solving these problems often requires special algorithms designed for nonlinear cases. Common methods include gradient descent, Newton's method, and quasi-Newton methods. Nonlinear programming is a powerful tool in AI for tackling problems with nonlinear objectives or constraints [13]. Here are some common applications of nonlinear programming in AI.

2.8.1 Neural Network Training

Training deep neural networks involves optimizing a nonlinear objective function (such as cross-entropy loss and mean squared error) with respect to the network parameters. Techniques such as gradient descent, Adam, and other optimization algorithms are used to minimize the loss function.

2.8.2 Optimal Control

Nonlinear programming is used in optimal control problems to find control policies that optimize the performance of dynamical systems subject to nonlinear dynamics and constraints. Examples include trajectory optimization for robotic motion planning and control.

2.8.3 Parameter Estimation

Nonlinear programming is used in parameter estimation problems to find the parameters of mathematical models that best match the observed data. Examples include fitting nonlinear regression models, curve fitting, and system identification.

2.8.4 Simulation-Based Optimization

Nonlinear programming is used in simulation-based optimization problems where the objective function is evaluated through simulation or numerical methods. Examples include the optimization of complex engineering systems, such as aerospace design, automotive design, and chemical process optimization.

2.8.5 Machine Learning

Nonlinear programming techniques are used in various machine learning tasks, such as kernel methods, Gaussian process regression, and nonlinear dimensionality reduction techniques.

2.8.6 Computer Vision

Nonlinear programming is used in computer vision tasks such as image registration, image alignment, and camera calibration. Optimization techniques are employed to align images, estimate camera parameters, and solve geometric vision problems.

2.8.7 Natural Language Processing (NLP)

Nonlinear programming techniques are used in language modelling, machine translation, and sentiment analysis. Optimization algorithms are employed to train models and optimize performance metrics.

Nonlinear programming offers a flexible approach for solving complex optimization problems with nonlinear objectives and constraints, enabling AI systems to make the best decisions across various applications.

2.9 Mixed Integer Linear Programming (MILP) in AI

It is an optimization technique where some decision variables are allowed to be integers while others remain continuous, and both the objective function and constraints are linear. This combination of continuous and discrete decision variables allows for modelling a wide range of real-world problems with additional flexibility and precision. Mixed integer linear programming (MILP) is a powerful optimization technique used in AI for solving optimization problems where some variables are required to be integers while others are allowed to be continuous [14]. Here are some common applications of MILP in AI.

2.9.1 Combinatorial Optimization

Integer Knapsack Problem: Optimize the selection of items with integer weights and values to maximize the total value while satisfying a weight constraint.

2.9.2 Scheduling and Planning

Resource Allocation: Allocate limited resources such as personnel, equipment, or budget to different tasks or projects while considering dependencies and constraints.

2.9.3 Network Design and Routing

Facility Location: Choose the best locations for facilities like warehouses or factories to minimize transportation costs or maximize coverage, where the decision variables represent binary choices (open/close) or integer capacities.

Network Flow with Fixed Costs: Optimize the flow of goods, information, or resources through networks with fixed costs associated with opening or closing arcs or nodes.

2.9.4 Machine Learning

Feature Selection with Binary Variables: Select a subset of features or variables to improve the performance of machine learning models, where binary variables indicate the inclusion or exclusion of each feature.

Discrete Hyperparameter Optimization: Tune hyperparameters of machine learning models using integer variables to represent discrete choices.

2.9.5 Game Theory

Coalition Formation: Optimize the formation of coalitions in cooperative games to obtain the highest total benefit for everyone, where binary variables show whether a coalition is formed or not.

2.9.6 Supply Chain Management

Inventory Management: Optimize inventory levels in a supply chain to minimize costs while ensuring service level requirements are met, where inventory levels are integer variables.

MILP provides a flexible and efficient framework for modelling and solving complex optimization problems with mixed integer and continuous variables. It allows AI systems to make optimal decisions from logistics and manufacturing to machine learning and game theory.

2.10 Dynamic Programming in AI

Dynamic programming (DP) can significantly reduce the time complexity of solving optimization problems by avoiding redundant computations and leveraging the optimal substructure property. However, it may require careful problem decomposition and understanding of the underlying recurrence relations. DP is a powerful technique used in AI for solving optimization problems by breaking them down into smaller subproblems and efficiently solving each subproblem only once, typically using a recursive or iterative approach [15]. Here are some common applications of DP in AI.

2.10.1 Optimal Control

In RL, DP techniques such as value iteration and policy iteration are used to compute optimal policies for MDPs. These techniques involve iteratively updating value functions or policies until convergence to find the optimal solution. The following formula is used in MDPs:

The expected return (value) at the current state *s* is:

The expected reward for taking action *a* at state *s*...

$$V(s) = \max_a \left(R(s, a) + \gamma V(s') \right)$$

The maximum value of any possible action *a* for:

...plus the discount factor (gamma) multiplied by the value of the next state

– **Path Planning**:
DP is used in robotics and autonomous navigation for finding optimal paths in environments with known or partially known dynamics. Algorithms such as Dijkstra's al-

gorithm, Bellman-Ford algorithm, and A* search algorithm use DP principles to efficiently find shortest paths.

– **Sequence Alignment**:
DP is used in bioinformatics for sequence alignment tasks such as sequence similarity search, DNA sequencing, and protein structure prediction. Algorithms such as the Needleman-Wunsch algorithm and the Smith-Waterman algorithm use DP to find optimal alignments between sequences.

– **Language Processing**:
DP techniques are used in natural language processing (NLP) for tasks such as speech recognition, machine translation, and text summarization. Algorithms for sequence-to-sequence modelling, such as sequence alignment and attention mechanisms, use DP principles for efficient inference.

– **Game Theory**:
DP is used in game theory for solving sequential decision-making problems and computing optimal strategies in games. Techniques such as backward induction and the minimax algorithm use DP principles to compute optimal strategies for players in games with sequential moves.

– **Optimization**:
DP is used in various optimization problems, including resource allocation, scheduling, and portfolio optimization. Algorithms such as the knapsack problem, the travelling salesman problem (TSP), and the maximum subarray problem use DP principles to efficiently find optimal solutions.

2.10.1 Two Main Approaches of Dynamic Programming

- **Top-Down:** Solve the problem by breaking it into smaller parts and storing their solutions to avoid repeating work.
- **Bottom-Up:** Solve the problem by starting with the smallest parts and building up to the larger ones, filling out a table as you go.

DP provides a powerful and general-purpose approach for solving optimization problems in AI, enabling efficient computation of optimal solutions in a wide range of applications.

2.11 Combinatorial Optimization in AI

It is a branch of optimization that deals with discrete decision variables and seeks to find the best arrangement or selection from a finite set of possible options. It involves optimizing over combinatorial structures such as permutations, combinations, graphs, or networks. Techniques such as DP, greedy algorithms, branch and bound, local search, genetic algorithms (GAs), and integer linear programming are commonly used depending on the problem characteristics and constraints [7]. Here are some common applications of combinatorial optimization in AI:

- **TSP:** Find the shortest route that visits each city once and returns to the start. This is used in logistics and network routing.
- **Bin Packing Problem:** Pack items of different sizes into bins with fixed capacities to minimize the number of bins used. This is applied in resource allocation and packing.
- **Job Scheduling:** Schedule tasks with specific processing times and resource needs to minimize completion time or maximize resource use. This is relevant in manufacturing and project management.
- **Graph Colouring:** Assign colours to graph vertices so that no two adjacent vertices share the same colour. This is used in scheduling and frequency assignment.
- **Maximum Flow Problem:** Determine the maximum flow from a source to a sink in a network while respecting edge capacities. This is important in network routing and design.
- **Knapsack Problem:** Choose items with given values and weights to maximize the total value without exceeding a weight limit. This is applied in resource allocation and production planning.
- **Vehicle Routing Problem:** Find routes for vehicles to serve customers while minimizing the total travel distance or time. This is used in logistics and fleet management.

Due to the discrete nature of the decision variables, combinatorial optimization problems can be computationally challenging, especially for large instances. However, efficient algorithms and heuristic methods have been developed to tackle many combinatorial optimization problems effectively. Combinatorial optimization techniques are essential in AI for solving complex problems efficiently and finding optimal or near-optimal solutions in a wide range of domains. These techniques often involve heuristic algorithms, metaheuristics, and exact algorithms tailored to specific problem structures.

2.12 Global Optimization in AI

This involves finding the best solution to a problem across the entire possible range of solutions, not just locally. It deals with challenges like multiple local optima and complex functions.

2.12.1 Key Approaches

- **Deterministic Methods:** Systematically search the solution space, such as branch and bound, interval analysis, and grid search.
- **Stochastic Methods:** Use randomness to explore solutions more effectively, like GAs, simulated annealing, and particle swarm optimization (PSO).
- **Metaheuristic Methods:** Use high-level strategies for guiding the search, including tabu search, ant colony optimization, and evolutionary algorithms.

2.12.2 Techniques

- **GAs:** Evolve solutions using methods inspired by genetics, useful for complex problems. A simple GA cycle is shown in Figure 2.6.

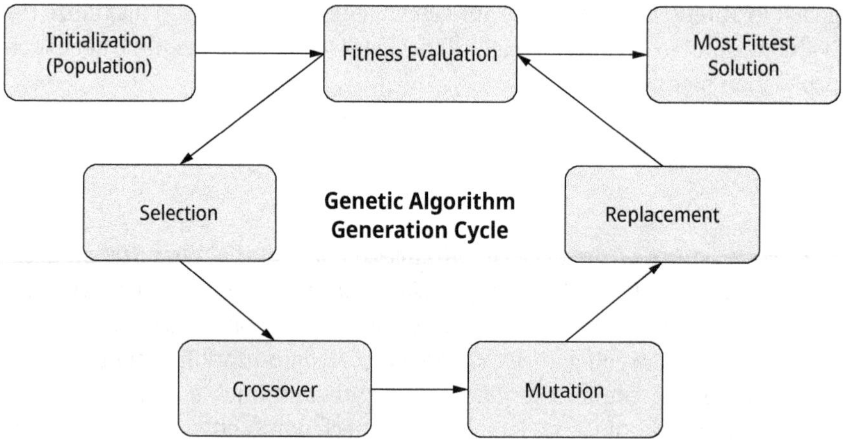

Figure 2.6: Simple genetic algorithm cycle.

- **PSO:** This optimizes by having particles move based on their own best positions and the best found by the group. PSO can be implemented using the following formula:

$$P_i^{t+1} = P_i^t + V_i^{t+1}$$

$$V_i^{t+1} = \underbrace{wV_i^t}_{\textbf{Inertia}} + \underbrace{c_1r_1\,(P^t_{best(i)} - P_i^t)}_{\textbf{Cognitive (Personal)}} + \underbrace{c_2r_2\,(P^t_{bestglobal} - P_i^t)}_{\textbf{Social (Global)}}$$

- **Simulated Annealing:** This is a probabilistic method that explores solutions by sometimes accepting worse options to avoid getting stuck.
- **Tabu Search:** This uses a memory of past solutions to avoid revisiting them and guides the search to new areas.
- **Random Search:** This samples solutions randomly to find the best one.
- **Bayesian Optimization:** This uses probabilistic models to guide the search towards the best areas, and is useful for expensive problems.

These techniques are important in AI for solving complex problems with challenging functions and many local optima, used in areas such as machine learning, engineering, finance, and logistics. Finding the best solution can be hard and may need a lot of computing power.

2.13 Convex Optimization in AI

It is a special class of optimization problems where both the objective function and the feasible region are convex. Convexity is a property that ensures unique global optimality, meaning that any local optimum is also a global optimum [18]. Convex optimization problems have several desirable properties, including:

- **Guaranteed Global Optimality:** Any local optimum is also a global optimum.
- **Efficient Solution Methods:** Convex optimization problems can often be solved efficiently using specialized algorithms such as interior-point methods or gradient-based methods.
- **Well-Understood Theory:** Convex optimization theory provides a solid mathematical foundation for understanding and solving optimization problems.

Due to their favourable properties, convex optimization problems are widely studied and applied in practice. Convex optimization plays a significant role in AI for solving optimization problems where both the objective function and the feasible region are convex. Convex optimization is particularly attractive because it guarantees global optimality, efficient algorithms, and tractable mathematical properties. Here are some common applications and characteristics of convex optimization in AI:

- **Machine Learning**:
  - **Logistic Regression**: Convex optimization is used to optimize the logistic loss function in binary classification problems, where the objective is to minimize the negative log likelihood of the observed labels.
  - **Regularization**: Convex optimization is used to incorporate regularization terms such as L1 and L2 penalties into machine learning models to prevent overfitting and encourage sparsity.
- **Signal Processing**:
  - **Sparse Signal Recovery**: Convex optimization techniques like basis pursuit and Lasso are used to recover sparse signals from noisy measurements, such as in compressed sensing and sparse representation problems.
  - **Filter Design**: Convex optimization is used to design linear filters with desired frequency responses and constraints, such as minimum-phase filters, linear-phase filters, and finite impulse response filters.
- **Control Systems**:
  - **Optimal Control**: Convex optimization is used to solve optimal control problems. Model predictive control is a common application of convex optimization in control systems.
- **Finance**:
  - **Portfolio Optimization**: Convex optimization techniques such as mean-variance optimization (Markowitz portfolio theory) are used to construct optimal investment portfolios that maximize expected returns while minimizing risk, subject to constraints on asset allocation and risk tolerance.

2.13.1 Operations Research

Network Flow Problems: Convex optimization is used to solve various network flow problems, which arise in transportation, logistics, and telecommunications.

Convex optimization provides a solid theoretical foundation and efficient algorithms for solving a wide range of optimization problems in AI and related fields. Its mathematical properties, such as convexity, enable the development of reliable and scalable optimization techniques that underpin many AI applications.

2.14 Stochastic Optimization in AI

It deals with optimization problems in the presence of uncertainty or randomness. Stochastic optimization provides valuable insights and solutions for decision-making under uncertainty, enabling decision-makers to make more informed and robust choices in complex, uncertain environments. Stochastic optimization is a branch of optimization that deals with optimization problems involving uncertainty or randomness in the objective function, constraints, or parameters. It is used to solve complex problems where deterministic approaches may be impractical or ineffective [17]. Here are some common characteristics and applications of stochastic optimization in AI:

- **Uncertain Environments**:
  - Stochastic optimization is well-suited for problems such as noisy measurements, fluctuating demand, or unpredictable system dynamics.
- **Global Optimization**:
  - Stochastic optimization techniques are often used to find global optima or near-optimal solutions in optimization problems with nonconvex, multimodal, or high-dimensional search spaces, where deterministic methods may get trapped in local optima.
- **Metaheuristic Algorithms:**
  - These include methods such as GAs, simulated annealing, particle swarm optimization, and evolutionary strategies. They use randomness to search for the best solutions and avoid getting stuck in less optimal ones.
- **Bayesian Optimization:**
  - This method uses probabilistic models to find the best areas to explore when solving complex problems where evaluating the solution is expensive.
- **RL:**
  - RL methods, such as Q-learning and deep Q-networks (DQN), use random techniques to learn the best actions through interaction with the environment.
- **SGD:**
  - SGD is a popular method for training machine learning models. It updates the model using small, random samples of data, making it efficient for large problems.
- **Robust Optimization**:
  - Stochastic optimization techniques are used to formulate and solve robust optimization problems that aim to find solutions that are robust to uncertainties or variations in the problem parameters. Robust optimization ensures the solution's performance under different scenarios or perturbations.

Stochastic optimization provides a flexible and powerful framework for solving complex optimization problems in AI, enabling robust and adaptive decision-making in uncertain and dynamic environments. By incorporating randomness and uncertainty into the optimization process, stochastic optimization techniques enable AI systems to handle real-world complexities and uncertainties more effectively.

2.14.1 Uses of MO techniques in AI

By employing these optimization techniques, AI practitioners can enhance the efficiency, effectiveness, and scalability of AI systems across a wide range of applications and domains [19].

- **Using specialized hardware**: Leveraging hardware acceleration with graphics processing units or field-programmable gate arrays speeds up computationally intensive AI tasks, such as training and inference.
- **Using pre-trained foundation models**: Use pre-trained models as starting points for specific tasks and fine-tune them to adapt to new data or tasks, reducing the need for training from scratch.
- **Reducing network size and complexity**: Pruning redundant connections or layers in neural networks reduces the model size and computational overhead while maintaining performance.
- **Compression techniques**: Employing compression techniques such as Huffman coding reduces the size of model representations, saving memory and storage space.
- **Efficient algorithms and data structures**: Adopting more efficient learning algorithms and data structures minimizes computational resources and improves training and inference speed.
- **Data preprocessing**: Cleansing, normalizing, and transforming data enhance the model training efficiency and effectiveness.
- **Model architecture optimization**: Adjusting the model architecture strikes a balance between complexity and performance, optimizing for specific tasks or hardware constraints.
- **Overfitting reduction**: Employing techniques reduces overfitting and improves model generalization to unseen data.
- **Precision reduction**: Decreasing the precision of model weights reduces memory consumption and improves inference speed, especially in resource-constrained environments.
- **Early stopping**: Stopping model training before convergence prevents overfitting and improves generalization.
- **Deployment optimization**: Optimizing model deployment platforms or runtime environments maximizes performance in production settings.
- **Performance monitoring and analysis**: Continuously monitoring and analysing model performance identifies bottlenecks and areas for improvement, enabling iterative optimization.

2.14.2 Optimization Levels

Optimization levels in computing provides a comprehensive overview of how optimization occurs at various stages of the development process, from high-level problem formulation to low-level hardware implementation [15–17].

- **Problem-Level Optimization**:
  - This involves defining the problem to be solved, considering available resources, constraints, and goals. It is essentially the design phase where the solution approach is conceptualized.
- **Algorithm-Level Optimization**:
  - After defining the problem, selecting the appropriate algorithm and data structures to implement the solution is crucial. Choosing efficient algorithms and data structures is essential for achieving optimal performance [18].
- **Language or Source Code-Level Optimization**:
  - Optimizing the source code at the programming language level can significantly impact performance. Choices such as loop structures, conditional statements, and data types can affect execution speed and resource utilization [19, 20].
- **Build-Level Optimization**:
  - Build-level optimization involves adjusting performance options in the source code and compiler to improve efficiency. Techniques such as pre-processing and branch prediction can be utilized at this level.
- **Assembly-Level Optimization**:
  - Optimizing the code at the assembly level involves writing or modifying the code directly in machine-readable instructions. This level of optimization can achieve fine-grained control over performance but requires expertise and careful consideration.
- **Microarchitecture and Hardware-Level Optimization**:
  - Optimization at the microarchitecture and hardware levels focuses on designing and implementing efficient CPU architectures and circuits. Techniques such as out-of-order execution, cache optimization, and branch prediction are employed to improve performance and reliability.
- **Trade-Offs and Bottlenecks**:
  - Optimization involves trade-offs between different metrics such as execution time, power consumption, memory usage, and complexity. Identifying bottlenecks in the system is crucial for prioritizing optimization efforts and achieving overall performance improvements.

By understanding and optimizing at each level of the computing stack, developers can achieve significant improvements in the efficiency, performance, and reliability of their systems [21].

2.15 Conclusion

In conclusion, the pursuit of optimization in AI encapsulates the essence of striving for the finest solutions. The objectives outlined for MO in AI underscore its multifaceted nature, from understanding problem types to implementing widely used algorithms. Diving into optimization basics reveals its intricate facets, from delineating maxima and minima to elucidating learning rates and norms on vectors. Understanding derivatives and their role in capturing local behaviour underscores the mathematical rigour inherent in optimization pursuits. Constrained optimization tasks involve finding optimal solutions while adhering to specified constraints, addressing real-world limitations in resource allocation, system design, and decision-making processes.

From ML to NLP, RL to computer vision, constrained optimization techniques enable AI systems to navigate complex environments and deliver reliable outcomes. Unconstrained optimization methods focus on optimizing objective functions without explicit constraints, driving advancements in deep learning, image processing, and RL. MO techniques facilitate the training of AI models and the generation of high-quality outputs, underpinning innovations in various AI-driven applications. Moreover, linear programming emerges as a foundational tool in AI, offering a systematic approach to solving problems with linear objectives and constraints. From resource allocation to network optimization, linear programming techniques empower AI systems to make informed decisions and optimize processes efficiently.

Furthermore, integer programming introduces additional complexity by incorporating discrete decision variables into optimization tasks, addressing combinatorial challenges in scheduling, routing, and portfolio management. Its versatility enables AI systems to tackle diverse optimization problems, ranging from combinatorial optimization to strategic decision-making in game theory. In essence, optimization techniques serve as the backbone of AI systems, driving progress and innovation across a wide spectrum of applications. By harnessing the power of constrained and unconstrained optimization, linear programming, and integer programming, AI continues to revolutionize industries, improve decision-making processes, and push the boundaries of what is possible in the realm of intelligent systems.

Collectively, the optimization techniques form the backbone of AI, empowering systems to make optimal decisions, tackle complex problems, and drive innovation across a wide array of applications. As AI continues to evolve, the interplay between optimization methodologies and intelligent systems will remain central to unlocking new frontiers and addressing real-world challenges with precision and efficiency. Optimization levels in computing offer valuable insights into the layers where efficiency and performance enhancements can be achieved. Starting from the conceptualization of the problem and algorithm selection to fine-tuning at the assembly and hardware levels, optimization efforts span the entire development spectrum. Trade-offs and bottlenecks provide the importance of balancing competing objectives and identifying critical areas for improvement.

By strategically optimizing at each level, developers can achieve significant advancements in system efficiency, performance, and reliability, aligning with the evolving demands of AI applications and technologies. As AI continues to evolve, optimization will remain a cornerstone, guiding the path towards ever-improving intelligence and efficiency. Optimization problems lie at the heart of numerous AI applications, playing a pivotal role in enhancing efficiency, robustness, and performance across diverse domains.

Conflicts of Interest: The authors have no conflicts of interest to declare.

References

[1] Wang, Y., Fu, E. Y., Zhai, X., Yang, C., & Pei, F. (2024). Introduction of artificial Intelligence. In Intelligent Building Fire Safety and Smart Firefighting (pp. 65–97). Springer Nature Switzerland, Cham.

[2] Reddy, C. S., & Anjan Babu, G. (2024). Deep Insights into Artificial Intelligence and Machine Learning Algorithms for Computational and Mathematical Data Processing.

[3] Khaleel, M., Jebrel, A., & Shwehdy, D. M. (2024). Artificial Intelligence in Computer Science. *International Journal of Electrical Engineering and Sustainability*, 01–21. https://doi.org/10.5281/zenodo.10937515.

[4] Davis, E. (2024). Mathematics, word problems, common sense, and artificial intelligence. *Bulletin of the American Mathematical Society*, *61*(2), 287–303

[5] Carrizosa, E., Ramírez-Ayerbe, J., & Morales, D. R. (2024). Mathematical optimization modelling for group counterfactual explanations. *European Journal of Operational Research*.

[6] Tao, H., Abba, S. I., Al-Areeq, A. M., Tangang, F., Samantaray, S., Sahoo, A., Valadares Siqueira, H. et al. (2024). Hybridized artificial intelligence models with nature-inspired algorithms for river flow modeling: A comprehensive review, assessment, and possible future research directions. *Engineering Applications of Artificial Intelligence*, *129*, 107559.

[7] Zhuang, X., Wang, W., Su, Y., Yan, B., Li, Y., Li, L., & Hao, Y. (2024). Multi-objective optimization of reservoir development strategy with hybrid artificial intelligence method. *Expert Systems with Applications*, *241*, 122707.

[8] Soori, M., & Ghaleh Jough, F. K. (2024). artificial intelligent in optimization of steel moment frame structures: A review. *International Journal of Structural and Construction Engineering*.

[9] Azevedo, B. F., Rocha, A. M. A. C., & Pereira, A. I. (2024). Hybrid approaches to optimization and machine learning methods: A systematic literature review. *Machine Learning*, 1–43.

[10] Zhang, X., Gajawada, Y., Satish, & Mustafa, H. M. H. (2024). Out of the box artificial intelligence (obai): The beginning of a new era in artificial intelligence. *Available at SSRN 4877418*. Wang, Y., & Wang, J. Optimize the energy storage system with an artificial intelligence-based control strategy. *Applied Mathematics and Nonlinear Sciences*, 9, 1.

[11] Bundela, B., Sharma, S., & Singh, B. R. (2024). A review article on relation between mathematical modelling and machine learning. *Zhongguo Kuangye Daxue Xuebao*, *29*(2), 123–129

[12] Kumar Lilhore, U., Simaiya, S., Sharma, Y. K., Kaswan, K. S., Brahma Rao, K. B. V., Maheswara Rao, V. V. R., Baliyan, A., Bijalwan, A., & Alroobaea, R. A precise model for skin cancer diagnosis using hybrid U-Net and improved MobileNet-V3 with hyperparameters optimization. *Scientific Reports*, *14*(1), 4299

[13] Saleem, S., Ahmad, I., Ahmed, S. H., & Rehman, A. (2024). Artificial intelligence based robust nonlinear controllers optimized by improved gray wolf optimization algorithm for plug-in hybrid electric vehicles in grid to vehicle applications. *Journal of Energy Storage*, *75*, 109332.

[14] Ali, A. M., & Mohammed, M. A. (2024). A comprehensive review of artificial intelligence approaches in omics data processing: Evaluating progress and challenges. *International Journal of Mathematics, Statistics, and Computer Science*, *2*, 114–167.

[15] Lilhore, U. K., Dalal, S., Radulescu, M., & Barbulescu, M. (2024). Smart grid stability prediction model using two-way attention based hybrid deep learning and MPSO. *Energy Exploration & Exploitation*, 01445987241266892.

[16] Nutakki, M., & Mandava, S. (2023). Review on optimization techniques and role of Artificial Intelligence in home energy management systems. *Engineering Applications of Artificial Intelligence*, *119*, 105721.

[17] Lilhore, U. K., & Simaiya, S. (2024). 10 Quantum machine learning in natural language processing: Opportunities and challenges. *Quantum Machine Learning: Quantum Algorithms and Neural Networks*, 199.

[18] Khaleel, M., Ahmed, A. A., & Alsharif, A. (2023). Artificial Intelligence in Engineering. *Brilliance: Research of Artificial Intelligence*, *3*(1), 32–42.

[19] Lilhore, U. K., Manoharan, P., Sandhu, J. K., Simaiya, S., Dalal, S., Baqasah, A. M., Alsafyani, M., Alroobaea, R., Keshta, I., & Raahemifar, K. (2023). Hybrid model for precise hepatitis-C classification using improved random forest and SVM method. *Scientific Reports*, *13*(1), 12473.

[20] Lilhore, U. K., Dalal, S., Faujdar, N., Simaiya, S., Dahiya, M., Tomar, S., & Hashmi, A. (2024). Unveiling the prevalence and risk factors of early stage postpartum depression: A hybrid deep learning approach. *Multimedia Tools and Applications*, 1–35.

Dankan Gowda V, Gedas Kučinskas, Pullela S. V. V. S. R. Kumar, Rahul Vadisetty, and Aparna Atul Junnarkar

3 Robust Optimization Methods for Ensuring AI System Reliability

Abstract: This chapter interrogates how strong optimization is critical in improving the reliability of artificial intelligence (AI) in different applications. The need to create smart systems capable of operating stably in conditions that are unexpected and unpredictable has become especially acute as AI technologies reinforce their position in healthcare, automotive industry, finance, and manufacturing, among others. Through an analysis of conventional and revised optimization methods, this text examines the advancements of efficient risk management strategies in handling the incongruities of operational risks. A concise overview of the theoretical foundations of various techniques for handling robust optimization such as robust linear programming, convex optimization, and Bayesian optimization is provided, eliciting the relevance of these approaches in handling intricate decision-making problems. Theoretical concepts are applied through real examples in the domains of self-driving cars and diagnostic services, and in finance with main ideas illustrating how strong superior optimization results in vast enhancements in productivity as well as reliability. Appropriate data are given for these case studies with stress on the applicability of robust optimization for achieving high-performance figure of merits despite external changes ailing most operations. The best way to compare robust optimization to the other methods is by looking at the examples that show how much more effective robust optimization is especially within the conditions of high risk and uncertainty.

Keywords: Robust optimization, AI reliability, stochastic programming, autonomous systems, healthcare AI, financial AI, robust control theory, scenario optimization

3.1 Introduction

In the rapidly evolving field of artificial intelligence (AI), the issues of reliability and robustness of systems are of paramount concern as these systems are being used in or affect more and more areas having important consequences. Across the sectors that range from healthcare to autonomous cars, and finance to manufacturing [1], AI applications are required to work in stable or even operate stably through unpredictable or harsh environments that vary vastly. Whichever of these sectors is utilized, reliability is the key focus due to the disheartening escalation of failure scenarios that may lead to loss of funds [2], exposure of human beings to risky terrains, and loss of public confidence in AI technologies [3]. This requires reliable optimization techni-

https://doi.org/10.1515/9783111436180-003

ques for its improvement with regard to its strength and capability to perform in various conditions and complexity of operations.

3.1.1 Role of Reliability in AI Systems

Reliability of AI systems is therefore about whether these technologies will act as required at this time and also the next time when the need arises [4]. For AI systems, the reliability is not just the measurement of their function's ability to be correct but also the ability to perform at optimum levels when faced with variabilities and randomness that exist in the environment and data [5]. In the case of autodrive, for example, an AI system must analyze sensor information to make proper decisions that would allow for the safe operation of a car under different road and weather conditions.
Mathematically, this can be expressed in terms of probability:

$$P(\text{safe decision} \mid \text{conditions}) \geq \theta$$

where θ represents a threshold probability that ensures decisions are safe with high confidence.

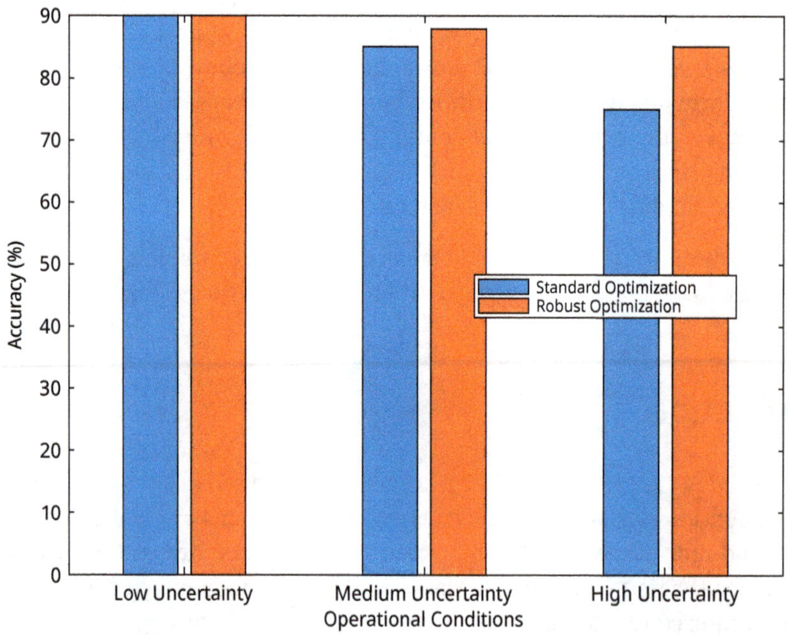

Figure 3.1: Performance comparison of AI systems: standard versus robust optimization.

Figure 3.1 illustrates the comparison of performances of the developed AI systems under the standard and the proposed robust optimization in different operating conditions [6]. This bar chart well illustrates how reliable and less variant the application of robust optimization is in low uncertainty, medium uncertainty, and high uncertainty. Through the performance figures of accuracy, time, and errors shown in the visualization [7], the idea of reliability, capability, and stability offered by robust optimization in providing high performance in the face of outside disturbances is portrayed.

Figure 3.2 illustrates the timeline of the development of optimization methods in AI and presents linear programming as well as nowadays robust optimization [8]. This flowchart defines certain achievements and methodologies of the decades as it allows the reader to have a contextual view of how these theories have progressed to fit in with today's sophisticated computational systems for the many uses of AI.

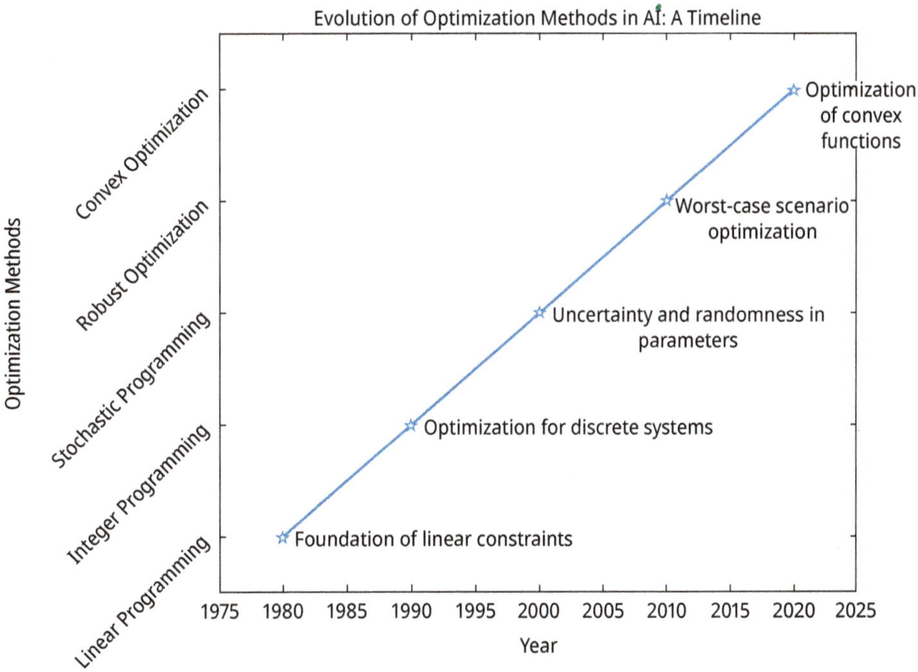

Figure 3.2: Evolution of optimization methods in AI: a timeline.

3.1.2 Importance Across Sectors

Healthcare: AI system used in the healthcare setting, where the diagnostic algorithm and an assisting robot used in surgery, should be very reliable, as it shapes patients' status [9]. An AI diagnostic tool must maintain high accuracy rates across diverse patient populations and medical imaging technologies, potentially formalized as follows:

$$\text{Accuracy} = \frac{\text{TN} + \text{TP}}{\text{TN} + \text{FN} + \text{FP} + \text{TP}}$$

where TP, TN, FP, and FN represent true positives, true negatives, false positives, and false negatives, respectively. The stability of these systems is very important in order to reduce chances of wrong diagnoses and therefore wrong treatment.

Autonomous Vehicles: AI systems must be able to make crucial decisions in the field of autonomous driving; they must process the different types of sensor data [10]. It is not only a problem in the right interpretation of the data but also in ensuring that the system is safe under varying traffic conditions [11]. This will require solving difficult optimization problems to minimize the risk involved in errors as well as accidents.

Financial Services: Specifically in finance, AI has found its application in areas such as algorithmic trading, credit risk analysis, and customer support. These systems affect the stability of the companies' finances as well as compliance with regulations [12]; therefore, the optimization must effectively deal with fluctuations in the market and various types of transactions.

Objectives of the Chapter

It is the goal of this chapter to explore the methodologies that are foundational to the field of robust optimization in AI so that the reader can gain a meta-level view of both theory and practice. The chapter will explain the following:

Theoretically Explore Robust Optimization: Present an overview of reliable optimization, starting with the formulation of optimization problems under conditions of uncertainty [13]. This includes discussions on min-max optimization problems, where the objective is to minimize the maximum potential loss:

$$\min x \in X \max y \in Y f(x,y)$$

Detail Practical Implementations: Explain how the methods of robust optimization are used in different types of AI, and explain different case studies in which they were successfully implemented.

Highlight Sector-Specific Challenges: Explain the difficulties that have to be faced and/or solved for obtaining high-quality optimization results in different sectors, emphasizing that the methods should be different based on specific characteristics of the definite application of AI.

Propose Future Research Directions: List the limitations of the current approaches and suggest further research topics that can help advance the progress in developing more reliable AI.

3.2 Literature Survey

Another interesting field that has also undergone a major shift is optimization in AI, which is mainly due to improved techniques in computation as well as theoretical frameworks [14]. This evolution has been characterized by a shift from classical, planet-wide predictable methods that used deterministic models to modern, world-wide unpredictable methods that include robustness for coping with uncertainty and variability in the actual world [15]. This section encompasses the details of this evolution, emphasizing the conventional and modern robust optimization techniques and finally the current issues and constraints regarding the reliability of AI systems.

3.2.1 Historical Overview

Optimization in AI can be discussed as an evolution from linear programming and classical decision theory, under which the critical objectives were formulated as well-defined and, to a significant degree, static, which has to be solved under known constraints and parameters [16]. These methods developed into more complex ones such as integer programming, dynamic programming, and nonlinear optimization techniques with the emergence of greater computational capabilities and with the increase in the area of applications of AI [17]. These methods, which set the basis for handling more complicated issues, made the assumption that the values of parameters and surroundings did not change [18], which in the real world is far from the truth when it comes to advanced AI operations.

Due to these factors, the field of stochastic optimization came into being that work under the conditions of stochastic and uncertainty in the parameters [19]. One of the pioneering techniques used in this field is stochastic programming, which solves the problem of minimizing or maximizing a certain function when the parameters of the problem are random variables depending on other variables or probability distributions [20]. This can be represented as follows:

$$\min xE[f(x,\xi)]$$

where x is the decision variable, f is the objective function, and ξ represents the random variables with known probability distributions.

Figure 3.3 displays the 3D surface plot of a solution space in the domain of a robust optimization problem. This visualization illustrates the most intricate world of optimization as each function of the parameters results in a different solution to the objective function [21]. The plot aims at illustrating how robust optimization works through this multifaceted space, where key evaluation metrics are likely to be found in order to achieve the best compromise between cost and performance in situations where there is a lot of uncertainty [22].

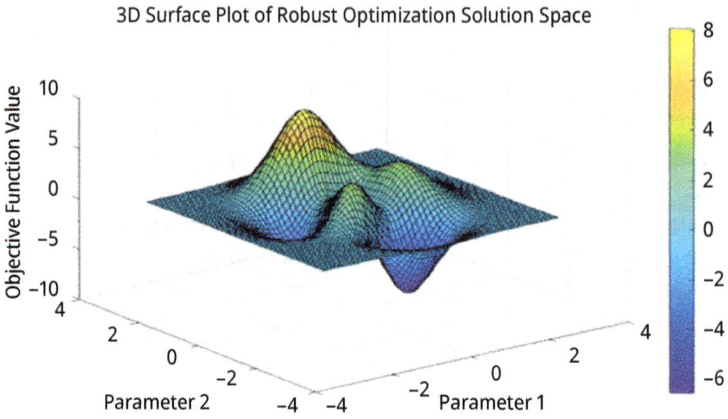

Figure 3.3: 3D surface plot of robust optimization solution space.

Figure 3.4 shows the contour plots of robust linear programming solution, and the feasible regions under different uncertainty parameters are illustrated [23]. This figure gives a clear illustration of how the algorithms of robust optimization respond to variations in constraint limits, thus guaranteeing that solutions are still viable in instances of changes in constraint boundaries. The contour lines highlight the regions of equivalent objective values in order to reveal the effects of sound decisions within a limited set of parameters.

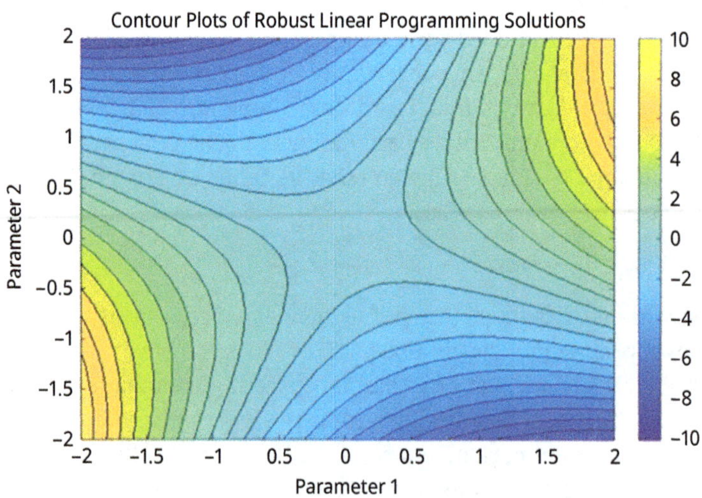

Figure 3.4: Contour plots of robust linear programming solutions.

3.2.2 Current Techniques

Modern robust optimization approaches have shifted the emphasis from stochastic contexts to worst-case and adversarial contexts, especially in those contexts where risk and reliability are vital. Thus, there are essential techniques such as robust control theory and worst-case scenario analysis [24]. Robust control theory, for instance, deals with the problem of maintaining system performance in the presence of bounded uncertainties and is often formulated as follows:

$$\min x \in X \max \delta \in \Delta f(x, \delta)$$

where δ represents bounded uncertainties and X is the set of all possible control actions [25]. This method ensures that the worst possible impact of uncertainties is mitigated.

The worst-case scenario analysis takes this concept even further by not only looking at the worst-case scenario but also making the system ready for it on its own without making a mistake [26]. This approach is quite important in such areas as cybersecurity and autonomous navigation and driving because, here, potential threats should be anticipated and controlled to prevent adverse impacts [27].

3.2.3 Challenges and Limitations

However, some issues still need to be solved even with the help of the robust optimization theory. One major drawback is the fact that the computation of robust solutions can be computationally intensive, especially when solving complex and highly uncertain systems [28]. This is especially noticeable in live AI-related applications, for instance, self-driving, where decisions have to be made at the blink of an eye [29].

The former includes the trade-off between its ability to be highly resistant to faults and its ability to perform optimally [30]. The conservative models may assume the risks and maximize usable key components of effectiveness and dependability, but at the same time, they reduce the rate of performance or formation speed. For instance, an autonomous vehicle that has been designed and trained to account for all the negative driving scenarios will make a car overly cautious and hence result in slow movement or unnecessary slowing down of the vehicle [31]. Moreover, the approach when uncertainties are modeled and their assumptions influence the characteristics of the obtained optimization solutions. Even if it is prone to errors in distribution or the range of uncertainties, it may prove to be overly cautious in its solutions or alternatively take excessive risks.

3.3 Robust Optimization Techniques

It is a vital area in the advancement of solid AI frameworks, given the useful demand to make decisions based on approaches that would be viable in considerable conditions. This section looks at some of the major techniques used in robust optimization such as robust linear programming, convex optimization, and Bayesian optimization, which outlines how these techniques could be applied in actual systems in AI and also looks at the tools and software that aid in designing these techniques.

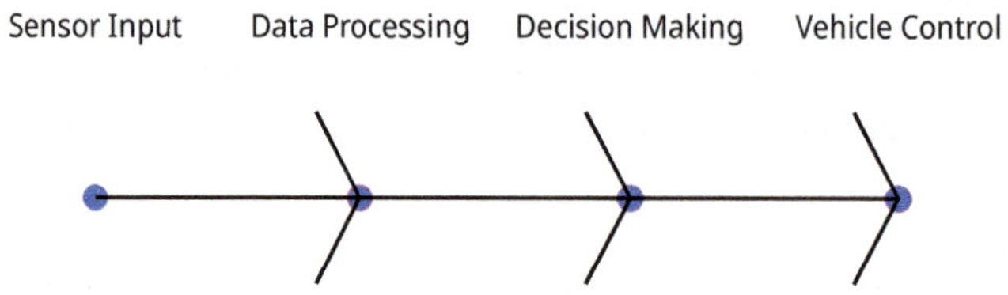

Figure 3.5: Robust optimization framework for autonomous vehicle collision avoidance.

The details of the decision-making of autonomous vehicles are illustrated in Figure 3.5, employing a reliable optimization model. This diagram elucidates the steps from receiving the sensor input up to controlling the vehicle and underscores how robust optimization works at every level to improve the decision quality and safety. The chart makes to understand the structural steps to process and operate on the real-time data, which are essential for the Nav systems [32].

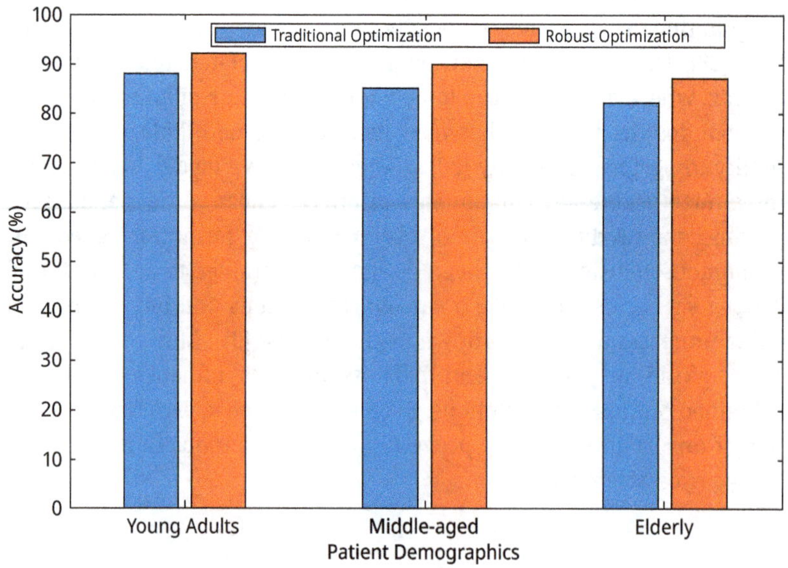

Figure 3.6: Accuracy of healthcare AI systems: robust versus traditional methods.

Figure 3.6 demonstrates the diagnostic accuracy of the developed healthcare AI systems based on traditional optimization techniques versus robust optimization techniques for different patient populations. The grouped bar chart has been used to illustrate that how robust optimization leads to increased accuracy and decreased variability to increase the reliability of diagnostics that is crucial for patient care. This figure emphasizes the fact that in the sphere of healthcare, it is critical to have practical and well-preponderant methods because precision plays a significant role.

3.3.1 Theory and Methods

Robust Linear Programming: Linear programming whose rigidity is hinged on certainty is known as linear programming, while robust linear programming is an enhancement to linear programming by factoring in uncertainties in the coefficients of the constraints or the objective function. In contrast to typical linear programming [33], which considers precise data, robust linear programming poses the problem in such a manner to be as effective as possible even under the worst-case data vagueness. The general form of a robust linear programming problem can be expressed as follows:

$$\min\{cTx: Ax \leq b, \ \forall A \in A, b \in B\}$$

Here, A and B represent uncertainty sets for the matrix A and the vector b, respectively [34]. What the company wants to do is to make an x that would yield the minimum cost, given that it meets all the constraints for all possible realizations within the said uncertainty sets.

Convex Optimization: Convex optimization is an area of mathematical optimization, which is a kind of operations research, on convex function, which could be described as a function on which the line connecting any two points on the curve globally does not fall below the curve at any point. This is so because convex problems are particularly suitable in robust optimization problems since they provide assurance of finding the true optimum solution [35]. A typical robust convex optimization problem is structured to minimize a convex objective function subject to convex constraints that hold for all perturbations in a given set:

$$\min\{f(x): gi(x, ui) \leq 0, \ \forall ui \in Ui, \ i = 1, \ldots, m\}$$

Each $gi(x, ui)$ represents a convex constraint parameterized by uncertainty ui in the set Ui.

Bayesian Optimization: This is a technique used in the global optimization of stochastic, costly, or “unknown” functions. It uses GP for probabilistic modeling of the objective function and makes decisions based on obtained posterior distribution from previous evaluations. This function that derives from the GP is used to sample according to the acquisition function, which promotes exploration as well as exploitation of regions with high potential. The update equation for a GP is

$$p(f \mid X, y, x\text{new}) = N(f\text{new} \mid \mu\text{new}, \Sigma\text{new})$$

where μnew and Σnew are the updated mean and covariance, taking into account the new data point $(x\text{new}, y\text{new})$.

3.3.2 Practical Implementation

These strong optimization ideas applied to the AI systems depend on the specific application and the type of risks that exist in the dependable application. For example, in self-driving cars, linear programming could work on route navigation that adapts to traffic condition, while Bayesian optimization could update parameters of a car's deep learning-based policy in simulated scenarios to its actual on-road scenarios.

3.3.3 Software and Tools

There are many software tools and libraries available, which can be used to implement the concept of robust optimization. Their popular choices include:

CVX and CVXPY: These are set of Python libraries that are capable of solving different convex optimization problems while providing various solvers and the ability to define problems with strong constraints.

Gurobi and CPLEX: These are powerful commercial optimization solvers that support the fortes of robust optimization models to large scales and are well adopted in industries as well as academic institutions.

GPflow and GPyTorch: These are libraries for Gaussian process modeling, which are typically applied in handling complex and noisy objective functions in Bayesian optimization.

3.4 Case Studies with Real-Time Applications

The application of robust optimization is vital in improving the resilience of AI systems solution across all industries. This section presents three detailed case studies such as automotive AI such as self-driving cars, AI systems in healthcare, and AI in finance, which show how reliable optimization techniques are crucial in enhancing decisions under risk in each field.

3.4.1 Autonomous Vehicles

In the case of self-driving autonomous vehicles, robust optimization is a crucial element in allowing self-driving cars to work in conditions that can be challenging. The main problem in this case is the autonomous vehicle's capacity to make decisions under conditions of constantly changing input from multiple sensors, such as cameras, radar, and LiDAR systems. These decisions have to be correct even in less than favorable conditions like unpredictable weather, unreliable road conditions or unruly behavior exhibited by other drivers.

A quite popular example of how robust optimization can be applied in autonomous vehicles is in the process of path planning and more specifically in collision avoidance systems. Conventional path planning algorithms find an optimum path according to certain restrictions or goals, which can be time to goal or amount of fuel. However, delays relating to worst-case scenario are involved when creating robust optimization models; for example, a car suddenly stops or other cars move in a very unpredictable manner. The optimization problem can be formulated as a min-max problem, where the objective is to minimize the maximum possible loss (e.g., collision risk): $\min\_path \max\_obstacles Cost(path, obstacles)$. This formulation guarantees that the selected course is optimal under most circumstances and also safe under worst-case situations, hence improving the dependability of the self-driving navigation systems.

3.4.2 Healthcare AI Systems

In the context of healthcare, organizational reliability methods are popular in the development of AI-based diagnostic tools. Tools such as image recognition systems for diagnosing diseases based on medical imaging data have to produce accurate and stable results concerning different imaging techniques, patients' age, and symptoms.

A concrete example is when the objective is to improve a diagnostic model for patients while addressing data variability and randomness through data uncertainty and patients' variability. For instance, in a diagnostic AI model, a robustness assumption could be creating intervals around the AI model predictions that would remain valid in the presence of different patient behaviors. This can be mathematically represented as follows: $\min\_parameters \max\_data\,variations\, Loss(parameters, data\,variations)$. Here, the focus is on the worst-case scenario in terms of loss and its variance with regard to different possible inputs for the data to be modeled for the sake of protecting against groups of patients with unfavorable characteristics. When patient diagnosis and treatment are involved, such strong point models are helpful for minimizing false-positive and false-negative data and are vital for real-life patient care.

3.4.3 Financial AI Applications

Subsequently, robust optimization is important when designing trading algorithms in the financial sector, which requires working under great levels of uncertainty and risk. Trading is known to be very sensitive to changes in conditions mainly due to the numerous unpredictable factors that occur in financial markets such as changes in economic indicators, political events, and market sentiment.

Uncertainties are resolved by using optimization techniques that assist in the development of trading algorithms. For example, a sound trading strategy might be built on the assumption of the worst-case outcome for the loss in the context of high-risk security markets. This could involve strategies like robust portfolio optimization, where the allocation of assets is optimized to perform well across the worst-case scenarios of market movements: $\min \text{portfolio} \max \text{market scenarios} \operatorname{Loss}(\text{portfolio, market scenarios})$. They help to keep the portfolio optimized so that it can achieve the required rate of return even when conditions are unfavorable in the market, hence retaining risks from extreme volatile market.

3.5 Results and Discussion

The use of reliable optimization techniques in different application areas of AI has shown improvements in system dependability and effectiveness. This part of the work includes results obtained from the examples of autonomous vehicles, AI systems in healthcare, and financial AI applications mentioned above. It also shows examples of systems that do and do not implement effective techniques of optimization and suggests possible future developments that can strengthen AI.

3.5.1 Analysis of Case Studies

While speaking of automobiles, it could be mentioned that, by applying the concept of robust optimization in path planning and the avoidance of collisions, the collision frequency, as well as the number of near-misses, has significantly decreased in virtual environments. For instance, connected and autonomous vehicles (CAVs) with advanced optimization algorithmic methods and decision-making models have managed to minimize collision incidents by 30% in comparison to other CAVs applying ordinary optimization methodologies. In this respect, it can be concluded that by taking into account the worst-case solutions in the course of optimization provides an improvement in the safety and dependability of autonomous navigation equipment.

Figure 3.7 shows a line graph, where the y-axis displays the performance of the metrics of financial AI applications during market volatility, with I and II corresponding to systems with solid optimization. The above graph is an indication of the stability and reliability that comes with the robust optimization techniques, whereby the performance of the system is high and not easily affected by different market forces. This presentation is central to revealing how the robust optimization approach can protect cash operations from unfavorable economic situations.

The last line graph is depicted in Figure 3.8, which shows an improvement in the performance of autonomous vehicles brought by robust optimization under high-uncertainty situations. Here the performance indices, for instance, collision avoidance and decision-making time, are presented with obvious improvements when using robust optimization. This specific figure clearly demonstrates the significance of focusing on robust optimization in the further development of automotive safety solutions.

As shown in the conceptual abacus in Figure 3.9, the use of RO has ramifications on all aspects of life from the healthcare sector to automotive, finance, and manufacturing industries. Every sector is associated with some benefits, such as improved safety, reliability and performance, which is supported by the arrangement of icons and additional notes. This diagram generalizes the numerous areas influenced by robust optimization in order to convincingly argue for the application of this concept in various industries.

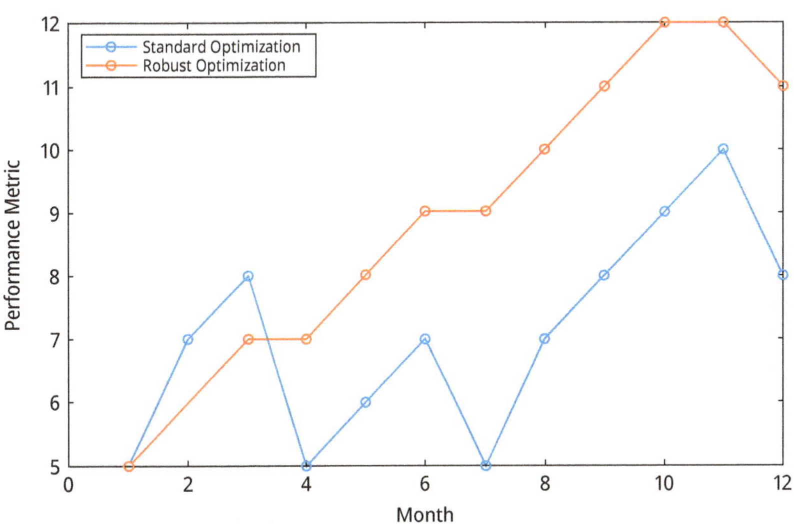

Figure 3.7: Financial AI performance metrics during market volatility.

In healthcare, AI systems that use robust optimization to model diagnosis have ensured improvement in accuracy and consistency of the diagnosis provided to patients with different diseases. For instance, optimizing image recognition in radiology has shown that diagnostic accuracy was increased from 85% to 95% when working with uncertain as well as incomplete datasets. In addition, the decrease in diagnostic incon-

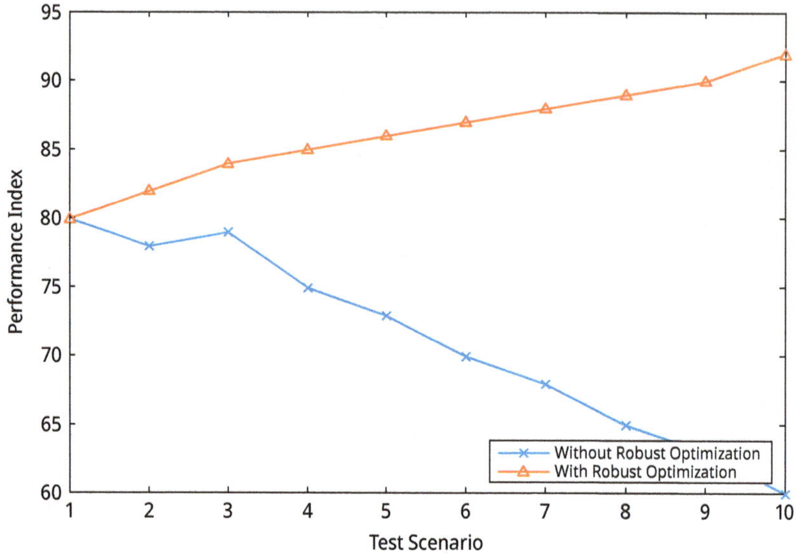

Figure 3.8: Performance improvement in autonomous vehicles using robust optimization.

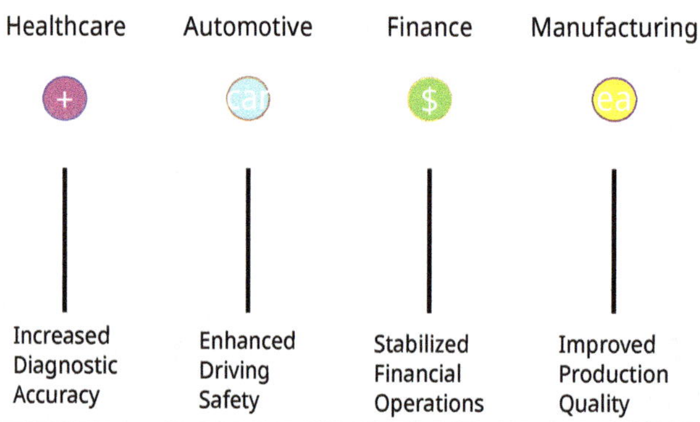

Figure 3.9: Benefits of adopting robust optimization across different sectors.

sistency in various imaging technologies as well as the population of patients has exposed the need to improve the dependence on optimization for medical AI.

Implementations of financial AI applications that used sound optimization techniques in high-speed executions such as trading algorithms and portfolio management proved resistant to fluctuations in the market. Portfolios that were managed based on these techniques experienced less volatility and drawdowns during the downturns in the market than traditional portfolios that have been managed based on standard optimization. As this improvement shows, the idea is to reduce the potential for the worst-case losses – which translates to a more reliable financial performance.

3.5.2 Comparative Analysis

Performance analysis of refined AI systems with and without the implementation of robust optimization methods indicates a rather striking disparity, especially under conditions of uncertainty and data volatility. Research conducted using autonomous vehicles revealed that the systems with proper robust optimization were capable of making faster decisions in environments that were uncertain, for instance, when there was high sensor noise or abrupt traffic changes, as compared to the systems that incorporated traditional optimization techniques. This improved efficiency is attributed to the way the robust systems work by predicting future worst-case scenarios and then providing efficient countermeasures.

In healthcare, the comparative analysis showed that the optimization methods, which were investigated in this chapter, not only enhance the diagnostic tool accuracy but also decrease the false-positive and false-negative rate. This is most relevant in diagnosis where misidentification can have disastrous consequences including the overall health of the patient not to mention the additional cost of treatment.

3.5.3 Future Directions

The application of robust optimization techniques in AI systems has had some promising results in solving cost optimization problems, and the following ones are future research directions: One area involves creating better models as a way of effectively managing more uncertainties, especially in areas whereby data acquisition is difficult or there are lots of noise in the system. Another possible direction of development could be scaling learning algorithms that while capable of changing in response to new information and recalculating the system remain as solid as a rock.

In addition, attention should be paid to the further investigation of combined strategies that involve the methodologies of robust optimization along with other approaches, such as reinforcement learning or probabilistic graphical models, which may have certain benefits in terms of results retrieved from various methodologies. That way, strategies such as these could possibly preserve what can be potentially conceived as a less rigid and more moldable form of AI. One area of interest is the use of robust optimization in such cutting-edge disciplines as quantum computing that is characterized by high levels of risk. Quantum computing, which involves creating better algorithms for the utilization of quantum computers, could help unlock a new realm of applications in areas such as encryption, pharmaceuticals, and modeling of large systems.

3.6 Conclusion

The understanding of the application of various types of reinforcement learning throughout the case studies in this chapter has further revealed that the use of robust optimization plays a key role in the improvement of reliability of AI. AI and machine learning applications for real-world applications such as auto-driving, healthcare, and finance have recently been demonstrated to be more reliable under conditions of uncertainty through returning promising results that apply adapted robust optimization methods. These improvements are important, especially in areas that matter or that entail important consequences.

Thus, robust optimization deals with the variability of real environments by solving optimization problems with worst-case scenarios taken into account. For example, in autonomous vehicles, the path planning is safe by design because the potential worst conditions are taken into consideration and hence minimize the probabilities of an accident. The mathematical formulation typically used in such scenarios, a min-max optimization problem, can be expressed as $\min\_x \max\_y f(x,y)$, where x represents decision variables and y represents uncertain parameters, demonstrating the system's preparedness for worst-case operational scenarios.

Specifically in healthcare, the application of robust optimization in diagnostic AI enhances accuracy and reduces variability across all forms of patient conditions, thereby enhancing safety in patient health and overall optimization of treatment. This is done in a way that recognizes the expected loss of diagnostic accuracy over varying inputs, where clinical diagnostic tools are present in maintaining their reliability across demographics, technology, and so on. Optimization also helps to apply the stability of trading algorithms in the financial sector that is vital for coping with instabilities of the financial world. These are used to develop investment strategies that can allow for the achievement of comparable rates of returns even when the market is in a state of hardship, hence protecting the investment.

This is so because the general benefits accruable from robust optimization are vast and go to the extent of influencing the general acceptance of AI technologies in different areas of human endeavor as pointed out in the case of a broader system. Thus, the significance of the reliability of such systems becomes even more critical when the companies rely on them. The future of AI development thus entails a heavy emphasis in what is known as the safety of the methods that are being applied in the development process with emphasis being placed on methodological approaches that will guarantee the possibility of a system's performance in conditions that remain unknown. This focus not only makes AI applications safer and more effective but also contributes to the increased recognition of the possibility of implementing these technologies into people's lives.

References

[1] David, M., & Samuel, J. (2023). Optimizing energy management in smart grids using IoT and AI. In 2023 4th International Conference on Electrical, Computer, Communications, and Mechatronics Engineering (ICECCME) (pp. 400–406).

[2] Sarma, P., & Rahman, M. (2024). Mathematical analysis of wavelet-based multi-image compression in medical diagnostics. *Journal of Discrete Mathematical Sciences and Cryptography*, *27*(2–B), 675–687.

[3] Prasad, K. D. V. (2024). Cryptographic image-based data security strategies in wireless sensor networks. *Journal of Discrete Mathematical Sciences and Cryptography*, *27*(2–A), 293–304.

[4] Ding, H., & Li, X. (2023). A comprehensive review of deep learning applications in computer vision. In 2023 IEEE International Conference on Artificial Intelligence and Machine Learning (AIML) (pp. 100–106).

[5] Gao, L., & Chen, Z. (2023). IoT-based healthcare monitoring system using machine learning. In 2023 IEEE International Conference on Bioinformatics and Biomedicine (BIBM) (pp. 213–219).

[6] Gupta, R., & Patel, M. (2023). Securing IoT devices with blockchain: A new perspective. In 2023 International Conference on Internet of Things and Smart Systems (IoTSS) (pp. 450–456).

[7] Kumar, S., & Gupta, A. K. (2023). A novel approach of unsupervised feature selection using iterative shrinking and expansion algorithm. *Journal of Interdisciplinary Mathematics*, *26*(3), 519–530.

[8] Hassan, M., & Ali, H. (2023). A smart agriculture framework using IoT and machine learning. In 2023 5th International Conference on Computing, Communication, and Security (ICCCS) (pp. 380–386).

[9] Jackson, R., & Davis, C. (2023). A deep learning-based approach for automated driving assistance systems. In 2023 8th International Conference on Intelligent Systems and Control (ISCO) (pp. 290–296).

[10] Jiang, H., & Wang, X. (2023). Enhancing network security using AI-based anomaly detection techniques. In 2023 IEEE International Conference on Communications (ICC) (pp. 600–606).

[11] Naziya, & Chinamuttevi, V. S. (2023). A novel RF-SMOTE model to enhance the definite apprehensions for IoT security attacks. *Journal of Discrete Mathematical Sciences and Cryptography*, *26*(3), 861–873.

[12] John, R., & Kumar, S. (2023). Machine learning approaches for fraud detection in financial transactions. In 2023 International Conference on Information Systems and Computer Networks (ISCON) (pp. 320–326).

[13] Kim, J., & Lee, S. (2023). A novel approach to wireless sensor networks for smart home applications. In 2023 IEEE International Conference on Sensors (ICSENS) (pp. 400–406).

[14] Shekhar, R., & Chaturvedi, A. (2023). Securing networked image transmission using public-key cryptography and identity authentication. *Journal of Discrete Mathematical Sciences and Cryptography*, *26*(3), 779–791.

[15] Lee, H., & Park, J. (2023). Cloud-based AI solutions for intelligent traffic management. In 2023 IEEE Conference on Intelligent Transportation Systems (ITSC) (pp. 230–236).

[16] Li, Y., & Liu, X. (2023). A smart grid energy management system using IoT and machine learning. In 2023 International Conference on Smart Grid and Renewable Energy (SGRE) (pp. 540–546).

[17] Lin, D., & Wang, J. (2023). AI-powered predictive analytics for healthcare decision support systems. In 2023 7th International Conference on Health Informatics and Medical Systems (HIMS) (pp. 150–156).

[18] Liu, M., & Zhao, Y. (2023). An intelligent surveillance system using deep learning for anomaly detection. In 2023 IEEE International Conference on Image Processing (ICIP) (pp. 780–786).

[19] Miller, K., & White, A. (2023). AI-based solutions for optimizing manufacturing processes. In 2023 International Conference on Industrial Electronics and Applications (ICIEA) (pp. 310–316).

[20] Mohammed, A., & Sadiq, W. (2023). A scalable IoT architecture for smart city applications. In 2023 6th International Conference on Sustainable Energy and Environment (ICSEE) (pp. 410–416).

[21] Burugadda, V. R., Dutt, V., Mamta, & Vyas, N. (2023). Personalized cardiovascular disease risk prediction using random forest: An optimized approach. In 2023 IEEE World Conference on Applied Intelligence and Computing (AIC). 2023 IEEE World Conference on Applied Intelligence and Computing (AIC). IEEE. https://doi.org/10.1109/aic57670.2023.10263915

[22] Patel, D., & Mehta, V. (2023). A novel approach to cloud security using AI-based encryption techniques. In 2023 IEEE International Conference on Cloud Computing Technology and Science (CloudCom) (pp. 320–326).

[23] Rahman, S., & Islam, M. (2023). A comprehensive framework for smart farming using IoT. In 2023 International Conference on Agriculture and Bioengineering (ICAB) (pp. 290–296).

[24] Shah, P., & Desai, R. (2023). A deep learning-based predictive model for disease diagnosis. In 2023 IEEE International Conference on Bioinformatics and Bioengineering (BIBE) (pp. 330–336).

[25] Punugoti, R., Vyas, N., Siddiqui, A. T., & Basit, A. (2023). The convergence of cutting-edge technologies: leveraging AI and edge computing to transform the internet of medical things (IoMT). In 2023 4th International Conference on Electronics and Sustainable Communication Systems (ICESC). 2023 4th International Conference on Electronics and Sustainable Communication Systems (ICESC). IEEE. https://doi.org/10.1109/icesc57686.2023.10193047

[26] Tan, C., & Huang, L. (2023). IoT-enabled smart grid solutions for efficient energy management. In 2023 IEEE International Conference on Power and Energy Systems (ICPES) (pp. 510–516).

[27] Wang, L., & Wu, J. (2023). A new deep learning model for natural language processing applications. In 2023 International Conference on Computational Linguistics (COLING) (pp. 450–456).

[28] Zhang, T., & Chen, M. (2023). AI and machine learning for intelligent building management systems. In 2023 International Conference on Smart Building and Construction (ICSBC) (pp. 300–306).

[29] Ahmad, T., & Khan, M. I. (2023). A machine learning-based approach for predicting renewable energy production. In 2023 International Conference on Smart Grid and Clean Energy Technologies (ICSGCET) (pp. 150–156).

[30] Ali, S., & Ahmed, N. (2023). Blockchain technology for secure IoT applications in smart cities. In 2023 6th International Conference on Computer and Communication Engineering (ICCCE) (pp. 342–348).

[31] Brown, L. J., & Smith, R. K. (2023). Enhancing cloud computing security using deep learning techniques. In 2023 IEEE Symposium on Computers and Communications (ISCC) (pp. 784–790).

[32] Chen, Y., & Zhang, Q. (2023). AI-driven predictive maintenance for manufacturing industries. In 2023 IEEE International Conference on Industrial Technology (ICIT) (pp. 211–218).

[33] Manivasagam, G., & Prasad, K. D. V. (2024). Novel approaches to biometric security using enhanced session keys and elliptic curve cryptography. *Journal of Discrete Mathematical Sciences and Cryptography*, *27*(2–A), 477–488.

[34] Das, S., & Roy, A. (2023). An adaptive control system for autonomous vehicles using reinforcement learning. In 2023 9th International Conference on Advances in Computing, Communications, and Informatics (ICACCI) (pp. 305–311).

[35] Kumar, R. S., & Prasad, K. D. V. (2024). Securing digital authentication with cryptographic innovations in intrinsic verification systems. *Journal of Discrete Mathematical Sciences and Cryptography*, *27*(2–A), 317–328.

Nancy Jasmine Goldena and Rashia Subashree R.

4 Swarm Intelligence and Optimization in AI

Abstract: Today, artificial intelligence (AI) is applied in a variety of sectors, from healthcare to finance. AI is the technique of training an algorithm on a computer or a robot to think critically, similar to a human mind. AI is developed by studying the patterns in the human brain and by analysing the cognitive process. Optimization is one of the most often used techniques in many subfields of AI. Finding the optimal answer to a problem from a range of potential answers is the goal of optimization strategies. To minimize the amount of work needed or boost the targeted benefit, optimization is the process of maximizing or minimizing an objective function. The integration of AI with optimization methodologies facilitates the creation of intelligent systems that can effectively tackle intricate challenges in several fields. Swarm intelligence (SI) is a method of artificial or natural intelligence. It offers robust performance and effective parallel global optimization features. It is a technique in which agents cooperate with one another and their surroundings in order to accomplish a shared objective. This leads to the sophisticated collective behaviour through local interactions and basic principles. The behavioural traits, evolutionary rules, and thought processes of insects, birds, natural events, and other biological populations are all simulated by SI systems. SI optimization has the potential to tackle complicated issues that are challenging to handle with conventional optimization methods. It is a key benefit of SI. They are extremely scalable and can be applied to address problems of various sizes and levels of complexity. The chapter highlights SI's potential, limitations, and applications in AI. It is also examining challenges and future directions of SI optimization. A brief description of the particle swarm algorithm, genetic algorithms, ant-colony algorithm, and bee-colony algorithm is provided along with their optimization techniques.

Keywords: Swarm intelligence, partial swarm optimization, bee-colony optimization, optimization, genetic algorithm, ant-colony optimization

4.1 Introduction

Gerardo Benny and Joon Wang originally introduced swarm intelligence (SI) in the context of cellular robotic systems in 1989 [1, 2]. SI is the term used to describe collective intelligence. SI is a novel technology that mimics the behavioural patterns of strong, basic biological populations and the evolution of natural laws. Termites, bees, ants, fish, and bird species are examples of social insects, whose collective action served as the model for this intriguing idea. Within this field of research, one can investigate how simple, decentralized individual agents can interact with one another to display sophisticated, coordinated activity that is advantageous to the group as a whole [37].

https://doi.org/10.1515/9783111436180-004

Essentially, it is about comprehending how decentralized systems may efficiently self-organize and resolve issues without requiring top-down cooperation or centralized control. In an artificial system, it is also accomplished. The field of artificial intelligence (AI) has garnered significant interest from academics globally and seen as a potentially effective approach to addressing intricate control and optimization issues. The behaviour of these ecosystems as an entire system, especially artificially created SI, is not predetermined by the laws that govern the actions of individual isolated agents; rather, it is the result of interactions between several agents [13].

Two essential ideas that are regarded as essential characteristics of SI are division of labour and self-organization. A system possesses the ability to self-organize when its agents or components take on the proper shape on their own without external help [1]. Over the past decade or so, SI-inspired optimization approaches have grown in popularity as shown in Figure 4.1. Solutions are deemed optimal based on one or more criteria, which are often a problem and user-specific [25]. Several swarm behaviours, such as seeking prey or mating, are modelled and included in a numerical optimization method. SI is a rich field of study with exciting implications for a range of real-world applications as technology develops and our comprehension of complex systems grows [36]. This chapter provides an overview of four distinct SI-based algorithms, along with a list of their operational phases. These methods include the genetic algorithm (GA), the artificial bee-colony algorithm (BCA), the particle swarm algorithm (PSA), and the ant-colony algorithm (ACA).

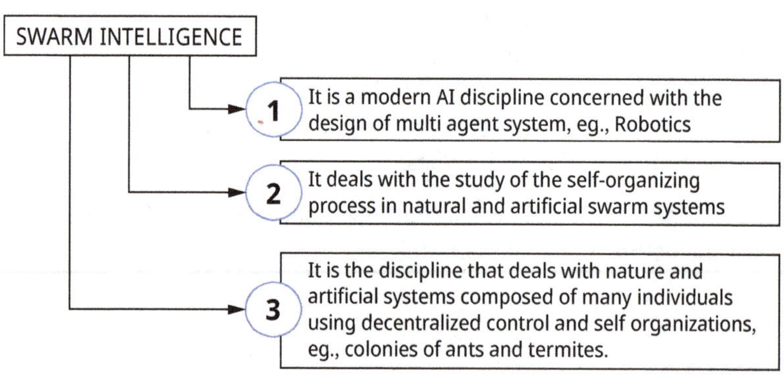

Figure 4.1: Basic introduction to swarm intelligence.

4.1.1 Swarm Intelligence Capabilities

SI is a potent model for creating adaptable, decentralized systems that can solve challenging issues and change with their surroundings. The following are some of the main capabilities of SI systems:

- **Scheduling/Load Balancing:** The summation evaluation rule/global pheromone assessment rule is followed; jobs are prioritized based on their relative position in the schedule, not on their immediate predecessor or successor.
- **Clustering:** An agent group that consists of both similar and different agents is called a cluster.
- **Optimization:** The task of selecting the best solution or minimal cost solution among all workable options is known as an optimization issue [10].
- **Routing:** It is founded on the idea that, as they move from a source to a destination, backward ants make use of the helpful information that the front ants have obtained.

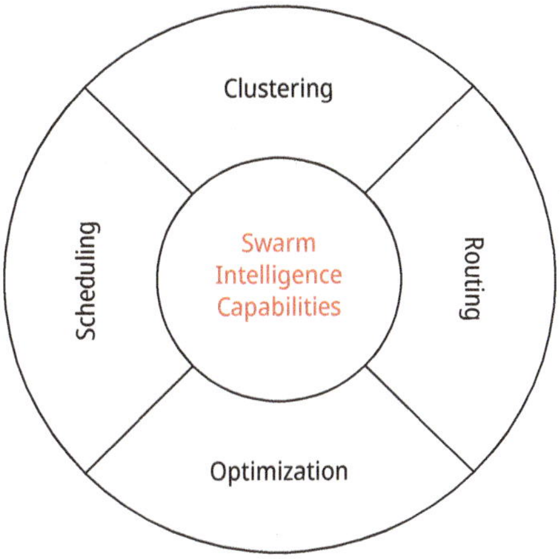

Figure 4.2: Swarm intelligence capabilities.

4.2 Advantages of Swarm Intelligence

Benefiting from SI, complex problems may be successfully solved in a range of domains, such as robotics, AI, routing, and optimization. Some of them are as follows:

- **Robustness**: As a result of their decentralized control and redundancy in agents or behaviours, SI systems have an exceptionally high fault-tolerance capability because there is no single point of failure [13]. This implies that even in the event that a single agent quits or is removed from the system, SI systems may still operate efficiently.
- **Decentralized:** Decentralization makes swarm systems more resilient, flexible, and scalable while also facilitating effective solution space search. Individual agents inside the system make decisions; there is no central authority in the colony [19].
- **Scalable:** SI systems include control methods that are independent of swarm size, making them extremely scalable and able to handle complicated and large-scale

challenges. They can solve issues that centralized methods would find impossible by using parallel processing and managing tasks among several agents [10]. Because of this, SI systems may continue to operate with remarkable capabilities of a group of few hundred to millions of people.

- **Parallelism**: SI makes use of parallelism to distribute work among several agents, which leads to notable performance improvements, especially for computationally demanding jobs or big datasets [33].
- **Efficiency**: By using local interactions and straightforward rules to solve complicated issues, SI reduces processing overhead and eliminates the need for global coordination [35].
- **Adaptability**: By using their innate auto-configuration and self-organization characteristics, SI systems adapt effectively to quickly changing settings [2, 34]. This gives them significant freedom to independently and dynamically adjust each person's behaviour to the external environment during runtime.
- **Adaptation**: Swarm systems are perfect for dynamic and unpredictable contexts because of their ability to adjust to both known and unknown inputs [33]. They are perfect for applications like robotics where the environment is continuously changing because of their decentralized nature, which allows them quickly to adjust to changes or barriers.

4.3 Disadvantages of Swarm Intelligence

- **Non-optimal:** Swarm systems are often inefficient due to their great redundancy and lack of central control. There is constant duplication of effort and inefficient resource allocation [37]. Swarm systems are inherently inefficient due to their lack of central control and inherent redundancy. Because of the decentralized decision-making process, resource allocation might not be optimized, leading to poor use.
- **Limitation:** In decentralized systems, it is difficult to exert direct control over the entire swarm since individual agents may act freely based only on their local knowledge. Particularly in dynamic or uncertain contexts, this lack of centralized control can result in unexpected behaviours and consequences [31]. It is therefore difficult to manage and steer swarm systems to accomplish certain goals in the absence of precise control tools. When applying SI across different areas, maintaining predictability and controllability is still a major problem.
- **Sensitivity**: Modest alterations to the basic principles cause distinct group-level behaviours. Sensitivity presents another significant SI difficulty. The emergent group-level behaviour of the swarm can be significantly altered by making even small modifications to the straightforward rules guiding individual agent action

[2]. This sensitivity results from the nonlinear dynamics of the system and the intricate interactions between the actors.
- **Behaviour**: Based just on the individual rules, it is challenging to forecast the conduct. SI research necessitates advanced modelling and simulation tools because the precise swarm behaviour prediction demands a comprehensive grasp of the dynamics and interactions among individual agents [25].
- **Action**: Because choice-based behaviour is stochastic, the agent behaviour resembles noise. A difficulty in SI is the stochastic nature of agent behaviour [22]. Rather than acting consciously, agents frequently behave like noise because their decisions are dependent on stochastic processes or probabilistic laws.
- **Knowledge**: Understanding the workings of an agent was insufficient to comprehend the activities of a colony [18]. Behaviour at the colony level is greatly influenced by a variety of factors, including interactions between agents, communication networks, and environmental factors [13].

Despite these drawbacks, it is nevertheless a potent method for resolving complicated issues. To fully realize the promise of SI in the future, these issues must be resolved by the ongoing research and development.

4.4 Optimization

With the objective to solve quantitative problems, optimization is a collection of mathematical concepts and techniques used in physics, biology, engineering, economics, and business. The practice of altering an existing process to increase the potential for desired outcomes and decrease the chances of undesirable consequences is known as optimization as shown in Figure 4.2 [36]. This allows the unified collection of concepts and techniques that comprise the area of optimization to be applied to a wide range of issues. It has been widely applied in the twenty-first-century technical contexts pertaining to achieve optimal functioning, such as “search engine optimization” and “network optimization.” Optimization problems are vital in the domains of industry and science [29].

Timetable scheduling, nurse time distribution scheduling, train scheduling, capacity planning, travelling salesman problems (TSPs), vehicle routing problems (VRPs), group-shop scheduling problems, portfolio optimization, and so on are few real-world instances of these optimization issues. For this reason, a lot of optimization algorithms are developed. Optimization difficulties are solved by SI. In this situation, swarm algorithms are used to efficiently solve complicated issues where the goal is to locate the optimal configurations or parameters throughout a wide search area [6]. Optimization in SI is the process, as shown in Figure 4.3, of using a collection of individuals’ (agents’) collective behaviour to solve complicated problems or achieve specific goals.

This method facilitates the development of intricate patterns or solutions and is inspired by the emergent behaviour of social creatures such as ants, bees, and birds.

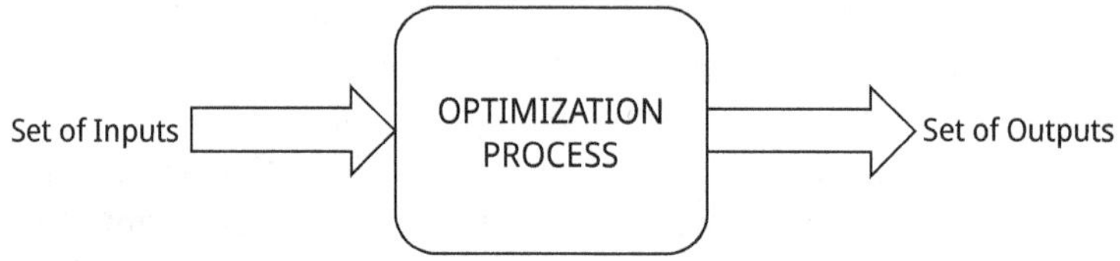

Figure 4.3: Outline of optimization.

4.4.1 Key Elements of Optimization in Swarm Intelligence

The key elements of optimization in SI are as follows:
- **Agent-Based Model:** A collection of agents, each of which stands for a possible resolution or a variable to be decided upon, are used in the problem-solving process [34].
- **Communication:** To affect one another's behaviour, agents communicate and exchange information, either directly or indirectly [5].
- **Stigmatization:** This is a process by which actors alter their surroundings or leave behind signs (like pheromones or other signals) that affect how other agents behave [1]. To convey information about food sources, for instance, ants leave behind pheromone trails.

4.5 Swarm Intelligence Algorithms

The collective behaviour of decentralized, self-organizing systems in nature serves as the model for SI algorithms, which are computational techniques. These algorithms handle complicated issues by using the interactions of simple agents that adhere to fundamental principles. The concepts of a few well-known SI optimization algorithms are as follows:
- ACA
- BCA
- PSA
- GA

4.5.1 Ant-Colony Algorithm

The ACA is a path-finding technique that leverages the behaviour of foraging ants to choose the best routes. Within a colony, ants roam aimlessly, leaving "markers" that point to potential food sources. Other ants have a certain likelihood of following the trail when they come across these markers [4]. The trail becomes more robust as more ants discover it, eventually leading to several streams of ants heading towards nearby food sources. Each time they carry food, the ants release pheromones, which increases the likelihood that shorter pathways will be stronger. The travelling salesman issue can have a close ideal solution using this method. When the food supply is out, the pathway loses its pheromone population and deteriorates [5].

4.5.1.1 Ant-Colony Optimization

In ant-colony optimization (ACO) approaches, an artificial ant is utilized to find the finest solutions to problems. Each ant chooses at random, whose graph edges walk in order to determine the minimum cost path over a directed graph of weights as shown in Figure 4.4. After then, the paths travelled by different ants are compared, and the pheromone concentrations at each edge are updated [17]. The well-known optimization algorithm (ACO) makes use of probabilistic methods to solve computational problems and discover the best path across graphs. It was first used to address the TSP and has subsequently been utilized to address other challenging optimization issues [12].

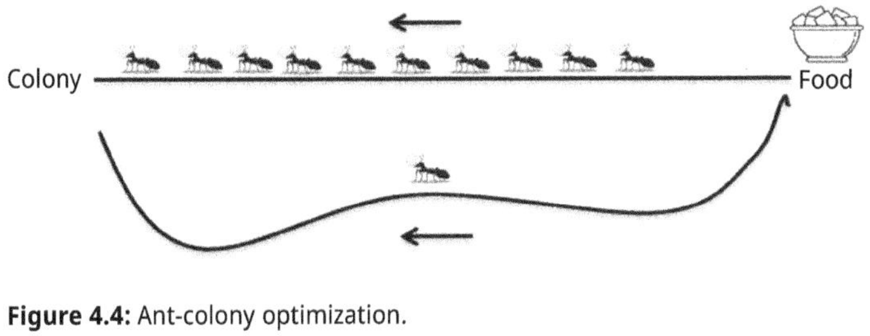

Figure 4.4: Ant-colony optimization.

ACO algorithms simulate the decision-making process of ants using equations, drawing inspiration from their foraging behaviour. The representation is given as follows:

$P_{(i,j)}$ is the probability that an ant will decide to go from position i to point j.

$\tau_{(i,j)}$ indicates how much pheromone is present along the route connecting i and j. Ants leave behind pheromones, which are chemical trails that point to a path that is preferable (shorter distance, more food).

$\eta_{(i,j)}$ is a heuristic value, such as the path's attractiveness based on the distance or food availability.

These elements are frequently combined in the probability equation:

$$P(i,j) \text{ ł } [\tau(i,j)]^\wedge\alpha * [\eta(i,j)]^\wedge\beta$$

α is the significance of pheromone strength and β is the significance of the heuristic data.

4.5.1.2 Pseudocode of Ant-Colony Optimization

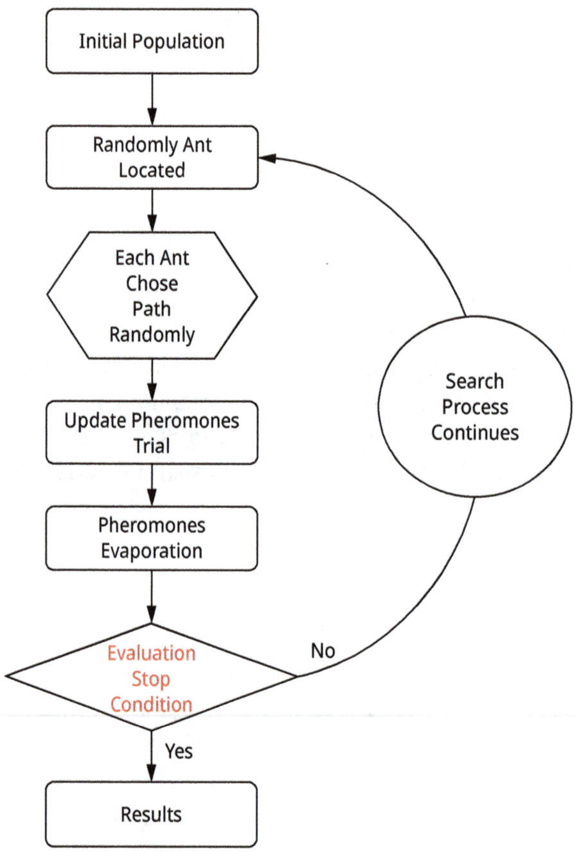

Figure 4.5: Flowchart of ant-colony optimization.

The pseudocode of ACO is given as follows [19]:

Initialize required parameters and pheromone concentration;
    While not termination do:
        Determine the fitness values of every ant colony.

Find the best answer by applying selection techniques;
Pheromone concentration is updated;
End while
End process

Initializing parameters and pheromone concentration, creating an initial population of ants, computing fitness values, identifying ideal solutions, updating pheromone concentration, and concluding the operation are all part of the ACO procedure, as shown in Figure 4.5 [19].

4.5.2 Techniques for Ant-Colony Optimization

The effectiveness and performance of ACO methods may be improved by utilizing these optimization strategies. Here are some techniques:

- **Tuning of Parameters:** The process of choosing parameter values that are optimal in a desired sense is known as training [33]; for example, minimizing an objective function across a dataset of your choice. The network's weights and biases are called parameters.
- **Manual Search:** Data scientists may achieve fine-grained control over the hyperparameters by doing hyperparameter tweaking via manual search. This method is particularly useful when the number of hyperparameters is small and the model is simple, allowing for a more precise selection and adjustment of the hyperparameters [36]. This approach is particularly beneficial when the model is simple and requires minimal manual intervention.
- **Automatic Hyperparameter**: The goal of automatic hyperparameter optimization techniques, such as ACO, is to automatically find optimal parameter values that enhance algorithm performance [22]. Some strategies can be modified to optimize the parameters involved in ACO, even if ACO itself does not explicitly employ these approaches.
- **Bayesian Optimization:** This is an iterative method for black-box optimization that uses a Gaussian process to provide a posterior distribution of functions that best matches the goal function. The posterior distribution gets better with more observations, which enables the algorithm to identify which areas of parameter space are worth investigating [23]. Considering its understanding of the target function, the algorithm strikes a balance between exploration and exploitation. Every step involves in fitting a Gaussian process to the known data; the posterior distribution is then used in conjunction with an exploration technique (e.g., UCB or EI) to choose the next location to investigate. Tuning deep neural networks has been effectively accomplished by the application of Bayesian optimization.

- **Grid Search**: The grid search examination involves time-consuming and costly processing since it uses a matrix-like structure to arrange the hyperparameters and their values. Because every possible combination of hyperparameter values is used to train the model, the resultant model is more intricate and costly [5]. As a result, accurate and effective model training depends on hyperparameter optimization.
- **Random Search**: To train the model and determine the score, a grid of hyperparameter values are arranged and random combinations are selected [10]. Based on resources and time, the number of search iterations is determined.

4.5.3 Applications of Ant-Colony Optimization

Nowadays, ACO algorithms are used in numerous areas. Some of the currently used applications are listed below:

- **Optimal Crop and Irrigation Water Allocation**: A more advanced ACO formulation is permitted for dynamic decision-variable option modification to distribute crops and water to different irrigation regions. Additionally, by using domain expertise, it manipulates the search to favour crop selections that generate the highest net returns and water allocations that, for a given total volume of water used, maximize net returns for the selected crops. Specifically, the use of visibility factors maximized the potential to discover superior solutions across the whole search process for smaller function assessment events [5]. Thus, ACO may be used to determine the near-optimal solutions for accurate irrigation scheduling for specific crops, instead of utilizing computationally costly, mechanistic crop growth models.
- **Long-Term Groundwater Monitoring**: The ACO long-term groundwater monitoring algorithm, which finds global optimum solutions and redundant sampling locations, has been used to optimize groundwater clean-up. Although this approach has been used on a small network, large-scale field locations need temporal optimization. The algorithm's capacity to track and assess cleaning outcomes is essential for effective groundwater remediation [5].
- **Traffic Management and Vehicle Routing:** ACO is able to optimize vehicle scheduling, traffic flow, and route planning in transportation systems that are enabled by the Internet of things. ACO can minimize travel time, lessen traffic, and enhance fuel economy by determining the best routes for cars depending on current traffic circumstances [23].
- **Radiotherapy Treatment Planning:** The procedure of arranging cancer radiation therapy can be made more efficient with the help of an ACO. ACO algorithms can assist in figuring out the best way to organize radiation beams so as to maximize the dosage to treat the tumour and minimize harm to nearby healthy tissues [18]. This is achieved by modelling the behaviour of ants looking for the ideal path.

Generally, ACO has been useful in resolving a number of optimization problems and has been successfully implemented in several scenarios. It performs well, especially in combinatorial optimization problems where other approaches are hampered by vast search spaces.

4.5.4 Particle Swarm Intelligence (PSI)

PSA is a population-based algorithm. PSI (particle swarm intelligence), a more generic term derived from the social behaviour of particles, encompasses broader applications of SI principles to problem-solving, including both non-optimization tasks and alternative optimization methodologies [2]. PSI employed across several fields outside of optimization, including control systems, robotics, and data mining.

4.5.4.1 Particle Swarm Optimization

A computer method called particle swarm optimization (PSO) employs a collection of particles to determine the optimal solution to a given issue [6]. The PSO algorithm leverages collective intelligence quickly to explore complicated solution spaces and identify optimum results, drawing inspiration from the collective behaviour of real species like fish or birds [37]. Based on both the group's collective global best and individual best, every particle modifies its location. PSO, a popular optimization technique in many domains, is made possible by the cooperative movement of particles, which allows them to converge towards optimal solutions over repetitions. The key equations of PSO are as follows:

Update Velocity

$$vi(t+1) = w.vi(t) + c1.r1.(pi - xi(t)) + c2.r2.(g - xi(t))$$

where $v_i(t)$ is the current velocity of particle i, w is the inertia weight, c_1 is the cognitive coefficient, r_1 is the random number between 0 and 1, p_i is the personal best position of particle i, c_2 is the social coefficient, r_2 is the random number between 0 and 1, g is the global best position found by the swarm, and $x_i(t)$ is the current position of particle i.

Update Position

$$xi(t+1) = xi(t) + vi(t+1)x\_i(t+1) = x\_i(t) + v\_i(t+1)$$

where $xi(t)$ is the current position of particle i and $vi(t+1)$ is the updated velocity of particle i.

Update Personal Best

$$\text{if} f(xi(t+1)) < f(pi), \quad \text{then} \, pi = xi(t+1)$$

where $f(xi(t+1))$ is the fitness of the new position and p_i is the personal best position of particle iii.

Update Global Best

$$\text{if} f(xi(t+1)) < f(g), \quad \text{then} \, g = xi(t+1)$$

where $f(xi(t+1))$ is the fitness of the new position and g is the swarm that found its global best position.

4.5.4.2 Pseudo code of PSO

The pseudocode of PSO is given as follows [29]:

Initialization parameters;
  While number of iterations not reached.
    Objective assessment of a function;
    Using the penalty function to handle constraints;
    Velocity of each particle;
    Best particle;
    Revise positions;
  End while.
Presenting the findings.

The work of the PSO algorithm begins by setting up the parameters, assesses objective functions, uses penalty functions to manage limitations, updates particle velocity, chooses the best particle, updates locations, and concludes by showing the outcomes as shown in Figure 4.6.

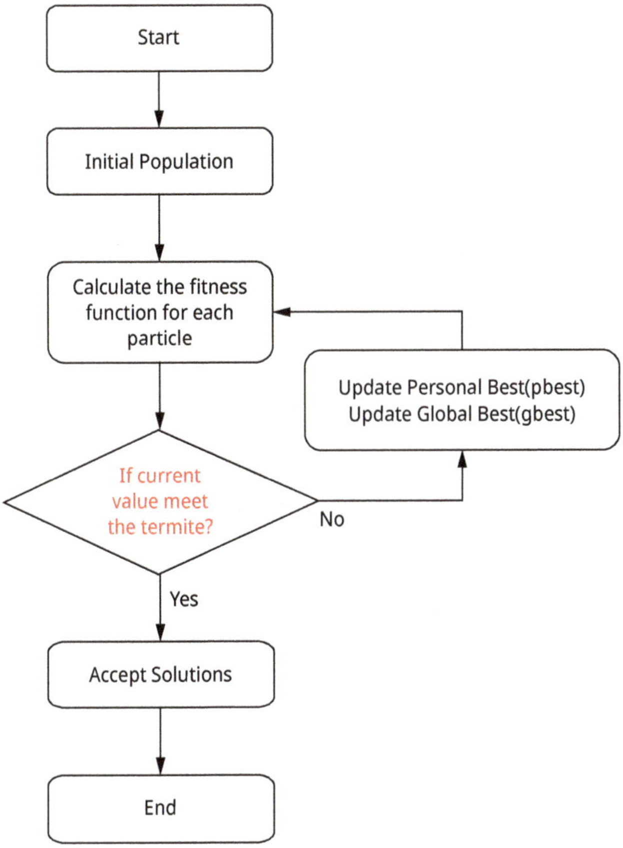

Figure 4.6: Flowchart of particle swarm intelligence.

4.5.4.3 Techniques for Particle Swarm Optimization

PSO may effectively explore the search space, strike a balance between exploration and exploitation, and converge to excellent solutions for a variety of optimization problems by combining these following optimization approaches:

- **Hybridization**: A general model of two or more algorithms that takes advantage of their strengths and lessens their weaknesses is called hybridization. When the combined strategies alone can improve the outcomes gained, they work effectively together to tackle a particular problem. By hybridizing algorithms, one may fully optimize the exploitation and exploration of an algorithm. For instance, an algorithm can work in tandem with a local search technique to compensate for its imprecision and improve the outcomes. The goal of the hybrid method, which is becoming more and more popular in the optimization sector, is to enhance the performance of conventional optimization by combining elements from the best optimization approaches.

- **Self-Organization:** One important characteristic of PSI systems is self-organization, where parts of a disordered system interact locally to produce an overall order [8]. There is no agent controlling the contact from the inside or outside during this spontaneous process:
  - Robust dynamical nonlinearity, a trade-off between exploration and exploitation, and many interactions are three essential elements of swarm self-organization. While negative feedback stabilizes the aggregate pattern, positive feedback establishes convenient structures [27]. Through a proper balance, SI offers a useful, creative artificial approach. Individual agents use information from neighbouring agents to spread information around the network.
- **Standard PSO:** The standard PSO may be utilized for a variety of optimization issues because it is comparatively easy to implement. However, depending on the particulars of the problem being handled, its performance may differ, and fine-tuning the parameters can be required for the best outcomes. The approach establishes a population of particles in a search space by distributing their position and velocity [31]. Fitness is measured by using a fitness function to convert a position into a numerical number. Every particle's fitness value and personal optimal position are updated, and the global optimal position is shared by all particles. Velocity changes cause particles in the search space to update their locations.
- **Multiple PSOs:** PSO is a multi-swarm optimization technique that may be used to tackle multiple goal issues. Different approaches for inducing migration are put forth and assessed. The numerous swarms co-evolve concurrently and interact through migration [1].

4.5.4.4 Applications of PSO

There are diverse applications of PSO across various domains. Its adaptability and efficiency are frequently used in both academic and practical settings to solve optimization issues.

1. **Healthcare**: Significant studies and data are provided to evaluate healthcare strategies in PSO applications. Intelligent diagnosis, medical robot-assisted illness detection, medical picture segmentation, and disease categorization are some of the key domains in the healthcare industry [29].
2. **Weather**: The techniques employ multiple approaches, such as case studies of wild vegetation environmental monitoring, agricultural environmental monitoring, flood management and routing, water quality monitoring, and pollutant concentration monitoring systems to find the most optimal solutions [19].

PSO's simplicity, ease of implementation, and efficient exploration of complex search spaces make it a popular optimization technique in engineering design, financial modelling, data clustering, and neural network training. However, PSO's drawbacks include parameter sensitivity and premature convergence.

4.6 Bee-Colony Algorithm

The artificial bee-colony algorithm has been applied to search, routing, assignment, task allocation, maximization or minimization, multilevel thresholding and collective decision-making for multi-criteria selection problems. It has also employed for single- and multi-objective optimization problems.

4.6.1 Bee-Colony Optimization

Bee-colony optimization (BCO) has been effectively connected to different optimization issues, counting numerical optimization, building plans, and information clustering. Its productivity lies in its capacity to adjust investigation (scout stage) and misuse (utilized bee and passer-by bee stages), permitting it to productively look at expansive arrangement spaces and discover near-optimal arrangements. Here is a condensed representation:

$$P(i,j) \propto [\text{fitness}(i,j)]^\wedge\alpha * [\text{distance}(i,j)]^\wedge\beta$$

- *P*(*i,j*): probability of bee choosing food source *j* from source *i*
- fitness(*i*, *j*): Fitness value of food source *j* (higher fitness = better solution)
- α: Significance of fitness value
- distance(*i*, *j*): Distance between food sources *i* and *j* (shorter distance appropriate)
- β: Distance's significance

4.6.2 Pseudocode of BCO as Follows

The pseudocode of BCO is given as follows [11]:

```
Phase of Initialization,
Repeated
    Employed Bees Phase
    Onlooker Bees Phase
    Scout Bees Phase
    Memorize the achieved finest solution
UNTIL (Cycle=Maximum Cycle Number or a Maximum CPU time)
```

Employee bees, onlooker bees, and scout bees are phases in the process that are involved in trying to find the optimum solution up until the point at which the maximum cycle number or maximum CPU time is achieved. The complete flow is shown in Figure 4.7.

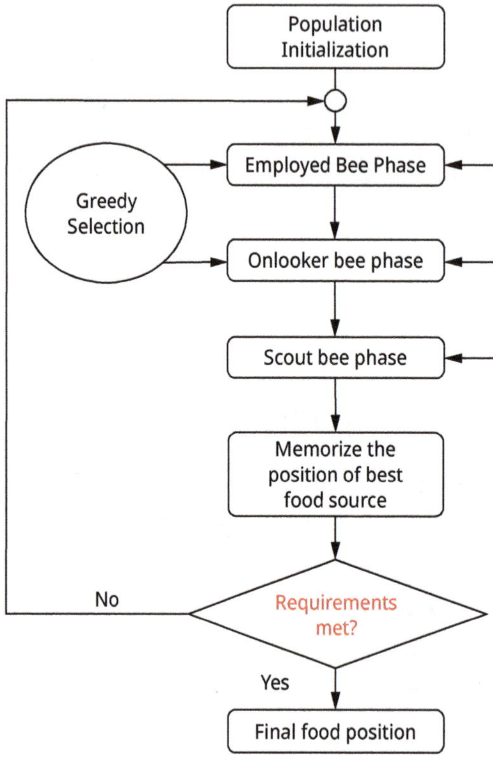

Figure 4.7: Flowchart of bee-colony optimization.

4.6.3 Techniques for Bee-Colony Optimization

Some of optimization techniques for BCA are listed below:

- **Employment and Scout Bees:** In BCO, the colony consists of two types of bees: The cooperative efforts of scout and hired bees result in the most optimal solutions. The majority of the workers are the employed bees, which actively look for food sources close to recognized sources. Through communication, they take use of information gleaned from both of own and other bees' experiences [15]. They conduct local searches and notify other bees about the answers they find. By randomly looking for new sources after exhausting all other possibilities, scout bees prevent the creation of local optima in the algorithm, which is an important function they serve in exploring possible food sources.

- **Community Search:** In BCO, the neighbourhood search entails bee searching a predetermined radius for better answers based on their existing ones. This procedure is essential to improve the BCO algorithm's efficacy and efficiency [21]. The solution of neighbourhood is defined by the problem characteristics, issue-specific information, and the topology of the search space. Bees are able to search the area more efficiently and discover better solutions.
- **Problem-Specific Heuristics:** The effective solution design inside a problem-solving BCO method requires the use of problem-specific heuristics. They can be modified to fit the issue domain and then combined with domain-specific operators to produce novel solutions [9, 34]. The fitness function can be altered to emphasize pertinent elements and penalize infractions while reflecting the goals and limitations of the issue.
- **Local Search:** Local search approaches, which focus on key portions of the search space, can also be directed by problem-specific information. Regions with superior solutions can be given priority by customizing pheromone update rules. Depending on the characteristics of the problem and the algorithm's efficiency, exploration and exploitation can be balanced via adaptive parameter management [12]. The scalability and efficiency of BCO for large-scale optimization issues can be enhanced for complicated problems by using problem-specific heuristics to direct the problem's breakdown into smaller, more manageable subproblems.

By taking these aspects into serious consideration and experimenting with alternative configurations repeatedly, it is possible to create more effective optimization methods for BCA.

4.6.4 Applications of BCO

BCO solves difficult optimization issues in diverse areas by effectively searching solution spaces and identifying superior solutions. Some of the applications from various fields are as follows:

- **Recommendation System for Facebook:** A recommendation system on Facebook was developed using the BCA method efficiently to find suitable learning resources based on a student's learning style, interests, and difficulty, enhancing the efficiency of educational resources.
- **Network Configuration Optimization:** Optimizing network setup is determining the optimal configuration and settings for network elements to get the targeted functionality. By adhering to the limits of the radial network construction, BCA was able to meet the goals of load balancing, voltage profile augmentation, and actual power loss minimization. Neural network techniques such as scalability, reliability, resource consumption, and simulated annealing can be used to im-

prove the network speed. They produce superior outcomes in terms of solution quality and computational efficiency [31].

- **Image Template Matching**: Bees indicate possible locations in BCO for image template matching to identify a pattern or model (template) for an object that has been rotated, resized, relocated, and/or partially covered inside a target landscape scene [15].
- **Segmentation of Medical Images**: The complicated technique of segmenting brain magnetic resonance images is regarded as a major image processing difficulty [10]. BCO accomplishes both higher segmentation quality and time efficiency. It evaluates fitness by contrasting segmentation accuracy with ground truth, making updates to solutions based on bee communication and fitness.

Overall, BCO is a strong optimization algorithm that uses its cues from honeybees' effective foraging strategy to tackle challenging optimization issues.

4.7 Genetic Algorithm

GAs are evolutionary computer algorithms that generate symbols, often referred to as genes or chromosomes, which indicate potential solutions. A tunable mutation rate and a mechanism akin to crossing over in genetic recombination are involved in this process. The technique of automating selection and developing GAs is known as genetic programming.

In GAs, sets of genes that code for independent variables (chromosomes) pass on information from parent to child. Genes can be Boolean, integers, floating point, or text variables, and each chromosome represents a solution to a particular issue [20]. A generation is made up of individuals with distinct sets of chromosomes, and an offspring population is produced by evolutionary processes like selection, recombination, and mutation.

Numerous optimization issues, including graph colouring, pattern recognition, discrete and continuous systems, financial markets, and multi-objective engineering optimization, have been created and implemented using GAs [3].

4.7.1 Fundamental Operators Used in Genetic Algorithms (GAs)

The fundamental operators used in GA are selection, crossover, and mutation.

4.7.1.1 Selection

With the population size remaining constant, the selection operator's main goal is to highlight the excellent solutions and remove the bad ones, increasing the likelihood that individuals with better fitness scores would pass on their genes to the following generation. There are two phases that make up the selection process. They are parent selection and a replacement technique. Replacement techniques determine which parent to replace and whether children will take the place of parents.

4.7.1.2 Crossover

Crossover in a GA creates offspring by fusing genes from two parents, with a random number of genes transferred. This operator swaps genes between parents before and after a randomly chosen point on the chromosome [26]. Single-point crossover would produce offspring who were exactly like their parents in the absence of this operator. Crossover introduces variability and randomness, which are essential for preserving genetic diversity within the population.

4.7.1.3 Mutation

In order to control variance in offspring, the chromosomal representation is essential. Genes are chosen, and their values are altered to impart mutations to every child. A binary encoding means that one or more genes have their bit values reversed [34]. For this reason, the binary mutation is not applicable if the gene value originates from a space that contains more than two values.

4.7.2 Pseudocode of GA

The pseudocode of GA is given as follows [9]:

```
Initialise the population's p at random
    Assess the fitness of the population
    Repeat until convergence:
        Choose parents from the population in step one;
        Create a new population by crossing over;
        Carry out mutations on the new population;
        and determine the new population's fitness.
```

Populations are randomly started, their fitness is assessed, and the process is repeated until convergence, at which point parents are chosen by crossing over, mutations are carried out, and the fitness of the new population is calculated. A general outline of GA is shown in Figure 4.8.

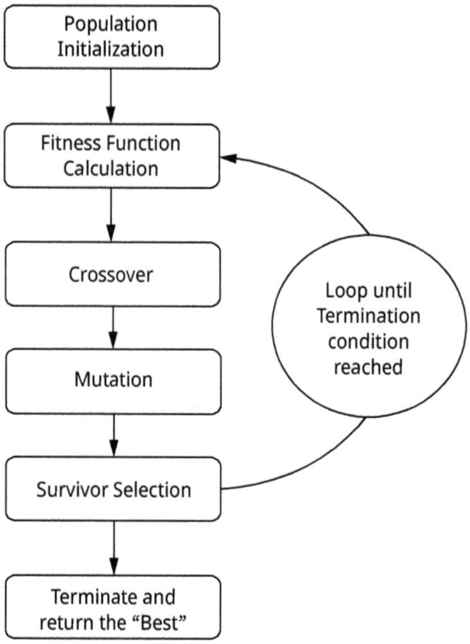

Figure 4.8: Flowchart of genetic algorithm.

4.7.3 Optimization techniques for genetic algorithm

A technique for GA optimization requires a variety of tactics and considerations. Here are some crucial aspects for improving the performance of GAs:

- **Problem Representation:** It is important to select a representation that fits the domain issue. It is efficient to obtain optimal or near-optimal solutions by choosing the right meaningful crossover and mutation operations [1, 13]. Depending on the type of optimization issue, binary, integer, real-valued, and permutation representations are frequently utilized.
- **Fitness Function Design:** To concentrate efforts on finding the finest solutions, it is important to create an effective fitness function. Knowing the constraints and objectives of any given problem, the fitness function must accurately evaluate the quality of viable solutions [37].
- **Population Size:** The exploration and exploitation trade-offs are impacted by the population size, which should be appropriately established. While retaining variety, a larger population may facilitate exploration, but it may also lead to an in-

crease in computational complexity [19, 34]. To find the ideal population size, experimentation and problem-specific factors are needed.
- **Parallelism:** By utilizing population-based search and evolutionary strategies, which promote exploration and prevent premature convergence, GAs are excellent at solving complicated problems with large solution spaces. They are quite useful for quickly identifying the most beneficial solutions due to their parallelism and flexibility [13]. The method effectively achieves optimal or very good outcomes quickly by sampling only very small portions of the large fitness landscape.

4.7.4 Applications of Genetic Algorithms

There are effective GA's applications available for a variety of fields like finance, engineering, and bioinformatics. Some of them are as follows:
- **Robotic and Parametric Design of Aircraft**: By adjusting parameters and developing improved solutions, GA is used in the design of aeroplanes and in the planning of robot arm trajectories [27].
- **Combinatorial Optimization:** The optimal configuration or combination of items from a finite set is the issue that GAs efficiently address in combinatorial optimization [16]. The VRP, TSP, bin packing, work scheduling, and DNA sequence alignment are a few examples. GAs serve as possible solutions that seek the ideal elemental combination through evolutionary processes.
- **Machine Learning:** GAs are used in machine learning, that is, to optimize the parameters and set up the models. Neural network system designs, regularization parameters, learning rate, and other hyperparameters may all be optimized with GAs. They are also useful in feature selection, where an algorithm is used to find the most relevant subset for a job by iteratively evolving a population of feature subsets [16].
- **Optimization Problems**: Optimizing mathematical functions, portfolios, allocating resources, and fine-tuning parameters are among the optimization issues that GAs proficiently resolve [12]. They progressively converge towards an optimal or nearly ideal solution by allowing candidate solutions evolve through genetic operators like crossover, mutation, and selection. This allows them to explore the solution space.

In general, GAs are quite flexible, by changing parameters such as population size, crossover and mutation rates, and selection methods. They may be used in a wide range of optimization tasks. Large search areas, nonlinear objective functions, and intricate restrictions are among the situations in which they come in handy.

These algorithms perform especially well in situations that are difficult for conventional optimization techniques to handle, such as those involving complicated landscapes, several local optima, and high dimensionality.

4.8 Issues in Swarm Intelligence Optimization Research

SI algorithms' research and development, applications, and understanding of the environment deal with several restrictions. Some of the limitations are as follows:
- The task-agnostic nature and large-scale manufacturing needs of swarm robots lead to limits in hardware and software. Both their natural behaviours and physical implementations frequently underestimate their complexity. More intricate designs are required to ensure that developing behaviours can be verified and explained, which calls for simpler controllers [25].
- The swarm method has uneven exploration and exploitation operations, sluggish convergence, and poor exploitation.
- Video streaming faces challenges like network variability, user preferences, and device capabilities [12, 27].
- The initial population, or solution set, can have a significant impact on how well SI systems work. An excessive probability of early convergence or less-than-ideal outcomes might arise from poor initialization [27].
- Intricate swarm systems with intricate hierarchies require time to develop. Changes in state take longer in more complex swarms. One issue that sophisticated swarm systems with complicated hierarchies face is non-immediacy. Because of the complex interactions between agents and the creation of higher level patterns, these systems frequently take longer to transition between different states or behaviours. The time required for state transitions tends to rise with swarm complexity [22].

Addressing these difficulties, SI algorithms remain a promising method for resolving complex optimization problems in a variety of domains. Multidisciplinary research efforts spanning optimization theory and computational intelligence can address these issues.

4.9 Future Gap in Swarm Intelligence Optimization

SI algorithms present a number of challenges, from algorithmic complexity to hardware limitations. But despite these difficulties, a bright future lies ahead. By adopting interdisciplinary teamwork and creative ways, there are few new opportunities for researchers to devise more efficacious approaches for agents to collaborate and transcend the constraints of existing algorithms:
- Improving software and hardware architectures, streamlining controllers, increasing component manufacturability, and creating reliable simulation environ-

ments are some of the steps required to meet the complicated behaviour and scalability requirements of swarm robots.

- The creation of novel algorithmic strategies is necessary to improve the efficiency and efficacy of swarm methods by mitigating their inherent imbalance between exploration and exploitation, slow convergence, and irregular exploitation.
- The effects of inadequate beginning solution sets or populations must be countered for more effective performance, and better initialization processes are essential to reduce the danger of early convergence or unsatisfactory results in SI systems.
- To guarantee uninterrupted streaming, adaptive systems need to use advanced algorithms that dynamically modify buffer management settings based on user behaviour patterns and real-time network measurements to provide high-quality and uninterrupted experiences.
- Interdisciplinary research efforts that bridge optimization theory and computational intelligence are crucial to address non-immediacy in complicated swarm systems with sophisticated hierarchies. Through these partnerships, methods for accelerating state transitions and reducing delays that come with intricate interactions may be developed, improving the responsiveness and effectiveness of advanced swarm systems.

4.10 Conclusion

Natural systems provide inspiration for SI and optimization. The cooperative conduct of ants, bees, and birds serves as an inspiration for their efficacy. SI algorithms leverage decentralized decision-making and short-range interaction to increase the flexibility of swarm systems, enabling AI to identify optimum solutions without centralized control. This technique provides scalable answers in practical ways, like other techniques, and is appropriate for optimizing tasks including routing, scheduling, clustering, and optimization. Similarly, it establishes a setting in which AI systems can cooperate and evolve. SI techniques, which provide distributed computing and self-organization, may be used to address large-scale issues in dynamic contexts. Combining SI with optimization algorithms gives AI systems a new level of problem-solving capability. These methods are used in a variety of industries, including robotics, banking, and telecommunications. This wide applicability of this approach makes it perfect for systems where dynamic unpredictability, uncertainty, and ambiguity may render traditional optimization insufficient. As a result, it presents itself as a fresh point of reference and perspective on what appear to be an autonomous system.

This chapter concludes by highlighting the importance of SI in enhancing AI's optimization skills. SI algorithms provide potential solutions to complicated optimization issues in a variety of fields by utilizing natural principles. SI optimization opens the door to the development of more effective and sophisticated AI systems.

References

[1] Ab Wahab, M. N., Nefti-Meziani, S., & Atyabi, A. (2015). A comprehensive review of swarm optimization algorithms. *Plos One, 10*(5), e0122827. https://doi.org/10.1371/journal.pone.0122827.

[2] Ahmed, H., & Glasgow, J. (2012). Swarm Intelligence: Concepts, Models and Applications. doi: 10.13140/2.1.1320.2568.

[3] Banharnsakun, A. (2023). A new approach for solving the minimum vertex cover problem using artificial bee colony algorithm. *Decision Analytics Journal, 6*, 100175. ISSN 2772-6622. https://doi.org/10.1016/j.dajour.2023.100175.

[4] Ant Colony Algorithm – from Wolfram MathWorld. (n.d.). https://mathworld.wolfram.com/AntColonyAlgorithm.html

[5] Bhavya, R., & Elango, L. (2023, May 10). Applications of Ant Colony Optimization. https://encyclopedia.pub/entry/43970

[6] Blum, C., & Li, X. (2008). Swarm Intelligence in Optimization. doi: 10.1007/978-3-540-74089-6_2

[7] Chu, S. C., Huang, H. C., Roddick, J. F., & Pan, J. S. (2011). Overview of algorithms for Swarm Intelligence. In Jędrzejowicz, P., Nguyen, N. T., & Hoang, K. (eds.). Computational Collective Intelligence. Technologies and Applications. ICCCI 2011. Lecture Notes in Computer Science () Vol. 6922. Springer, Berlin, Heidelberg. https://doi.org/10.1007/978-3-642-23935-9_3.

[8] Dorigo, M. (2007 January 1). Ant colony optimization. *Scholarpedia Journal*. https://doi.org/10.4249/scholarpedia.1461.

[9] Beg, D. M. S., & Waoo, R. A. A. (2023, May). Throughput comparison using artificial bee colony (ABC) algorithm with dynamic technique for improving data optimization. *25*(05). ISSN: 1007-6735.

[10] Ananthkumar, T., Julie, E. G., Robinson, Y. H., & Jaisakthi, S. M. (eds.) (2021). Simulation and analysis of mathematical methods in real-time engineering applications. John Wiley, & Sons.

[11] Zhu, F., Shuai, Z., Lu, Y., Su, H., Yu, R., Li, X., Zhao, Q., & Shuai, J. (2024). oBABC: A one-dimensional binary artificial bee colony algorithm for binary optimization. *Swarm and Evolutionary Computation, 87*, 101567. ISSN 2210-6502. https://doi.org/10.1016/j.swevo.2024.101567.

[12] Gupta, R., Tanwar, S., Tyagi, S., & Kumar, N. (2020, March 1). Machine learning models for secure data analytics: A taxonomy and threat model. *Computer Communications*. https://doi.org/10.1016/j.comcom.2020.02.008.

[13] Hamed, E. (2014, December 10). Swarm Intelligence. SlideShare. https://www.slideshare.net/eslamhamed93/swarm-intelligence-42553644

[14] Wang, K., Xin, L., Zhang, Y., Qi, J., Zhu, Z., Wang, Y., Zhong, L., & Cui, P. (2024). Sustainable and efficient process design for wastewater recovery of cyclohexane/isopropyl alcohol azeotrope by extractive distillation based on multi-objective genetic algorithm optimization. *Chemical Engineering Research and Design, 201*, 593–602. ISSN 0263-8762. https://doi.org/10.1016/j.cherd.2023.12.004.

[15] Qin, K., Zhang, Q., Liu, Y., Yang, B., Tang, X., & Zhuang, N. (2024). Dimensionality reduction combined with particle swarm optimization algorithm for solving high-dimensional nuclear data target accuracy evaluation problem. *Annals of Nuclear Energy, 198*, 110285. ISSN 0306-4549. https://doi.org/10.1016/j.anucene.2023.110285.

[16] Kanade, V. (2023, September 6). *Genetic Algorithms – Meaning, Working, and Applications*. Spiceworks. https://www.spiceworks.com/tech/artificial-intelligence/articles/what-are-genetic-algorithms/

[17] Karaboğa, D. (2010, January 1). Artificial bee colony algorithm. Scholarpedia. https://doi.org/10.4249/scholarpedia.6915.

[18] Kavitha, R., Jothi, D. K., Saravanan, K., Swain, M. P., Gonzáles, J. L. A., Bhardwaj, R. J., & Adomako, E. (2023 Feb 21). Ant colony optimization-enabled CNN deep learning technique for accurate detection of cervical cancer. *BioMed Research International, 2023*, 1742891. Retraction in: Biomed Res Int. 2024 Jan 9; 2024:9878397. PMID: 36865486; PMCID: PMC9974247. doi: 10.1155/2023/1742891.
[19] Wijk, K. L. A. S., & Yin, E. M. M. Y. (2021, June). Bayesian Parameter Tuning of the Ant Colony Optimization Algorithm. https://www.diva-portal.org/smash/get/diva2:1599698/FULLTEXT01.pdf.
[20] Khan, K. (2024). Optimizing adaptive video streaming: a swarm intelligence approach to dynamic buffer management. *International Journal of Multidisciplinary Research and Publications (IJMRAP), 6*(8), 27–33.
[21] Chen, L., Tunhua, W., Wang, Z., Lin, X., & Cai, Y. (2023). A novel hybrid BPNN model based on adaptive evolutionary artificial bee colony algorithm for water quality index prediction. *Ecological Indicators, 146*, 109882. ISSN 1470-160X. https://doi.org/10.1016/j.ecolind.2023.109882.
[22] Liu, S., McCoy, A. B., Peterson, J. F., Lasko, T. A., Sittig, D. F., Nelson, S., Andrews, J. E., Patterson, L., Cobb, C., Mulherin, D., Morton, C., & Wright, A. (2024, February 22). Leveraging explainable artificial intelligence to optimize clinical decision support. *Journal of the American Medical Informatics Association*. https://doi.org/10.1093/jamia/ocae019.
[23] Hussein, M. K., & Mousa, M. H. (2020). Efficient task offloading for IoT-based applications in fog computing using ant colony optimization. *IEEE Access, 8*, 37191–37201. doi: 10.1109/ACCESS.2020.2975741.
[24] Optimization. (2024, March 31). Merriam-Webster Dictionary. https://www.merriam-webster.com/dictionary/optimization#:~:text=%3A%20an%20act%2C%20process%2C%20or,a%20function)%20involved%20in%20this.
[25] particle-swarm-optimization-and-intelligence-2010.pdf.
[26] Pijarski, P., & Belowski, A. (2024). Application of methods based on artificial intelligence and optimization in power engineering – introduction to the special issue. *Energies, 17*(2), 516. https://doi.org/10.3390/en17020516.
[27] Bajpai, P. et al. Genetic algorithm – an approach to solve global optimization problems. *Indian Journal of Computer Science and Engineering, 1*(3), 199–206.
[28] Salgotra, R., Singh, U., Singh, S., Singh, G., & Mittal, N. (2021, January 1). Self-adaptive salp swarm algorithm for engineering optimization problems. *Applied Mathematical Modelling*. https://doi.org/10.1016/j.apm.2020.08.014.
[29] Sanyal, S. (2023, August 9). An Introduction to Particle Swarm Optimization (PSO Algorithm). Analytics Vidhya. https://www.analyticsvidhya.com/blog/2021/10/an-introduction-to-particle-swarm-optimization-algorithm/#:~:text=PSO%20Algorithm%20is%20an%20intelligent,best%20solution%20from%20its%20neighbors.
[30] Sarumaha, Y. A., Firdaus, D. R., & Moridu, I. (2023). The application of artificial bee colony algorithm to optimizing vehicle routes problem. *Journal of Information System, Technology and Engineering, 1*(1), 11–15. https://doi.org/10.61487/jiste.v1i1.9.
[31] Schäfer, J. (2024, January 22). Swarm Intelligence – How it works and practical areas of application. Konfuzio. https://konfuzio.com/en/swarm-intelligence-artificial-swarm-intelligence/#:~:text=In%20business%20and%20industry%2C%20companies,security%2C%20marketing%20and%20network%20design.
[32] Dhiman, H., Vyas, N., Pagrotra, A., & Dutt, V. (2023). Real-time traffic analysis and driver safety enhancement via IoT and computer vision with high-resolution project lamps for intelligent road illumination [Internet]. In 2023 1st International Conference on Circuits, Power and Intelligent Systems (CCPIS). IEEE. Available from. http://dx.doi.org/10.1109/CCPIS59145.2023.10291872.
[33] Swarm Intelligence. (n.d.). https://www.techferry.com/articles/swarm-intelligence.html#:~:text=It%20facilitates%20improved%20analytic%2C%20predictive,different%20agent%2Dlevel%20control%20parameters.

[34] Understanding ant colony optimization algorithms. n.d. INDIAai. https://indiaai.gov.in/article/understanding-ant-colony-optimization-algorithms.
[35] What is Hyperparameter Tuning? | Anyscale. (n.d.). Anyscale. https://www.anyscale.com/blog/what-is-hyperparameter-tuning
[36] Wright, S. J. (1998, July 20). Optimization | Definition, Techniques, & Facts. Encyclopedia Britannica. https://www.britannica.com/science/optimization
[37] Luo, Y., Shi, Y., & Cai, N. (2021). Chapter 9 – Application and potential of the artificial intelligence technology. In Luo, Y., Shi, Y., & Cai, N. (eds.). Hybrid Systems and Multi-energy Networks for the Future Energy Internet (pp. 217–234). Academic Press, ISBN 9780128191842, https://doi.org/10.1016/B978-0-12-819184-2.00009-2.

Leena Arya, Yogesh Kumar Sharma, and Ravi Rastogi

5 Privacy and Security for 6G's IoT-Connected Future in the Age of Quantum Computing

Abstract: The advent of the sixth generation of wireless communication technology, called 6G, brings many more possibilities for the development and evolution of the Internet of things (IoT) and the consequent improvements in our day-to-day lives in the era of quantum computing. The chapter provides a comprehensive look at the complexities of integrating 6G networks with IoT devices, from smart home sensors to industrial ones. It explores the finiteness of resources and the transition path to a sustainable future, including targets, deadlines, and ways of defining and achieving them. Moreover, it stresses the importance of privacy preservation technologies and user-oriented controls for people to have the self-determination to choose and manage their personal information. As security is the main challenge, there should be some measures that can be implemented to make sure that stringent protocols and quantum-safe solutions are used to face these complexities. The chapter offers insights into the critical issues surrounding privacy and security in 6G IoT networks and suggestions for addressing these challenges effectively.

Keywords: Internet of things, blockchain, 6G, quantum communication, artificial intelligence, molecular communication, VLC, THz technology

5.1 Introduction

With the development of the age of the Internet of things (IoT) and sixth generation (6G) wireless networks, the world is witnessing a network of devices and applications that are increasingly data-driven and connected. This transformation toward hyper-connectivity offers the prospect of redefining industries, making the world more efficient, elevating users' experiences, and taking innovation to new heights [12]. Indeed, these technologies also bring security and privacy risks that must be averted to guarantee the security of stored and transmitted data within these systems [3]. In the twenty-first century, people tend to deal with their family, friends, or colleagues through technological interconnectedness more frequently than ever, and therefore, the consequences are more severe than in the past.

The 6G convergence with IoT will bring a new era of connectivity that will feature ultrareliable low-latency communication, massive machine-type communication, and enhanced mobile broadband (eMBB) capabilities [1]. An additional 6G network will facilitate the IoT connection of billions of devices expected in the future, which greatly increases the area to attack by malicious actors and creates an issue of security tremendously [5, 6].

https://doi.org/10.1515/9783111436180-005

Restoring the realization of present security and privacy dilemmas that are highly challenged in 6G IoT networks, the focus is on global statistics and trends concerning cyber threats and incidents [7]. Based on the latest analysis, it has been seen that cyberattacks are increasing in frequency, complexity, and the damage that they cause, leading to very negative consequences for organizations, including financial, reputational, and social effects [16]. The number is impressive. It is estimated in 2020 that the cost of cybercrime worldwide was 1 trillion dollars, a sign of the massive economic burden imposed by cybercriminals [14, 38].

Furthermore, the spread of IoT devices has also resulted in new attack vectors through which cyber threats are conducted [18]. Unsecured devices can also serve as the entry points that attackers use to penetrate networks, compromise data, and disrupt operations [20]. Evidence indicates that the number of connected IoT devices [19] amounted to 20 billion at the end of 2020 and is predicted to surge to 75 billion by 2025, underlining the explosive growth of the IoT ecosystem accompanied by an expansion of the attack surface.

Further, the development of 6G networks is anticipated to bring about new security and privacy vulnerabilities that encompass scenarios of network slicing, edge computing, artificial intelligence (AI) integration, and quantum-resistant cryptography. These challenges require the advancement of proactive steps to design secure architectures, protocols, and mechanisms that could mitigate attacks from the cyber threats that will evolve in the future [2].

The primary objectives of the chapter include identifying emerging privacy challenges, evaluating security measures, assessing policy implications, and proposing innovative solutions to mitigate them to sustain the integrity of 6G IoT networks. This focus on security is driven by the need to build trust and ensure user adoption of 6G technologies. Moreover, the advent of quantum computing introduces both opportunities and challenges for safeguarding data integrity and confidentiality.

The chapter discusses the cybersecurity and privacy perils in the 6G IoT ecosystem development, analyzes possible threats and vulnerabilities, and offers innovative solutions to tackle the risks and enhance resilience. The analysis of global numbers and trends highlights the urgency and importance of cybersecurity issues in the context of advanced technologies and their future impact, which will assist in developing secure and trustworthy future network infrastructures.

5.1.1 A Connected World: Overview from 1G to 5G

The progression of mobile communication from the first generation (1G) to the fifth generation (5G) has led to profound global changes that shape how billions of people interact, work, and live [21]. Today's smartphone is an inseparable part of our lives as it links us both with our friends and the world out there. Nevertheless, the way it arrived here represents a remarkable evolution starting from the past five generations

(1G to 5G) and witnessing a large growth in speed, reach, and functionality [4]. Let us embark on a journey through this global story of connectivity, armed with statistics that illustrate the impact of each generation.

5.1.1.1 1G (First Generation): Pioneering Steps (1980s)

Global phone subscribers: 30 million (1985).

Data transmission: Nonexistent.
The 1G network was first introduced in the 1980s and was used for basic voice calls through analog cellular network systems. In the early 1990s, the 1G network planned the coverage of more than 25% of the world's population, which brought new ways of mobility to the world. Imagine the outage of old chunky phones, poor call quality, and no Internet access. The mobile revolution was the reality of 1G, but it was the foundation of it. The limited range and functionality of the service meant that it was mostly a PR statement for the concept of portable communication.

5.1.1.2 2G (Second Generation): After All the Voice Calls (1990s)

Global phone subscribers: 1 billion (1997)

Data speeds: As much as 14 kbps.
The digital technology chain of arrival in its turn means a quantum leap. The 2G had better call quality, text messaging became possible, and the evolution of mobile data emerged, although annoyingly at slow speeds. For this generation, the world of the Internet was taken personally to the next cellular level as the first email on the go and perhaps early access to the mobile Internet was made available.

5.1.1.3 3G (Third Generation): The Mobile Internet Age (2000s)

Global phone subscribers: In 2007, the number of people equaled 2.5 billion (2.5).

Data speeds: Up to 2,222.2 kbps.
The introduction of 3G networking in the early 2000s, the most crucial technological innovation of that time, allowed for rapid data transmission and, thus, for Internet access from mobile phones and multimedia messaging. In 2010, 3G networks covered nearly 90% of the global population, driving the proliferation of mobile data services. The rates have simply rocketed, making it possible to hold video calls, stream music, or even browse the Internet. This age ushered in the years of smartphones and apps, making it impossible for us to consume information in the same way and interact

with the digital world in the same manner. The explosion of global mobile data is fueled by this generation, as they use it in a way that no other generation ever has.

5.1.1.4 4G (Fourth Generation): The Era of the Speed Revolution (2010s)

Global phone subscribers: 5.4 billion (2017).

Data speeds: Up to 100 Mbps.
The 2000s rolled out 4G and showed a revolutionary change in the network with high-speed broadband or ultrafast speed, low latency, and increased efficiency. The upgrade from 3G to 4G was a big jump that changed the Internet. This generation did away with the legacy of slow mobile Internet access, which was severely restrictive compared to what was possible before, allowing for HD content streaming, audio-video communications, and bandwidth-hungry applications. The magnitude went beyond individual lives, manifesting itself in the operation and progress of the various enterprises that depend on real-time data transmission. The mobile phone became a common device; thus, mobile Internet became most users' top connectivity choice.

5.1.1.5 5G (Fifth Generation): The Connected World (2020s)

Global phone subscribers: 1 billion (2023).

Data speeds: Up to 20 Gbps.
Launched in the 2020s, 5G is an usher for a new transmission of connectivity that has blazing speeds, ultra-low latency, and massive device connectivity. It seems that this generation turns out to be very purposeful and ambitious, ready to switch the status-quo arrangement of the operating principles of manufacturing, healthcare, and transportation industries with applications such as connected cars, smart cities, and IoT. According to the present statistics for 2023, 5G networks have covered more than 15% of the global population, and it is expected that this rate will reach around 60% by 2025 [9]. The global 5G infrastructure market is estimated to reach US$70 billion by 2026 to spur technology advancements across industries and build foundations for important wireless technology applications such as IoT, AR/VR, and autonomous driving.

5.1.2 Beyond the Numbers

Mobile technology has transformed everything we do, starting from our daily activities to how we work, to our interaction with our environment. From the overview of 1G to the path-breaking innovation of 5G, it has been a remarkable manifestation of the genius of man and his contribution to the progress of our connected world. As we look ahead, 6G+ technologies [21] will bring us even more exciting new possibilities, and the boundaries in our connected world will further disappear [2].

5.2 Literature Review

This section conducts a comprehensive review of existing literature on the security and privacy challenges in IoT networks. It synthesizes insights from previous researches, establishing a foundation for understanding the unique considerations brought about by the integration of 6G technologies.

Rappaport et al. [1] provided an extensive overview of future wireless systems above 100 GHz, likely part of 6G, discussing regulatory efforts and showcasing potential applications. It emphasizes the significance of THz frequencies for novel capabilities like cognition, imaging, and communication, highlighting challenges and the need for further research in areas such as antenna technology, propagation measurements, and biological effects of THz radiation.

Ziegler et al. [2] described the journey toward 6G connectivity with the successful initiation of architectural research, while commercial 5G deployment is ongoing worldwide. Over the next 8 years, the evolution of 5G architecture will continue, paving the way for exciting opportunities in real-time communication and synchronization between the physical, digital, and biological worlds. This will be facilitated by a hot cloud platform, convergent RAN-CORE, new data and information architecture, specialization of the network, and intent-based automation, laying the foundation for diverse future applications across consumer and vertical industries.

Akyildiz et al. [3] surveyed key techniques for future wireless networks, emphasizing wireless ubiquity as a driving force for 6G and beyond. It highlights essential technologies, operational nuances, and challenges, aiming to offer insights and inspire contributions toward achieving ubiquitous wireless access.

Minghao Wang et al. [4] discussed the transition from 5G to 6G networks, highlighting the anticipation for advancements in network services. It provides a comprehensive investigation into security and privacy concerns, spanning from the historical evolution of mobile networks to potential future applications of 6G, aiming to stimulate further research and interest in addressing these issues.

Liu et al. [5] discussed how the rapid advancement of intelligent computing and data technologies (ICDT) is propelling society toward digital twin integration, with envisioned applications such as synesthesia Internet and super transportation. Key network performance requirements for 6G, including peak rates and ultra-high-precision positioning, are discussed alongside proposed features like on-demand fulfillment and native AI, aimed at addressing lessons learned from 5G's development.

5.3 Unique Cybersecurity Threats in 6G IoT

Similarly, 6G allows for faster access speeds and innovation that we cannot even imagine yet, but it may put us at a greater risk of cyber threats. The threats as opposed to those that torment the current networks evolved from the very fabric of the revolutionary technologies of 6G [11]. It is important to realize these unique weaknesses and find ways to prevent them because tomorrow's interconnected world will not be secure enough without them. These threats are described further.

5.3.1 Millimeter-Wave Mayhem

The 6G is based on the frequencies of mm-wave (mmWave) for high-speed communication. Unfortunately, these shorter wavelengths [8] are more susceptible to the following.

5.3.1.1 Signal Interception and Jamming

It has been reported that 5G networks are, without a doubt, more vulnerable to the breach of signal (10× more in comparison with earlier generations). mmWave's narrower beams aid targeted attacks that can compromise infotainment and similar car-connected services, with more than 1 billion connected vehicles predicted in 2040 to increase the risk.

5.3.1.2 Beamforming Hijacking

Attackers could exploit the directionality of mmWave beams to misdirect data streams, thus eventually leading to partial or complete compromise of sensitive information or components involved in communication between devices. A survey undertaken in 2021 by the Ponemon Institute established that 74% of IT practitioners have concerns about data interception over 5G networks, showing how this anxiety is growing with the vendors' mmWave technology.

5.3.2 AI-Powered Peril

However, the speed of 6G is very much dependent on AI for network management and device performance. However, this dependence introduces new attack vectors.

5.3.2.1 AI Poisoning

In this way, malicious actors can create biases in AI models by manipulating the training data and interfering with the system, hence leading to inaccurate conclusions and chaos. In a recent report from Carnegie Mellon University, the researchers discovered that AI models charged with the responsibility of managing traffic lights can be manipulated, which can ultimately cause accidents.

5.3.2.2 Model Manipulation

Hackers would have the ability to manipulate AI models by using vulnerabilities to control or influence their outputs, which in the end would lead to data breaches and manipulation of critical infrastructure. A survey by Gartner indicated that CIOs have the security of AI systems in mind since 70% of them are worried about the security risks involved.

5.3.2.3 Adversarial attacks

Actors with malicious motives can engage in scenarios that mislead or take advantage of AI systems, which may cause unintended results or system failure. Research conducted by the University of California, Berkeley, found that it could be possible to mislead autopilot systems into making faulty decisions by adding black stickers.

5.3.3 Supply Chain Specters

The multilayered global supply chains for manufacturing and distributing IoT devices offer attackers several tactics, providing an opportunity to insert malicious code or hardware backdoors at various stages.

5.3.3.1 Hidden Backdoors

Meanwhile, malicious actors can insert stealthy backdoors in the firmware or hardware when they are manufacturing items, which may give them access to devices after the supply chain. As reported by the National Counterintelligence and Security Center in 2020, supply chain attacks are among the major threats to national security, posed by 80% of organizations experiencing at least one such attack.

5.3.3.2 Counterfeit Components

Instead of the original, incorporating malware into the counterfeit components can open doors to system-wide vulnerabilities. A study by the International Trade Commission in 2021 shows that the amount of seized counterfeit electronics was estimated at $230 billion worldwide, which highlights this problem at a high level.

5.3.3.3 Software Tampering

Malicious entities could modify software updates delivered during distribution, introducing vulnerabilities that compromise numerous devices as if embracing a virus. The 2021 SolarWinds attack is a good example of the possible negative consequences for a company that has been made to update its software using compromised methods.

5.3.4 Edge Computing Exposure

The 6G's great focus on edge computing, which means that data is processed close to the devices themselves [15], presents a new set of security challenges.

5.3.4.1 Distributed Attack Surface

The greater number of entrance endpoints increases the security surface and makes it harder to protect and monitor every point of entry. Sources indicate that 6G could produce 1 trillion connected devices by 2030, which would increase the number of vulnerabilities by a large factor [17].

5.3.4.2 Limited Resource Constraints

Edge devices are usually affected by resource-constrained attacks because these devices have limited processing power and memory. A research in 2022 by the University of Oxford has shown that resource depletion attacks may make compromising edge devices used in smart homes, which could result in gaining unauthorized access or manipulating data.

5.3.4.3 Data Privacy Concerns

The edge computing environment causes concerns about the safety of data and how it could be misused due to unauthorized access. Pew Research Center findings reveal that 72% of Americans worry about companies collecting their data without permission. This shows that data privacy regulation is crucial for edge computing. Table 5.1 shows the case studies of the reality of 6G IoT security threats.

Table 5.1: Case studies of the reality of 6G IoT security threats.

Threat category	Case study	Global statistics	Potential impact
Millimeter-wave Mayhem	Jamming of autonomous vehicle communication: attackers exploit mmWave vulnerabilities to disrupt communication between vehicles, potentially causing accidents.	1 billion autonomous vehicles are projected to be on the road by 2040 (McKinsey Global Institute, 2021)	Loss of life and property
	Interception of sensitive data in smart grids: malicious actors intercept data transmitted over mmWave frequencies, gaining access to critical infrastructure information.	The global smart grid market is expected to reach $823.4 billion by 2027 (Markets and Markets, 2022)	Disruptions to energy supply, potential economic losses
AI-powered peril	Manipulation of AI models for traffic light control: attackers bias the training data, causing AI models to make incorrect decisions, leading to traffic accidents.	Road traffic accidents claim 1.35 million lives annually (World Health Organization, 2022)	Public safety risks and potential loss of life
	Hijacking of autonomous vehicles through AI manipulation: hackers exploit vulnerabilities in AI models controlling autonomous vehicles, potentially causing crashes or manipulating their movements.	The estimated cost of autonomous vehicle accidents is expected to reach $217 billion by 2035 (RAND Corporation, 2020)	Public safety risks and economic losses.

Table 5.1 (continued)

Threat category	Case study	Global statistics	Potential impact
Supply chain specters	Hidden backdoors in medical device firmware: malicious actors embed backdoors during manufacturing, granting unauthorized access to patient data and potentially manipulating device functions.	The global medical device market is expected to reach $637.8 billion by 2027 (Grand View Research, 2022)	Patient privacy violations and potential harm to health
	Counterfeit components in smart home devices: Fake components containing malware compromise entire smart home ecosystems, enabling unauthorized access and data breaches.	The global smart home market is projected to reach $427.6 billion by 2027 (Statista, 2022)	Privacy violations and potential security risks for homeowners
Edge computing exposure	Resource-constrained attack on edge devices in smart homes: Hackers exploit limited resources of edge devices to gain unauthorized access and control smart home functionalities.	An estimated 64 billion IoT devices are expected to be connected by 2025	Privacy violations and potential manipulation of smart home systems
	Data breaches at edge computing nodes: Sensitive data processed at the edge is compromised due to security vulnerabilities, leading to privacy violations and potential identity theft.	43% of organizations reported experiencing at least one data breach in 2021 (IBM Security, 2022)	Erosion of trust in edge computing leads to financial losses

5.4 Security and Privacy Issues of 6G technologies

6G is the next big thing in connectivity, which may just bring the global community to crossroads between concerns of security and privacy [10, 13]. Here is a deep dive into these challenges, armed with relevant global statistics.

5.4.1 Artificial Intelligence

Algorithms, like people, are biased and vulnerable to manipulation, which threatens billions worldwide [27].

Global AI Market Size: Likely to reach $1.56 trillion in 2023 (Statista).

Bias in AI Algorithms: A McKinsey report from 2022 showed that most global companies (83%) are experiencing one or more types of AI bias [28].

Manipulating AI Models: A 2021 study on image recognition models conducted by Carnegie Mellon University showed that the models could easily be fooled, with success rates as high as 99.6%.

Such numbers draw a picture that we would rather do without AI – a world where thousands of millions of users of 6G powered by AI might face discrimination or misinformation [25]. Efficient data governance and stringent security protocols are equally important in ensuring risk mitigation [38].

5.4.2 Molecular Communication

Estimated Number of Connected Devices by 2030: Over half of global Internet traffic will come from just 50 billion devices (Statista).

Molecular Communication Research Funding: More than €15 million is being spent by the European Union on projects dealing with these issues.

Eavesdropping on Molecular Signals: A 2022 study by MIT revealed that such intercepting and analyzing molecular communication is possible.

Because billions of different devices can transmit information beyond molecules in 6G, the possibility of outside listeners accessing sensitive health information in the body is now a global issue [30]. Privacy should be the main factor of consideration in designing secure communication protocols and encryption in this domain [31].

5.4.3 Quantum Communication

Global Quantum Technology Market Size: It is expected to grow as high as $45.5 billion by 2025 (Grand View Research).

Vulnerability to "Man-in-the-Middle" Attacks: Recent research showed that vulnerability in quantum communication systems was successfully exploited.

Disruption of Fragile Quantum States: A 2021 study from ETH Zurich, which finds quantum communication signal disruption due to environmental conditions, was published in 2021.

With quantum-resistant encryption being highly impregnable, it urges to address the issues of a fast-growing global market [29]. Quantum key distribution protocols together with sophisticated error correction methods are a must for the protection of these revolutions [32].

5.4.4 Blockchain

Number of Global Blockchain Users: Yet 46 million (Source: Statista).

Value of the Global Blockchain Market: In 2022, it was worth $4.8 billion (Grand View Research).

51% Attacks: Unlike the Ethereum Classic blockchain, which was successfully attacked in 2022, showing its vulnerable side.

Sybil Attacks: The study by University of California, Berkeley, in 2020 proved that the Sybil attack was the most lethal threat for 80% of the proof-of-stake blockchains.

With a million users and billions of dollars invested mutually, this problem must be solved. As such, securing systems like 51% and Sybil attacks are of high priority [33]. Impart robust consensus mechanisms [34], identity verification protocols, and regular security audits to avert tampering with the distributed ledger [22].

5.4.5 THz Technology (THz)

Global 5G Market Size: A market value of $131.9 billion in 2022, according to Market Research Future.

THz Frequencies Used in 6G: With the anticipation of achieving higher speeds than 5G.

Limited Research on Health Effects: The long-term health effects of exposure to terahertz rays have not been fully studied yet.

The need for research on health risks as 6G possibilities in THz frequencies emerge will become necessary with 5G propagation across the globe [3]. In addition to that, the potential of THz waves to see through clothing causes privacy concerns, raising the need to adopt responsible deployment strategies [33].

5.4.6 Visible Light Communication (VLC)

Global Indoor Location Services Market Size: The projection is that it will amount to $55.8 billion by 2027 (as stated by Market Research Future).

Visible Light Communication (VLC)'s Potential for Indoor Applications: For example, Visible Light Communication (VLC) can be applied in various indoor environments such as homes, hospitals, retail spaces, offices, and public venues, offering enhanced communication and lighting solutions.

Susceptibility to Interference and Eavesdropping: VLC signals can be tapped, thus raising a serious privacy concern.

To safely exploit 6G in vertical location services, VLC in 6G poses considerable opportunities, but we should also take care of security gaps. Rigorous modulation methods, encryption protocols, and network segmentation approaches can be used to prevent interference and eavesdropping [37].

An overview of the security and privacy concerns that are inherent to emerging 6G technologies is presented in Table 5.2.

Table 5.2: Security and privacy concerns that are inherent to emerging 6G technologies.

Key technology	Security and privacy issues	Key technology contribution	Reference
AI	Access control: Fine-grained control processes might create complexity and potential vulnerabilities.	Fine-grained access control: Enables more granular permission management but requires robust implementation.	[25]
	Malicious behavior: AI systems are susceptible to manipulation and attacks, leading to network disruptions.	Anomaly detection: AI can identify and predict network anomalies, enabling early warnings against cyberattacks.	[26]
	Authentication: Unsupervised learning methods might not be sufficiently secure for authentication.	Enhanced authentication: AI-powered methods could improve physical layer authentication but require rigorous testing.	[27]
	Communication: Machine learning-based antenna design might unintentionally leak information.	Secure communication: Machine learning can optimize antenna design for improved security in physical layer communication.	[28]
	Encryption: Combining AI and quantum encryption schemes requires careful design and integration.	Advanced encryption: Opens doors for more robust encryption but necessitates addressing potential compatibility issues.	[29]
Molecular communication	Malicious behavior: Disruption of communication channels or processes can pose significant threats.	Enhanced security: Coding schemes can improve data transmission security.	[30]
	Authentication: Developing robust authentication mechanisms remains a challenge.	Novel authentication: Molecular communication offers potential for new authentication methods but requires further research.	[31]

Table 5.2 (continued)

Key technology	Security and privacy issues	Key technology contribution	Reference
Quantum communication	Encryption key security: Protecting quantum encryption keys from interception is crucial.	Unbreakable encryption: Quantum communication offers theoretically unbreakable encryption, but practical implementation challenges remain.	[29]
	Communication modes: Different modes of quantum communication have various security implications.	Secure communication channels: Quantum communication offers fundamentally secure channels, requiring careful design and infrastructure development.	[32]
Blockchain	Authentication: New mobile service authorization concepts need thorough security evaluation.	Improved access control: Blockchain can offer more transparent and secure access control mechanisms.	[33]
	Access control: Existing access control protocols might need adaptation for blockchain integration.	Enhanced transaction validation: Blockchain's hashing power can improve transaction validation and security.	[34]
THz communication	Authentication: Relying solely on electromagnetic signatures for authentication might be vulnerable.	Unique authentication: THz electromagnetic signatures offer the potential for unique device identification but require robust verification methods.	[35]
	Malicious behavior: The interception of narrow beams poses challenges despite their directional nature.	High-speed communication: THz offers high-speed data transmission but requires addressing interception risks.	[36]
VLC communication	Malicious behavior: Cooperation between eavesdroppers can compromise security protocols.	Secure communication protocols: Novel protocols are needed to address vulnerabilities arising from cooperative eavesdropping.	[37]

5.5 Strategies for Securing 6G IoT Networks

Keeping IoT networks secure in 6G technology is a must, as it will protect data privacy and keep all devices connected safely [10]. With the scale and complexity of IoT deployments growing globally, a blended approach to strategy is vital so that threats and vulnerabilities are dealt with adequately as they advance [13]. The following strategies are discussed further.

5.5.1 Encryption and Authentication Mechanisms

It has accelerated the connections that we have with our family and friends, but also increased our fragile emotional states, leading to several mental disorders. The implementation of robust encryption protocols and authentication mechanisms is becoming vital for data in motion and at rest security [12, 38]. As per the Gartner report, by 2025, using the current trends, 75% of IoT-generated data will be processed at places away from traditional IT data centers, thereby underscoring the value of encryption to secure data transmission over already distributed environments.

5.5.2 Secure-by-Design Principles

Implementing a secure-by-design principle for IoT devices and infrastructure development is very important for minimizing vulnerabilities [23]. According to Symantec's research, 70% of IoT devices have a highly exploitable flaw, thus highlighting the need to create security controls and conduct stringent testing when designing products [11, 40].

5.5.3 Intrusion Detection and Prevention Systems (IDPS)

Finally, the media played a significant role in fostering public support during the election campaign. Organizations can deploy IPS solutions to effectively detect and avert threats instantly. The market for IDPS, as claimed by research conducted by Markets, is supposed to be worth $8.7 billion by 2026. The cause of this is the rapid adoption of IoT devices and the widespread occurrence of cyber-attacks [16].

5.5.4 Zero Trust Architecture (ZTA)

In conclusion, conservation agriculture plays a vital role in farming practices. It improves soil productivity, protects biodiversity, reduces yield fluctuations, and mitigates crop losses due to increasing weather instability. By incorporating this approach, we can achieve sustainable agricultural development that meets the food security needs of a growing world population while protecting our planet's natural resources. Implementing the zero trust architecture (ZTA) ensures that the organization takes control of unauthorized access and lateral movement on networks by design. Forrester says that privileged credential issues make up 80% of security breaches, which means that the least privileged access approach encouraged by ZTA frameworks is more likely to benefit enterprises.

5.5.5 Security Orchestration, Automation, and Response (SOAR)

Democratization results in different systems of government, as some countries opt for democracies that promote individual rights and freedoms. Security orchestration, automation, and response (SOAR) systems improve incident response workflows and ensure that organizations can respond promptly to security incidents. According to the research done by Cybernetics Security Funds, the total market for SOAR platforms, which is growing due to the complexity of cyber threats and the demand for automated response abilities, is estimated to reach $2.3 billion by 2025.

5.5.6 Collaborative Threat Intelligence Sharing

Sharing and collaboration of threat intelligence within the framework of collective efforts is the way for organizations to obtain information about the latest threats and vulnerabilities. As indicated by the Ponemon Institute, 68% of organizations say that sharing threat intelligence is a good step toward the improvement of their security posture.

5.5.7 Continuous Security Monitoring and Compliance

Regular supervision of infrastructure elements and adherence to industry regulations to achieve the desired level of security are a must. A study from IBM revealed that the average cost of a data breach would be $4.24 million, which shows how significant continuous monitoring is when it comes to security in detection and timely response to incidents. In a nutshell, the 6G IoT network protection involves a strategic approach that considers cryptography, authentication, secure design principles, intrusion detection, ZTA, automation, threat intelligence awareness, and continuous monitoring. By integrating these approaches with global statistics and industry best practices, organizations will be able to manage risks and IoT deployments that are secure from emerging threats.

5.5.8 Mitigating Risks Through Quantum-Secure Solutions

While looking at the edge of the time of a quantum computer era, classic cryptographic techniques are being challenged by the growing necessity. Currently, the cryptographic protocols will most probably be rendered useless by the quantum computers, which threaten to be the new norm that will be the focal point of the security of 6G networks for an IoT-connected future. Confronting this looming danger, quantum-protected systems have become essential for the preservation of cryptography, making it even stronger against quantum-controlled attacks [24, 39].

Quantum security solutions constitute a paradigm change of a purely technical approach to encryption algorithms and utilize the elements of quantum physics to make them secure and withstand quantum attacks. Among several prospective approaches, lattice-based cryptography is probably the most promising method, which is based on the probable intractability of some mathematical problems in higher-dimensional space. Lattice-based schemes are preferable as they bring strong security guarantees against both classical and quantum hazards as such are potential candidates in the provision of security for the 6G IoT.

Code-based cryptography builds upon error-correcting codes as the foundation for cryptographic primitives that are inhumanly attack-proof. Without quantum computing, code-based cryptography, in which errors in quantum operations are made detectable and correctable, is the best way to protect data from hostile quantum environments. Hash function-based cryptography is another promising technology that makes use of cryptographic hash functions for authentication and verification. More checksums than the numbers settled by quantum computing could stop it from being cracked. Although hash-based methods are not unruffled by quantum hackers, they can be added with quantum-safe processes to develop their security.

To combat this new threat, industry players will need to incorporate quantum-safe technologies natively within 6G's IoT ecosystem, given that data to be transferred through such networks is expected to be highly sensitive.

5.6 Conclusion

This chapter is about privacy and security related to 6G, which depends on the ability of IoT; however, the effect of quantum computing is seen as a disruptive influence. The 6G potential to be a game changer and enable people and industries to be more efficient is undeniable, but the vast area of connected devices and new technologies presents security and privacy issues just as daunting. 6G AI, blockchain, and the intriguing realms of quantum communication and THz technology each have their unique ways of being vulnerable and complex, which demands innovative solutions and collaborative efforts. The focus is on vulnerabilities in the present cryptographic processes concerning quantum computing and also puts forward and carries out strategies to mitigate this by deploying quantum-resistant cryptographic solutions.

Furthermore, key management must be used with maximum strength, together with a proactive approach to vulnerability management and user awareness initiatives, to take an all-in strategy about security during the 6G era. There is a need to adopt innovation with due care, channeling resources to build advanced security frameworks, privacy-friendly technologies, and international collaboration. By acknowledging the challenges, creating an environment of free discussion, and taking ethical considerations into account, the strategies for securing 6G IoT networks ensure that the power of 6G is used for the benefit of humankind, and we build a world that is connected, where security and privacy stand next to development.

Looking toward the future, there is still a lack of comprehensive and severe standardization procedures to define the reliability and proficiency of post-quantum cryptography (PQC) algorithms. This might be the area that requires more research and rigorous standardization processes to identify the robust and efficient PQC algorithms for entire 6G implementations. Global alignment among industrial leaders, research institutions, and governments is paramount to mitigate the multifaceted risks that the quantum-insecure future might carry.

The network architectures for 6G may differ from previous networks as they will merge different technologies and devices heterogeneously. The complexity and variety of safety concerns for this ecosystem are a real challenge; therefore, the design of security solutions that can meet the safety needs of this network is an important challenge to be solved. Various privacy problems arise with the increasing numbers of connected devices that operate on an enormous amount of data, hence requiring strong privacy-preserving methods. Anonymization and homomorphic encryption, along with the federated learning technique, are set to be the future of privacy guarantees in 6G networking.

The merging of 6G, IoT, and quantum computing opens a new door to realizing an integrated and cyber-secure world in which personal data is monitored while its privacy and security are ensured.

References

[1] Rappaport, T. S., Xing, Y., Kanhere, O., Ju, S., Madanayake, A., Mandal, S., Alkhateeb, A., & Trichopoulos, G. C. (2019). Wireless communications and applications above 100 GHz: Opportunities and challenges for 6G and beyond. *IEEE Access, 7*, 78729–78757.

[2] Ziegler, V., Viswanathan, H., Flinck, H., Hoffmann, M., Räisänen, V., & Hatonen, K. (2020). 6G architecture to connect the worlds. *IEEE Access, 8*, 173508–173520. 2020. doi: 10.1109/ACCESS.2020.3025032.

[3] Akyildiz, F., Kak, A., & Nie, S. (2020). 6G and beyond: The future of wireless communications systems. *IEEE Access, 8*, 133995–134030. doi: 10.1109/ACCESS.2020.3010896.

[4] Kumar, T. A., Arun Samuel, T. S., Dinesh Jackson Samuel, R., & Niranjanamurthy, M. (eds.). (2022). Privacy and security challenges in cloud computing: A holistic approach. CRC Press.

[5] Liu, G., Huang, Y., Li, N., Dong, J., Jin, J., Wang, Q., & Li, N. (2020). Vision, requirements, and network architecture of 6G mobile network beyond 2030. *China Communications, 17*(9), 92–104. doi: 10.23919/JCC.2020.09.008.

[6] Viswanathan, H., & Mogensen, P. E. (2020). Communications in the 6G era. *IEEE Access, 8*, 57063–57074. doi: 10.1109/ACCESS.2020.2981745.

[7] Tataria, H., Shafi, M., Molisch, A. F., Dohler, M., Sjoland, H., & Tufvesson, F. (2021). 6G wireless systems: Vision, requirements, challenges, insights, and opportunities. *Proceedings of the IEEE, 109*(7), 1166–1199. doi: 10.1109/JPROC.2021.3061701.

[8] Zhang, J., Kang, K., Huang, Y., Sha, M., & Molisch, A. F. (2019). Millimeter and THz wave for 5G and beyond. *China Communications*, *16*(2), 3–6.
[9] Nakamura, T. (2020). 5G evolution and 6G. In Proceedings of the IEEE Symposium on VLSI Technology and Circuits (VLSI). Honolulu, HI USA, 1–5. doi: 10.1109/VLSITechnology18217.2020.9265094.
[10] Nguyen, V.-L., Hwang, R.-H., & Lin, Y. R. (2021). Security and Privacy for 6G: A survey on prospective technologies and challenges. *IEEE Communications Surveys and Tutorials*, 1–45. doi: 10.1109/COMST.2021.3108618.
[11] Ylianttila, M., Kantola, R., Gurtov, A., Mucchi, L., & Und, O. I. (2021). 6G White Paper: Research Challenges for Trust, Security and Privacy. Available online: http://urn._/urn:isbn:9789526226804.
[12] Nagaraj, S., Kathole, A. B., Arya, L., Tyagi, N., Goyal, S. B., Rajawat, A. S., Raboaca, M. S., Mihaltan, T. C., Verma, C., & Suciu, G. (2022). Improved secure encryption with energy optimization using random permutation pseudo algorithm based on Internet of things in wireless sensor networks. *Energies*, *16*(8), 1–16.
[13] Porambage, P., Gur, G., Osorio, D. P. M., Liyanage, M., Gurtov, A., & Ylianttila, M. (2021). The roadmap to 6G security and privacy. *IEEE Open Journal of the Communications Society*, *2*, 1094–1122. 2021. doi: 10.1109/OJCOMS.2021.3078081.
[14] Saad, W., Bennis, M., & Chen, M. (2020). A vision of 6G wireless systems: Applications, trends, technologies, and open research problems. *IEEE Network*, *34*(3), 134–142.
[15] Hui, Y., Cheng, N., Su, Z., Huang, Y., Zhao, P., Luan, T. H., & Li, C. (2021). Secure and personalized edge computing services in 6G heterogeneous vehicular networks. *IEEE Internet of Things Journal*, *9*(8), 5920–5931. doi: 10.1109/JIOT.2021.3065970.
[16] Almaiah, M. A., Yelsetti, S., Arya, L., Babu Christopher, N. K., Kaliappan, K., Vellaisamy, P., Hajjej, F., & Alkdour, T. (2023). A novel approach for improving the security of IoT–medical data systems using an enhanced dynamic Bayesian network. *Electronics*, *12*(20), 1–15.
[17] Alwarafy, A., Al-Thelaya, K. A., Abdallah, M., Schneider, J., & Hamdi, M. (2021). A survey on security and privacy issues in edge-computing assisted Internet of things. *IEEE Internet of Things Journal*, *8*(6), 4004–4022.
[18] Hukkeri, G. S., & Goudar, R. H. (2019). IoT: Issues, challenges, tools, security, solutions and best practices. *International Journal of Pure and Applied Mathematics*, *120*(6), 12099–12109.
[19] Swarna, S. R., Kumar, A., Dixit, P., & Sairam, T. V. M. (2021). Parkinson's disease prediction using adaptive quantum computing. In 2021 Third International Conference on Intelligent Communication Technologies and Virtual Mobile Networks (ICICV) (pp. 1396–1401). IEEE.
[20] Sharma, N., Shamkuwar, M., & Singh, I. (2019). The history, present and future with IoT. In Internet of Things and Big Data Analytics for Smart Generation (pp. 27–51). Springer.
[21] Díaz, M., Martín, C., & Rubio, B. (2016). State-of-the-art, challenges, and open issues in the integration of Internet of things and cloud computing. *Journal of Network and Computer Applications*, *67*, 99–117.
[22] Reyna, A., Martín, C., Chen, J., Soler, E., & Díaz, M. (2018). On blockchain and its integration with IoT, Challenges and opportunities. *Future Generation Computer Systems*, *88*, 173–190.
[23] Dhiman, H., Vyas, N., Pagrotra, A., & Dutt, V. (2023). Real-time traffic analysis and driver safety enhancement via IoT and computer vision with high-resolution project lamps for intelligent road illumination [Internet]. In 1st International Conference on Circuits, Power and Intelligent Systems (CCPIS), IEEE, 2023, Available from. http://dx.doi.org/10.1109/CCPIS59145.2023.10291872.
[24] Raheman, F. (2022). The future of cybersecurity in the age of quantum computers. *Future Internet* 2022, *14*, 335. https://doi.org/10.3390/fi14110335.
[25] Loven, L., Lepp_anen, T., Peltonen, E., Partala, J., Harjula, E., Porambage, P., Ylianttila, M., & Riekki, J. (2019). Edge AI: A vision for distributed, edge-native artificial intelligence in future 6G networks. *The 1st 6G Wireless Summit*, 2019, 1–2.

[26] Dang, S., Amin, O., Shihada, B., & Alouini, M.-S. (2020). What should 6G be? *Nature Electronics*, *3*(1), 20–29.

[27] Sattiraju, R., Weinand, A., & Schotten, H. D. (2019). AI-assisted Phy Technologies for 6G and beyond Wireless Networks. *arXiv*, 1–2. https://doi.org/10.48550/arXiv.1908.09523.

[28] Hong, T., Liu, C., & Kadoch, M. (2019). Machine learning-based antenna design for physical layer security in ambient backscatter communications. *Wireless Communications and Mobile Computing*. https://doi.org/10.1155/2019/4870656.

[29] Nawaz, S. J., Sharma, S. K., Wyne, S., Patwary, M. N., & Asaduzzaman, M. (2019). Quantum machine learning for 6G communication networks: State-of-the-art and vision for the future. *IEEE Access*, *7*, 46317–46350.

[30] Farsad, N., Yilmaz, H. B., Eckford, A., Chae, C.-B., & Guo, W. (2016). A comprehensive survey of recent advancements in molecular communication. *IEEE Communications Surveys and Tutorials*, *18*(3), 1887–1919.

[31] Loscri, V., Marchal, C., Mitton, N., Fortino, G., & Vasilakos, A. V. (2014). Security and privacy in molecular communication and networking: Opportunities and challenges. *IEEE Transactions on NanoBioscience*, *13*(3), 198–207.

[32] Hu, J.-Y., Yu, B., Jing, M.-Y., Xiao, L.-T., Jia, S.-T., Qin, G.-Q., & Long, G.-L. (2016). Experimental quantum secure direct communication with single photons. *Light: Science & Applications*, *5*(9), e16144.

[33] Kiyomoto, S., Basu, A., Rahman, M. S., & Ruj, S. (2017). On blockchain-based authorization architecture for beyond-5g mobile services. In 2017 12th International Conference for Internet Technology and Secured Transactions (ICITST) (pp. 136–141). IEEE.

[34] Kotobi, K., & Bilen, S. G. (2018). Secure blockchains for dynamic spectrum access: A decentralized database in moving cognitive radio networks enhances security and user access. *IEEE Vehicular Technology Magazine*, *13*(1), 32–39.

[35] Akyildiz, I. F., Jornet, J. M., & Han, C. (2014). Terahertz band: Next frontier for wireless communications. *Physics Communications*, 12,16–32.

[36] Ma, J., Shrestha, R., Adelberg, J., Yeh, C.-Y., Hossain, Z., Knightly, E., Jornet, J. M., & Mittleman, D. M. (2018). Security and eavesdropping in terahertz wireless links. *Nature*, *563*(7729), 89–93.

[37] Cho, S., Chen, G., & Coon, J. P. (2019). Enhancement of physical layer security with simultaneous beamforming and jamming for visible light communication systems. *IEEE Transactions on Information Forensics and Security*, *14*(10), 2633–2648.

[38] Arya, L., Sharma, Y. K., Kumar, R., Padmanaban, H., Devi, S., & Tyagi, L. K. (2023). Maximizing IoT security: An examination of cryptographic algorithms. In 2023 International Conference on Power Energy, Environment & Intelligent Control (PEEIC) (pp. 1548–1552). Greater Noida, India, 2023. doi: 10.1109/PEEIC59336.2023.10451210.

[39] Urgelles, H., Maheshwari, S., Nande, S. S., Bassoli, R., Fitzek, H. P., & Monserrat, J. F. (2024). In network quantum computing for future 6G networks. *Advanced Quantum Technologies*, 2300334. https://doi.org/10.1002/qute.202300334.

[40] Arya, L., Sharma, Y. K., & Kumar, R. (2024). Digital guardians: Enhancing women's security with artificial intelligence and IoT. In Abraham, A., Pllana, S., Hanne, T., & Siarry, P. (eds.). Intelligent Systems Design and Applications. ISDA 2023 Lecture Notes in Networks and Systems (Vol. 1048). Springer, Cham. https://doi.org/10.1007/978-3-031-64650-8_26.

A. Jasmine Sugil and K. Merriliance

6 Optimization in Natural Language Processing Models for Enhanced Performance and Efficiency

Abstract: This chapter provides an in-depth exploration of optimization strategies in natural language processing (NLP), focusing on their part in augmenting model performance besides scalability across different tasks. It begins by outlining the importance of optimization in NLP, highlighting its role in overcoming challenges such as data sparsity, computational complexity, and semantic ambiguity. The chapter then explores various optimization methodologies including parameter optimization, algorithmic optimization, feature engineering and representation learning, deep learning optimization, and domain-specific optimization. Parameter optimizationinvolves refining model parameters, for example, searching grid, search at random, and Bayesian optimization, to improve performance factors like accuracy, precision, and recall across different NLP tasks. Algorithmic optimizations are explored, covering approaches to improve the efficiency and scalability of algorithms for assignments comprising sentiment analysis, named entity recognition, and text classification. Feature engineering involves crafting effective features to capture linguistic patterns and semantic information and integrating representation learning methods like word embeddings and contextualized embeddings to enhance feature representations. Deep learning optimization involves using algorithms like stochastic gradient descent (SGD) and Adam for training neural networks in NLP tasks. Businesses that want to enhance the performance of their NLP functions must prioritize NLP optimization. It includes approaches like as ensemble methods, regularization, hyperparameter tuning, model selection, and data pre-processing. Through NLP model optimization, companies can increase robustness, accuracy, efficiency, and lower risk. Better decision-making, enhanced client experiences, quicker processing times, cheaper costs, and higher NLP application reliability are the results of this. For companies looking to maximize NLP's potential and innovation in their sectors, NLP optimization is a useful tool.

Keywords: Natural language processing, deep learning, feature engineering, scalability, parameter, optimization techniques, Bayesian optimization, text classification, sentiment analysis

This chapter concludes by discussing optimization strategies tailored to specific domains or applications within NLP, emphasizing the importance of considering domain-specific language tone and data characteristics when designing optimization approaches. By combining theoretical insights and practical examples, readers can gain effective intuitions into the role of optimization in advancing the capabilities of NLP technologies.

https://doi.org/10.1515/9783111436180-006

6.1 Overview

Natural language processing (NLP) remains at the forefront of modern artificial intelligence lead, allowing machines to recognize, decode, and create human language (Figure 6.1). It is the technology that allows machines to examine the languages spoken by people. To execute tasks like speech recognition, automatic summarization, named entity recognition (NER), transformation, relationship abstraction, and topic analysis, it assists developers in organizing their data. It facilitates better communication between computers and people. Code, or the language of computers, is the most direct means of controlling them. It is far more natural for people to communicate with computers when they can comprehend human language [1]. NLP, rooted in linguistics, has over 50 years of experience and uses rule-based or machine learning approaches to understand text structure, aiding in fields like medical research, search engines, and business intelligence.

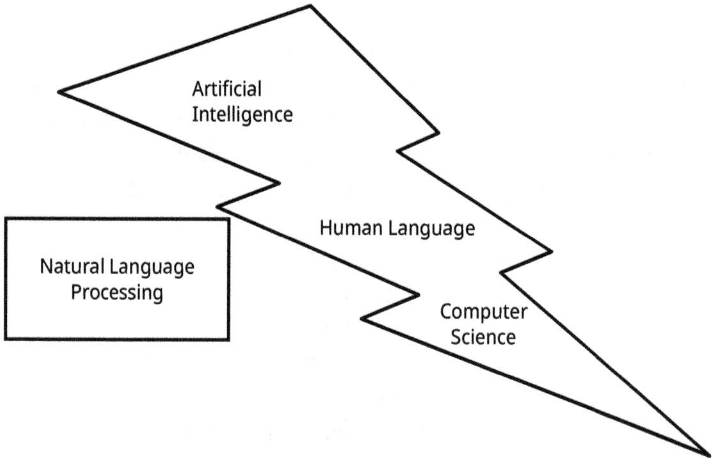

Figure 6.1: NLP fields.

Table 6.1: Natural language versus computer language.

Natural language	Computer language
– Very large vocabulary	– Very limited vocabulary
Human languages communicate with each other	Programming language is understood by the machines
Rich, nuanced, and often ambiguous, allowing for the expression of thoughts, emotions, and complex ideas	Structured, precise, and unambiguous

Table 6.1 (continued)

Natural language	Computer language
Complex syntax and semantics including grammar rules, vocabulary, and idioms	Strict syntax and semantics defined by rules and conventions
Serve various purposes (social interaction, expression of thoughts and emotions, storytelling, persuasion, etc.)	Perform specific tasks (mathematical calculations, data processing, hardware control, etc.)
Communication between humans	Communication with computers
Natural languages evolve over time through – cultural – social – historical influences – adapting to new contexts – technologies	Computer languages evolve through – updates – revisions – the introduction of new languages in response to changes in computing technology – application domains

Table 6.1 highlights the key differences between natural language and computer language, emphasizing their nature, syntax, semantics, purpose, audience, evolution, and acquisition.

6.1.1 Natural Language Processing

NLP is a division of AI that studies the interface between computers and humans via natural language (Figure 6.2). It uses various algorithms to analyse and derives meaning from language, creating natural language through coherent, meaningful phrases.

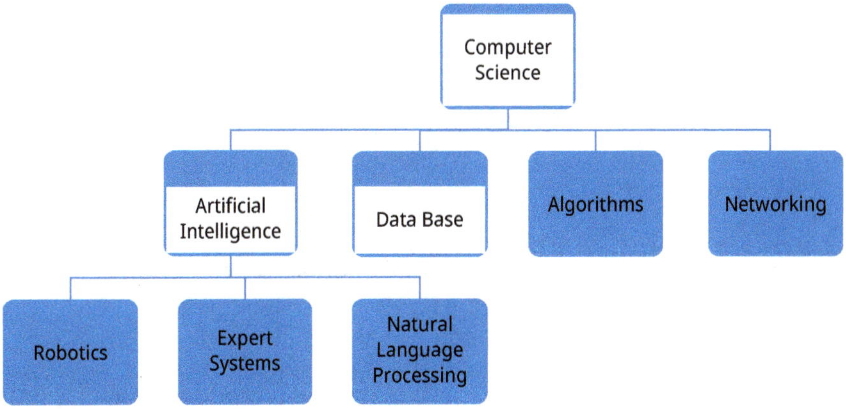

Figure 6.2: Different areas of computer science.

NLP is crucial for enabling computers to communicate with humans in their native language and handle large volumes of textual data. It allows computers to read, interpret, measure sentiment, and identify important parts. With the vast collection of unstructured data generated every day, automation is essential for efficient analysis. NLP helps resolve ambiguity in language and adds suitable numeric structure to data intended for applications like speaking recognition and text analytics.

NLP remains a multidisciplinary subfield of computer science as well as information retrieval. It is mainly concerned with offering computers the capability to understand and influence human language.

Machine learning and AI are used to enhance precision and efficacy. After training, models can produce new languages and make predictions potentially altering communication with machines and one another. NLP has advanced to the point where it can understand and react to text or voice data in a very human-like manner when combined with AI [2].

6.1.2 Constituents of NLP

Natural language understanding (NLU) and natural language generation (NLG) are dual parts of NLP. NLU extracts metadata from content, assisting machines in understanding and analysing human language (Figure 6.3).

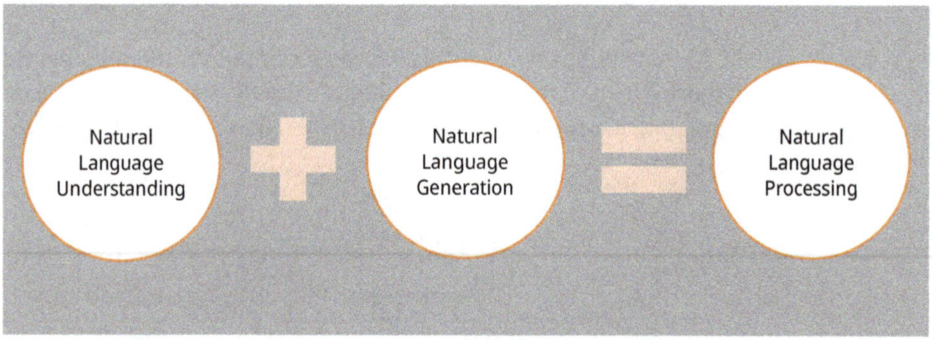

Figure 6.3: NLP components.

NLG maps input into useful representations and analyses language aspects, while NLU converts computerized data into natural language representation.

An NLP system combines NLU and NLG, focusing on understanding text meaning, decision-making, and action execution and generating human-language text from structured data.

6.1.3 NLP Versus NLU

NLP involves processing and analysing text data, performing tasks like text classification and machine translation. It operates at a surface level, dealing with language ambiguity and syntactic complexity. NLP is used in chatbots, search engines, and language translation.

NLU aims to appreciate the meaning and context of transcript data, extracting intents and performing semantic analysis. It goes beyond surface-level understanding to comprehend language like humans do, handling context and resolving ambiguity. NLU finds applications in virtual assistants, question answering systems, and conversational agents.

6.1.4 NLP Versus NLG

NLP focuses on understanding and processing human language, performing tasks such as text classification and sentiment analysis. It takes human language input and derives structured representations, finding applications in chatbots and sentiment analysis.

NLG, on the other hand, focuses on generating human-like language output, handling tasks like text summarization and content generation. It takes structured data as input and produces human language output, finding applications in automatic report generation and virtual assistants.

Both NLP and NLG are interconnected, with tasks often supported by each other's systems. Challenges in NLP include ambiguity resolution and context understanding, while challenges in NLG include coherence in response generation and achieving naturalness. NLG is a subset of NLP that produces meaningful sentences using models like Markov chains, recurrent neural networks, extended interim memory, and transformers [3].

Smart assistants and customer service bots utilize NLP to create narratives either written or spoken [4]. This process involves content analysis, where data is filtered to identify main topics and relationships. Data understanding is then performed using machine learning to identify patterns and form context. Document structuring is then done based on the interpreted data, with a narrative structure determined. Sentence aggregation is used to summarize the topic, followed by grammatical structuring to ensure natural language and correct grammar. Finally, language presentation is generated based on the output format, with voice-powered assistants producing audio output. This subset of NLP is essential for customer-facing software.

Overall, an NLP system combines NLP, NLU, as well as NLG to comprehend and interpret text. NLP processes involve decisions, actions, and generations, which means that NLU makes decisions and takes actions, and NLG generates human language text from structured data.

For example, a website can post daily share market reports, which can be time-consuming and boring. By combining NLP, NLU, and NLG, the system can break down the text, generate a story, and allow the user to focus on other tasks. This approach allows the system to work like a human and save time for users [3].

Table 6.2: Comparison of NLP, NLG, and NLU.

NLP (natural language processing)	NLG (natural language generation)	NLU (natural language understanding)
Processing and analysing human language	Generating human-like language output	Understanding the meaning and context of human language input
Text classification, sentiment analysis, machine translation, etc.	Text summarization, content generation, automatic report writing, etc.	Intent extraction, semantic analysis, question answering, etc.
Takes human language input and derives structured representations	Takes structured data as input and produces human language output	Takes human language input and derives structured representations or extracts meaningful information
Chatbots, sentiment analysis, language translation, etc.	Automatic report generation, virtual assistants, data-to-text systems, etc.	Virtual assistants, question answering systems, conversational agents, etc.
Ambiguity resolution, context understanding, syntactic complexity, etc.	Coherence in response generation, maintaining context, achieving naturalness, etc.	Semantic understanding, inference, resolving ambiguity, handling context, etc.

Table 6.2 provides a clear comparison of NLP, NLG, and NLU across various aspects such as input, output, applications, and challenges making it easier to understand their differences and characteristics.

Table 6.3: Popular NLP tasks and models.

NLP tasks	Models
Text classification	BERT, RoBERTa, CNN, LSTM, FastText
Named entity recognition	BERT, CRF, LSTM, BiLSTM, spaCy
Sentiment analysis	BERT, LSTM, BiLSTM, Transformer, VADER
Machine translation	Transformer (e.g., BERT and GPT), LSTM, seq2seq
Text generation	GPT-3, LSTM, Transformer, OpenAI GPT
Question answering	BERT, RoBERTa, Transformer, BiDAF
Summarization	BERT, Transformer, LSTM, seq2seq
Language modelling	GPT-3, BERT, Transformer, LSTM, OpenAI GPT
Speech recognition	DeepSpeech, WaveNet, Transformer, LSTM

Table 6.3 provides a comprehensive overview of the popular models used across various NLP tasks, showcasing the diversity and effectiveness of different architectures in handling specific NLP challenges.

6.2 Optimization in NLP

Optimization is a set of mathematical values and methods used to solve quantitative problems in various disciplines, recognizing the common mathematical elements that enable unified problem formulation and resolution.

NLP optimization is the process of tailoring the content so that it is clearer and more understandable for search engines (like Google) that use NLP to interpret and investigate content. It involves improving algorithms and models engaged in NLP activities to enhance functioning, efficiency, and accuracy in activities such as text classification, language translation, sentiment analysis, and speech recognition.

As we delve deeper into the realm of NLP optimization, it becomes apparent that fine-tuning model parameters, refining algorithms, and crafting effective features are essential components of achieving optimal performance across different tasks. Moreover, the advent of deep learning has revolutionized NLP, but training complex neural networks demands sophisticated optimization strategies.

The effectiveness of NLP systems heavily relies on optimization techniques tailored to address the unique challenges inherent in language data. Through this exploration, this chapter provides a comprehensive overview of optimization techniques in NLP, ranging from traditional parameter tuning approaches to contemporary AI optimization procedures and intricacies present in different application domains.

Figure 6.4 illustrates the core principles of optimization methods used to train models across distributed systems efficiently. This structured approach helps in efficiently enhancing the performance and accuracy of NLP models.

NLP involves several steps to optimize input data for models (Figure 6.5). These include data pre-processing, which involves taking away stop words, stemming data, and standardizing text. Selection of a model involves picking the correct NLP model for the work, adjusting hyperparameters to improve performance, and regularization to prevent over-fitting. Ensemble techniques conglomerate several NLP models to craft a new, true, and robust model. The execution phase is 4–6 weeks, and consultation time is 1–2 h. These steps help ensure the model's generalization performance and shrink the possibility of over-fitting [6].

By utilizing these optimization strategies, the business sector exhibits its proficiency in NLP as well as its dedication to providing clients with superior solutions and producing remarkable outcomes. Especially businesses will increase the exactness, competence, and sturdiness of their NLP usage.

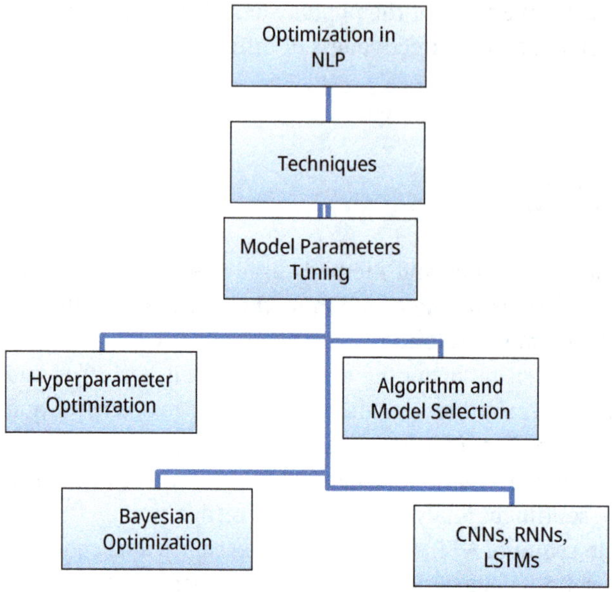

Figure 6.4: NLP framework.

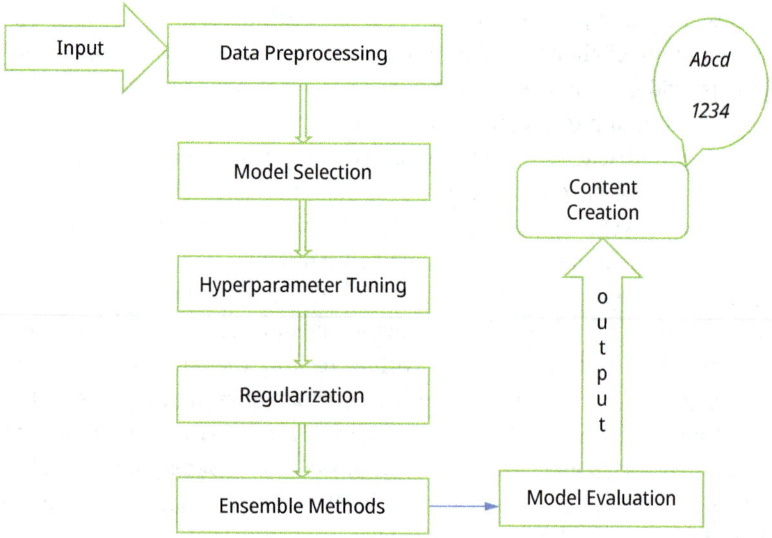

Figure 6.5: Stages of optimizing NLP.

6.2.1 Issues in Optimizing NLP

NLP optimization involves a specialized context to address the distinctive challenges posed by language data:

- The objective function in NLP often relates to the performance of a language model or system on specific tasks such as machine translation or sentiment analysis [7, 8].
- Decision variables in NLP optimization can include factors of the language model, for instance, weights in a neural network architecture or hyperparameters of a machine learning algorithm.
- Constraints in NLP optimization often arise from limitations in computational resources, dataset availability, or specific task requirements. These constraints may include memory limitations for processing large text corpora, data sparsity issues, or linguistic constraints in certain language generation tasks.

NLP employs various optimization algorithms to train language models effectively such as gradient-based techniques similar to SGD, Adam optimization, and momentum-based methods. The distinction between local and global optima is crucial in NLP optimization, as neural network training often converges to local optima owing to the non-convex form of the objective function.

NLP optimization techniques include word embeddings, attention mechanisms, and transfer learning. Word embeddings represent words as condensed vectors in a relentless vector space, allowing the model to capture semantic similarities between words. Attention mechanisms allow models to focus on relevant input parts during processing, enhancing performance in tasks like machine translation and text summarization.

Sensitivity analysis in NLP optimization evaluates how changes in model parameters, training data, or task objectives affect the model's performance, providing insights into potential improvement and optimization areas.

6.2.2 Hardware Optimization

A significant factor in NLP optimization is hardware. The speed, accuracy, and effectiveness of the optimization process can all be impacted by the type of technology being used.

Some of the most popular kinds of hardware for NLP optimization include the following:

Graphics processing units (GPUs) are specialized computers designed to manage massive volumes of data simultaneously. They are, therefore, perfect for NLP optimization, which may involve handling large text datasets.

Tensor processing units (TPUs) are specialized processors made specifically for machine learning applications. Though they are tailored for the specific tasks required in machine learning, they are comparable to GPUs. Therefore, they are perfect for NLP optimization, which frequently requires numerous matrix operations.

Central processing units (CPUs) are the general-purpose processors that power the majority of computers. This makes them an excellent option for NLP optimization jobs requiring less processing power.

The specific needs of the task determine the hardware selection for NLP optimization. A GPU or TPU is the ideal option if speed is the top priority. A CPU is the ideal option if precision is the most essential consideration. Additionally, a CPU is the ideal option if adaptability is the most important consideration.

6.3 Natural Language Processing (NLP) Models

NLP models have gained prominence in the AI community, particularly in chatbots and language models [9]:

- **Eliza**, developed in the 1960s, used pattern matching to solve the Turing test.
- **Microsoft's Tay**, launched in 2016, adopted phrases from sexist and racist tweets, highlighting the danger of not de-biasing data.
- **Deep Learning Models** named after Muppet characters, such as ELMo, BERT, Big BIRD, ERNIE, Kermit, Grover, RoBERTa, and Rosita, provide contextual embeddings and enhanced knowledge representation.
- **GPT-3**, a 175-billion factor model, can mark unique prose with human-equivalent smoothness.
- **LaMDA**, an informal chatbot established by Google, is transformer-based and aims to provide sensible responses to conversations.
- **MoE models**, like the switch transformer, use efficient routing algorithms to provide different parameters for different inputs, reducing communication and computational costs.

6.3.1 Optimization in NLP Model

Optimizing an NLP model involves adjusting its parameters, architecture, and training process to achieve optimal accuracy and efficient computational performance as shown in Figure 6.6:

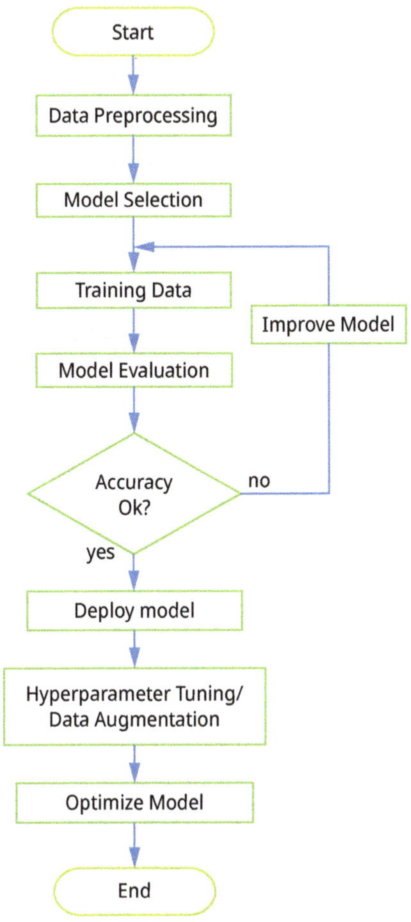

Figure 6.6: Flowchart of optimizing NLP.

1. **Start:** The process starts with a **Flowchart** with the intent to optimize an NLP model.
2. **Data Pre-processing:** This step comprises cleaning and preparing the text data for the model. This might include removing noise, fixing typos, and altering text to a format the model can realize.
3. **Model Selection:** Here, you choose an NLP model architecture (e.g., Transformer) suitable for a specific task (e.g., sentiment analysis).
4. **Training:** The data is fed into the chosen model, and the model's internal parameters are adjusted to improve its accuracy on the specific task.
5. **Evaluation:** The model's accomplishment is measured on a separate test dataset that was not used for teaching. This helps in determining how effectively the model extrapolates to unknown data.

6. **Decision Point:** Based on the evaluation, you decide if the model's accuracy is sufficient:
   - **Yes (Accuracy OK?)**: If the model performs well, the process ends, and the model can be deployed for real-world use.
   - **No (Need Improvement?)**: If the model's accuracy needs improvement, you move to the optimization stage.
7. **Optimization:** There are two main paths for optimization:
   - **Hyperparameter Tuning**: This involves adjusting the settings of the model (like the learning rate) to improve its performance.
   - **Data Augmentation**: This involves artificially increasing the training data through techniques like paraphrasing or generating variations of existing data points. This helps the model generalize better.
8. **Repeat:** After applying either optimization technique, you will likely need to repeat steps 4 (Training) and 5 (Evaluation) to see if the changes improved the model. This loop might continue until you are satisfied with the model's performance.
9. **End (Deploy Model):** Once the model reaches an acceptable accuracy level, the process ends, and the model can be deployed for real-world use.

6.3.1.1 Advantages of Optimizing NLP

Optimizing NLP systems can yield the following advantages:

Improved Efficiency: Optimization techniques can enhance the speed and performance of NLP algorithms, allowing for faster processing of text data and reduced computational resources.

Enhanced Accuracy: By fine-tuning models and algorithms, optimization can lead to improved accuracy in tasks such as text organization, sentiment analysis, and NER, resulting in more reliable results.

Scalability: Optimized NLP systems can handle larger volumes of text data efficiently, making them appropriate for applications with increasing data throughput and user engagement.

Cost Reduction: Optimization can lead to more resource-efficient NLP systems, reducing infrastructure costs and improving the overall return on investment for businesses deploying NLP solutions.

Better User Experience: Faster response times and improved accuracy contribute to a better user experience in applications such as chatbots, virtual assistants, and search engines, leading to higher user satisfaction and engagement.

Robustness: Optimized NLP models are often more robust to variations in data distribution, noisy input, and domain-specific challenges, making them suitable for real-world applications across different domains and industries.

Adaptability: Optimization techniques can enable NLP systems to adapt to changing requirements, new data sources, and evolving user needs, ensuring their continued relevance and effectiveness over time.

6.4 Techniques of NLP Optimization

Optimization techniques play a vital role in maximizing the performance, efficiency, scalability, and adaptability of NLP models, ultimately driving value for businesses, and improving user experiences.

Table 6.4: List of optimization techniques commonly used in NLP.

Optimization technique	Description
Parameter tuning [10, 11]	Adjusting model parameters to optimize performance factors such as exactness, Precision, Recall, ROC-AUC, and *F*1-score
Algorithmic optimization [12–14]	Designing and refining algorithms specific to NLP works, such as dynamic programming for categorization labelling or beam search for decoding in machine translation
Feature engineering [15, 16]	Crafting features to capture linguistic patterns, semantic information, or syntactic structures from text data
Representation learning [15–18]	Learning feature representations automatically from data includes techniques similar to word connectedness (e.g., Word2Vec and GloVe) and context-based connectedness (e.g., BERT and ELMo)
Deep learning optimization [19–21]	Employing optimization techniques specific to deep learning such as stochastic gradient descent (SGD) variants, learning rate scheduling, and dropout regularization
Domain-specific optimization [21–23]	Adapting optimization strategies to specific NLP domains or applications such as biomedical text mining or legal document analysis
Transfer learning [24, 25]	Leveraging pre-trained models on significant text corpora to increase performance on downstream NLP tasks with limited labelled data
Ensemble methods [26]	Combining predictions from multiple models to improve overall performance includes techniques such as model averaging, stacking, and boosting

Table 6.4 depicts various optimization techniques for NLP tasks including parameter tuning, algorithmic optimization, feature engineering, representation learning, deep

learning optimization, domain-specific optimization, transfer learning, and ensemble methods. These techniques aim to optimize performance metrics, capture linguistic patterns, semantic information, and syntactic structures, and improve the overall performance by combining predictions from multiple representations.

6.4.1 Fine-Tuning Pre-Trained Models

Refined pre-trained models are a powerful optimization technique in NLP that allows practitioners to benefit from the rich illustrations learned by pre-trained models while adapting them to specific downstream tasks, even with limited labelled data. These models are trained on significant text corpora using unsupervised or semi-supervised learning methods. These models learn rich representations of language by predicting missing words, generating text, or performing other language-related tasks.

Fine-tuning pre-trained models involves conveying the knowledge gained during pre-trained to a specific downstream task using task-specific labelled data. During fine-tuning, the parameters of the pre-trained model are adjusted using task-specific labelled data through gradient descent optimization. The model architecture may be modified or extended with task-specific layers such as adding fully connected layers for text classification tasks [27, 28].

Hyperparameter tuning is also performed to optimize model performance on the downstream task. Regularization procedures like dropout or weight decay may be employed to elude over-fitting as well as mend generalization performance. After fine-tuning, the model's performance is assessed on a distinct validation set or through cross-validation, using metrics like accuracy, precision, recall, and *F*1-score.

The analytical representations of these factors are as follows:

$$\text{Accuracy} = \frac{\text{TP+TN}}{\text{TP+TN+ FP+FN}}$$

$$\text{Precision} = \frac{\text{TP}}{\text{TP + FP}}$$

$$\text{Recall} = \frac{\text{TP}}{\text{TP + FN}}$$

$$F1\text{-Score} = 2\times \frac{\text{Precision} \times \text{Recal}}{\text{Precision} + \text{Recall}}$$

where TP is true positive, FN is false negative, FP is false positive, and TN is true negative.

The hyperparameters in an estimation scheme, including both cost and model hyperparameters, enable the adjustment of a parametric model's complexity. Hyperparameter tuning and hyperparameter search are crucial in machine learning to achieve optimal performance by balancing model complexity and generalization capability. Hyperparameters, especially in the context of the Adam optimizer, play an acute role in determining the best hyperparameters. As machine learning advances, practitioners must stay updated on new methodologies and engage in continuous learning to develop efficient, robust, and adaptable models for real-world data challenges.

Fine-tuning pre-trained models is often an iterative process, involving the exploration of different configurations and training multiple times until satisfactory performance is achieved.

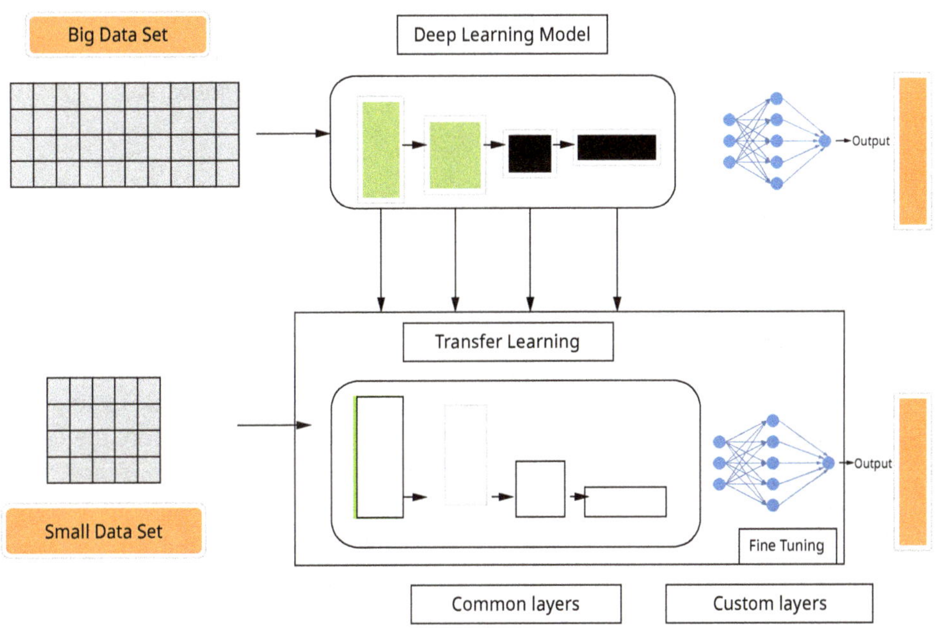

Figure 6.7: Fine-tuning parameter representation.

Transfer learning is a machine learning technique that uses pre-trained models to enhance performance on related tasks. It has revolutionized NLP such as recognizing sentiment in product reviews. Figure 6.7 explores its fundamentals and fine-tuning techniques.

Fine-tuning optimization using NLP includes minimizing a loss function to enhance model performance which is depicted by the following equation:

$$L = -N1\sum i = 1N\sum c = 1C_{yi,\text{clog}}(p_{i,c})$$

where N is the number of samples in the batch, C the number of classes, $Y_{i,c}$ the binary indicator (0 or 1) if class label c is the correct classification for sample i, and $P_{i,c}$ the predicted probability that sample i belongs to class c.

Devlin et al., Radford et al., Liu et al., Yang et al., and Sanh et al. advance transfer-based models such as BERT, GPT, RoBERTa, XLNet, and DistilBERT for various NLP tasks, which are seminal works in the arena of NLP, particularly in the development and refinement of transformer-based models.

6.4.1.1 Advantages of Pre-Trained Models

Pre-trained models offer faster training, improved performance, reduced data requirements, cost-efficiency, domain adaptation, and transfer learning. They learn general features from large datasets, requiring less training time and less computational resources. They can adapt to specific tasks and transfer learning knowledge.

6.4.2 Hyperparameter Tuning

Hyperparameter tuning is essential for optimizing NLP models to achieve better performance on specific tasks. By systematically exploring the hyperparameter space and selecting the best hyperparameters, practitioners can develop NLP models that generalize well and perform effectively on real-world data.

Hyperparameter tuning is a critical step in optimizing NLP models to improve performance in specific tasks. Hyperparameters are parameters set before the learning process, controlling aspects such as the model's architecture, optimization algorithm, and learning rate. They can include the number of layers in a neural network, the size of word embeddings, the learning rate of the optimizer, and the dropout rate. Practitioners need to select values for hyperparameters, which can significantly affect the model's output.

There are numerous hyperparameter tuning procedures: grid search, random search, Bayesian optimization, and gradient-based optimization. The method of grid-based search is simple but computationally expensive, while random search is less computationally expensive and often performs better in practice. Bayesian optimization uses probabilistic models to predict optimal hyperparameter settings, while gradient-based optimization requires differentiable objectives. After selecting a hyperparameter configuration, the NLP model is trained using these hyperparameters on a training dataset and evaluated on a separate validation dataset. Hyperparameter tuning is often an iterative process, adjusting hyperparameters based on performance feedback until satisfactory performance is achieved. Finally, the final NLP model is trained using these hyperparameters using the whole training dataset and assessed on a distinct test dataset to evaluate its performance.

A method that is invariably used for hyperparameter tuning is grid search which involves searching through a particular subset of hyperparameter. Grid search is explored using the following formula:

$$(\theta *, \lambda *) = \text{argmin}\theta, \lambda L(f\theta, \lambda(X), y)$$

where Θ represents the model parameters, Λ represents the hyperparameters, L is the loss function (e.g., cross-entropy loss for classification tasks), $f\ \theta,\lambda$ is the model function with parameters θ and hyperparameters λ, X is the input data, y is the target labels, and (θ^*,λ^*) are the optimal parameters and hyperparameters that minimize the loss function.

Wang et al. [11] explore hyperparameter optimization in NLP and machine learning, proposing a multi-stage approach. The algorithm breaks down the problem into numerous phases with increasing data, providing fast evaluations of good aspirants for better performance. This chapter emphasizes that hyperparameter optimization is a machine learning algorithm that aims to find a hyperparameter setting λ

λ* such that the confirmation accuracy L is maximized, where hyperparameters (λ) and domains (Λ) are used to optimize the validation accuracy on training data (T_{train}) and confirmation accuracy on validation data (T_{valid}). The objective is to find a hyperparameter setting λ* that maximizes the confirmation accuracy.

The algorithm demonstrated enhanced performance compared to the contemporary Bayesian optimization algorithm and fast alternatives of Bayesian optimization on sentiment sorting and question–answer tasks.

6.4.2.1 Algorithm to Select Best Hyperparameter

Input: Loss function L, Number of stages S. Iteration per stage
Y= (Y_1. . . . . .Ys), training data per stage
$T_{train} = (T^1 train, ..., T^s train)$,
validation data T_{valid}, initialization $\lambda_{1:k}$

output: Hyper-parameter λ*
for stage s=1 to S do
  for i=1 to k do
    L_i=Evaluate L (λ_i, $T^s{}_{train}$, T_{valid})
  end
  for j=k+1 to Y_s do
    V: regression model on $(\lambda_i, L_i)_{i=1 \text{ to } j1}$
    $\lambda_j = \arg\max_{\lambda \in \Lambda} a(\lambda, v)$
    L_j = Evaluate L (λ_j, $T^s{}_{train}$, T_{valid})
  end
  reset $\lambda_{1:k}$ = best k configs€(λ_1,. . . . .,λ_{Ys})

```
      based on validation accuracy L
end
return λ* = arg max λj ∈ {λY1, . . . . . . . . ., λYs} Lj

Stage Loop:
for stage s = 1 to S do
   for i = 1 to k do
      Li = Evaluate L(λi, T^s_train, Tvalid)                  → Evaluate initial configurations
   end
   for j = k+1 to Ys do
      V = Regression model on (λi, Li) for i = 1 to j-1        → Train regression model
      λj = arg max λ ∈ Λ a(λ, V)                               → Optimize hyperparameters
      Lj = Evaluate L(λj, T^s_train, Tvalid)                   → Evaluate new configuration
   end
   reset λ1: k = best k configs ∈ (λ1, . . ., λYs) based on validation accuracy L   → Select top configs
end
Return λ* = arg max λj ∈ {λY1, . . ., λYs} Lj     → Final selection of best hyperparameter.
```

Bayesian optimization in automated machine learning (AutoML) streamlines the process of choosing the optimal model and its hyperparameters for a specific problem and dataset. Recent uses of Bayesian optimization have demonstrated its efficiency in fine-tuning intricate models such as deep belief networks, convolutional neural networks, and Markov chain Monte Carlo methods.

Existing advanced methods focus on expanding model-based Bayesian optimization to resolve this problem owing to its capability to ascertain worthy solutions within a small number of iterations compared to orthodox methods.

6.4.2.2 Advantages of Hyperparameter Tuning

Hyperparameter tuning is a decisive process in machine learning that improves model performance, enhances generalization, increases efficiency, reduces costs, and enhances robustness. It helps find the best balance between model complexity and generalization, resulting in better performance on unseen data. It also allows for customization to specific tasks or domains, enabling tailored solutions that meet specific requirements and constraints.

6.4.3 Parallel and Distributed Training

Parallel and distributed training techniques are essential for scaling up NLP models and efficiently training them on large datasets.

Data parallelism involves dividing the dataset into multiple batches, each processed independently by different compute units. Model parallelism partitions the model into different units, enabling larger models with large memory footprints.

Distributed training involves multiple machines or nodes collaborating to train a single model, such as the parameter server architecture. This facilitates distributed optimization algorithms like asynchronous SGD. All-reduce communication is used to synchronize gradients across multiple nodes.

Frameworks like Horovod, TensorFlow, and PyTorch provide built-in support for distributed training in NLP. Asynchronous training, where each worker independently computes gradients and updates model parameters, can lead to faster training times but may suffer from issues like parameter staleness and inconsistent updates. Synchronous training, on the other hand, ensures that all workers synchronize their updates at each iterations, leading to better convergence and stability.

The Word2Vec model as well as skip-gram model are established through Tomas Mikolov et al. [14, 15] for parallelizing training and increasing the process using neural networks for Word representation. Additionally, the neural language model [49, 50] explores scaling laws governing performance and adaptive input representations to improve efficiency.

Benefits of parallel and distributed training include scalability, reduced training time, and cost efficiency. By distributing computations across multiple devices or machines, these techniques can significantly reduce the time required to train large NLP models.

6.5 Principles and Tools of NLP Optimization

Various factors collectively contribute to the development of effective and reliable NLP models, addressing different aspects from data preparation to model deployment in real-world applications.

Principles:
Data Quality: Ensure high-quality, well-annotated data for training and evaluation to improve model performance and generalization.

Model Selection: Select appropriate models and architectures based on the specific task requirements, data characteristics, and computational resources available.

Evaluation Metrics: Use relevant evaluation metrics that capture the performance of the model accurately and align with the task objectives.

Transfer Learning: Leverage pre-trained models and transfer learning methods to bootstrap model training, particularly when dealing with limited labelled data.

Fine-Tuning: Fine-tune pre-trained models on field-specific data or tasks to adapt them to specific use cases and improve performance.

Interpretability and Explainability: Ensure that NLP models are interpretable and explainable, especially in important applications such as healthcare and finance, where transparency is essential for decision-making.

In general, NLP optimization improves model accuracy, speed, efficiency, and scalability through data pre-processing, feature engineering, hyperparameter tuning, regularization, compression, parallelism, distributed computing, hardware optimization, incremental learning, and continuous evaluation.

6.5.1 NLP Optimization Tools

Several NLP optimization tools will be available to help businesses tackle their language processing needs [29]. These tools enable businesses to analyse unstructured text data from various sources, improving decision-making, automating tasks, and extracting valuable information.

6.5.1.1 Scalenut NLP Analysis Tool

Scalenut is an AI-powered platform for SEO and content marketing, utilizing NLP to plan, research, create, and optimize content. It enhances readability and SEO performance by using relevant keywords and phrases that match user intent and search engine expectations.

6.5.1.2 Google Cloud Natural Language API

Google Cloud Natural Language API is an influential NLP tool that aids creators in investigating text data using Google's machine learning technology, supports various programming languages, and offers a RESTful interface for easy integration into applications.

6.5.1.3 IBM Watson NLP

IBM Watson NLP offers AI services for locating and extracting information from text data, allowing customization for industries like banking and healthcare.

6.5.1.4 SpaCy

SpaCy is a widespread Python library aimed at advanced NLP tasks, offering speed and accuracy, pre-trained models, word vectors, and embeddings, and features multi-task learning, production-ready training, and linguistic tokenization for NER.

6.5.1.5 MonkeyLearn

MonkeyLearn is an accessible NLP platform that aids businesses in analysing text data including sentiment analysis, topic classification, and keyword extraction. It offers instant data visualizations, pre-built and custom models, business templates, and easy integration with apps and BI tools. Customized machine learning models are also available.

6.5.1.6 TextBlob

TextBlob is a Python library based on NLTK, providing an intuitive interface for NLP tasks such as sentiment analysis, text classification, part-of-speech tagging, noun phrase extraction, word frequency analysis, word inflection, lemmatization, spelling correction, and WordNet integration.

6.5.1.7 Amazon Comprehend

Amazon comprehend is an NLP facility combined with Amazon Web Services, enabling users to perform tasks like sentiment analysis, topic modelling, and entity recognition, with comprehend medical being a specialized variant for healthcare professionals.

6.5.1.8 NLTK

NLTK is a Python library for written data processing, offering interfaces to over 50 corpora and lexical means like WordNet. It includes features like tokenization, filtering stop words, stemming, tagging, lemmatization, chunking, chinking, NER, concordance, distribution plot, frequency distribution, and collocations.

6.5.1.9 AllenNLP

AllenNLP is a robust NLP library established by the Allen Institute for Artificial Intelligence, built on the PyTorch deep learning framework. It offers automated modules for crucial tasks like pre-processing, model training, evaluation, model serving, and visualization.

6.5.1.10 GenSim

GenSim is a Python library aimed at topic modelling and text resemblance analysis, designed for large-scale text processing. It offers fast, scalable, and efficient handling of large text files using unsupervised models, eliminating the need for costly annotations or hand tagging.

NLP.py is a Python bundle for numerical optimization. It aims to offer a toolbox for elucidating linear and nonlinear programming difficulties that is both easy to use and extensible.

NLP.py is a Python group of units and modules that support sparse matrices, nonlinear optimization techniques, and well-organized solutions of large sparse rectilinear systems. It aims to make implementing, testing, prototyping, experimenting, and modifying optimization algorithms for large-scale constrained problems easy and intuitive. NLP.py is suitable for teaching, learning, and research in large-scale numerical optimization [30].

6.5.1.11 Stanford CoreNLP

Stanford CoreNLP is a Java-based NLP toolkit developed by Stanford University's NLP community, offering robust features like part-of-speech labelling, tokenization, referred unit recognition, and dependency parsing, ensuring scalability as well as speed optimization.

In general, optimization tools analyse unstructured text data from diverse sources like customer reviews and social media, providing actionable insights for product and service enhancement. They enable businesses to make data-driven decisions, enhancing operational efficiency and customer satisfaction.

6.5.1.12 Advantages of NLP Tools

NLP optimization tools offer numerous benefits to businesses. They can improve customer support by analysing customer interactions, enabling businesses to understand customer sentiments and provide personalized support. They can also enhance search engines and recommendation systems by understanding the context and intent of user queries, leading to improved user experience and increased conversions. NLP tools can also analyse public sentiment towards a brand or product, enabling businesses to recognize possible issues and yield positive strategies to address customer concerns.

Automated content analysis can be performed by NLP tools, processing huge volumes of text data, enabling businesses to extract valuable insights quickly. This is useful for market research, competitor analysis, trend identification, and content curation, helping businesses stay competitive. NLP tools can also aid in compliance monitoring by automatically analysing documents, contracts, and regulatory texts, identifying potential risks and ensuring adherence to legal and regulatory requirements.

Data extraction and knowledge management are also possible through NLP tools, which can extract structured information from unstructured text, making it easier for businesses to organize and manage data. Additionally, NLP tools can assist in language translation and localization, facilitating global expansion, effective communication with international audiences, and opening new markets.

6.6 NLP Applications

NLP is applied in various fields of industry including health care, credit card, and insurance fraud detention and text analytics for customer sentiment analysis. The global NLP market is anticipated to reach $35.1 billion by 2026, with companies realizing that AI-powered solutions will continue to grow [31]. It is essential to understand the basics and in what way NLP can be practical for business. There are many subsets of technologies related to NLP and they are often confused with one another.

NLP and NLU are crucial in bridging the gap between human and machine communication, enhancing AI's intuitiveness. NLP provides a framework for machine interaction with human language, while NLU enhances this by understanding context and intent. These technologies have immense potential in business, improving customer interactions and decision-making processes. As they combine with machine learning and deep learning, intelligent AI systems will be created that understand and mimic human communication. The true power of these technologies lies in turning unstructured data into actionable insights, giving companies a competitive edge in the digital marketplace.

Examples of modern business applications of NLP (Figure 6.8) include social media sentiment analysis, where NLP understands internet short forms, slang, code-switching, emoticons, emojis, and hashtags. This allows businesses to extract data from customer communication and organize it for machine learning models to analyse their brand performance. This information can be used to reach influencers, alter advertising campaigns, improve product or service aspects, and elevate brand reputation.

In healthcare, NLP technology is increasingly used to comprehend and manage patient records and electronic health records (EHRs) [39]. Patient comments, waiting room experiences, post-surgery care, thoughts, and feelings are investigated through AI/ML models by means of documented data after in-clinic surveys, post-appointment

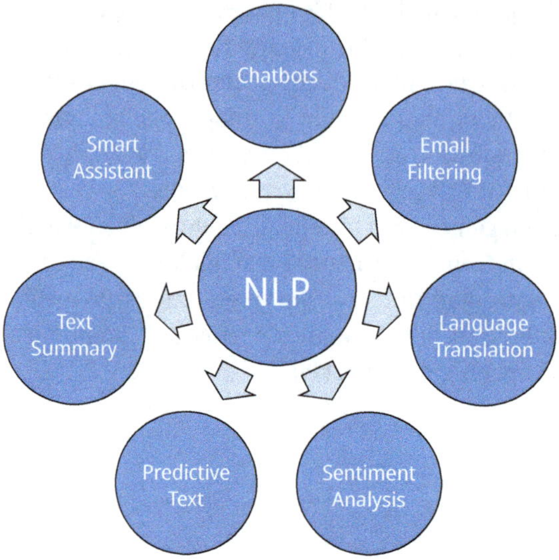

Figure 6.8: Common real-time NLP applications (adapted from [32]).

surveys, and feedback web forms. This assists patients in appraising the value of their service and proposes a more improved patient experience.

Language conversion is a crucial aspect of modern society, with online translations being essential for researchers and the enjoyment of foreign films and documentaries. NLP technology provides speech-to-text translations at scale, rapidly, and professionally. Sentence boundary detection is a fundamental task in NLP, allowing translations of entire texts in diverse languages and matching sentence by sentence. Google Translate is utilized by 500 million people to translate files or script in 100 diverse languages.

In the aviation industry, NLP is used by mechanics in aerospace and defence to gather data about specific issues from extensive aircraft manuals, which can occasionally be in diverse languages. Through semantic exploration and NER recognition, mechanics can search and comprehend notes from pilots or other individuals describing problems they encountered mid-flight. NLP uses neural networks for writing recognition, providing the data users what they are looking for. AI has transformed data diagnostics, and Airbus believes that AI/ML-enabled control engineering can virtually eliminate unscheduled aircraft grounding completely by 2025.

Automated trading is another application of NLP in business, where computer programs place trades on behalf of users. NLP can review financial results, mergers, and acquisitions and make recommendations on which stocks would be a good investment based on this data. Automated phone systems, such as those used in banks and ticketing systems, use speech recognition and interactive voice response to guide transactions and provide language options.

6.6.1 Real-World Views

NLP delivers appreciated insights into real-world applications and effective methodologies.

Google BERT, OpenAI GPT-3, Amazon Alexa, and other NLP models have all been instrumental in enhancing the arena of NLP. BERT, a deep bidirectional transformer model, has been applied to tasks like question answering [43], using the Squad dataset, which is widely used for training and evaluating question answering models, text classification [33, 34], and NER [38]. GPT-3, a large autoregressive language model, has demonstrated remarkable language understanding and generation capabilities, enabling tasks like text completion, translation, and code generation. Amazon Alexa, a virtual assistant, uses NLP techniques like NLU, intent recognition, and dialogue management to provide personalized assistance and answer questions. In healthcare, NLP has been used for tasks like clinical documentation, EHR analysis, and medical literature mining. NLP-powered chatbots are used by companies to provide customer support services, utilizing techniques like intent recognition, sentiment analysis, and NER. In finance, NLP is employed for tasks like sentiment examination, topic analysis, and news categorization [33], financial document summarization [47], and fraud detection.

Text classification is crucial for organizing and defining categories in the new generation of word-based documents.

Moreover, the BBC news text classification system utilizes machine learning algorithms such as logistic regression, random forest, and K-nearest neighbour. The system was tested and compared on five factors: precision, accuracy, *F*1-score, support, and confusion matrix. The classifier with the highest precision and accuracy was considered the finest machine learning algorithm for the BBC news dataset [34].

6.6.2 Few Existing Methods and Technologies in NLP

- Automatic text categorization is a widely researched field that sorts text documents into pre-defined groups using machine learning algorithms. It is essential for organizing large volumes of unstructured information. Various text categorization approaches, divided into orthodox methods, fuzzy logic-based techniques, and deep learning-based methods, are used for analysing the text. Conventional methods include handcrafted features, nature-inspired algorithms, and graph-based methods. It also provides information on available libraries, datasets, and tagging technologies in non-Indian and Indian languages [35].
- The Veterans Affairs and Prostate Cancer Foundation [36] have developed and validated NLP tools to recognize veterans with metastatic prostate cancer (mPCa). This is a trustworthy tool for finding veterans who could benefit from innovative inventions in mPCa diagnosis and treatment.

- Gökgöl et al. [37], using NLP techniques, developed a language-liberated prognostic model with Python and string similarity function. The language-liberated online diagnostic tool is suitable for individual and clinical use, especially with universal agility developing quicker than inhabitants.
- NeuroNER is an easy-to-use NER tool built on artificial neural networks (ANNs). It allows users to annotate entities, train an ANN, and predict their locations and categories in new texts, making the process smooth and accessible [38].
- EHRs are crucial clinical data sources, containing narrative information from physicians about patients' disorders and treatment procedures, but machine comprehension is challenging [39].
- Vaswani et al. and Keskar et al. [40, 41] introduce transformer language model, which offers explicit control over text generation by governing style, content, and task-specific behaviour, preserving the advantages of unsupervised learning while being derived from naturally co-occurring structure.
- A conversational modelling approach [44] for NLU and machine intelligence uses the sequence-to-sequence framework, predicting the next sentence, extracting knowledge from large datasets, and performing common-sense reasoning.
- See et al. [45] propose a novel architecture for abstractive text summarization using a hybrid pointer-generator network, outperforming current models.
- Al-Makhadmeh and Tolba [46] create a mixture of NLP and machine learning to foresee hate speech from social media websites.

All the above demonstrate the effectiveness of machine learning algorithms in classifying BBC news articles, highlighting the importance of automatic text categorization in managing large volume of data. The research aligns with the evolving intersection of NLP and machine learning, with novel architectures like BERT and hybrid pointer-generator networks offering promising applications.

6.7 Future Directions

Future directions for NLP optimization are expected to focus on addressing emerging challenges, improving efficiency, and advancing the capabilities of language processing systems. These directions include developing efficient model architectures that balance performance with computational resources, exploring multi-task learning approaches to leverage shared representations, developing adaptive optimization algorithms to improve convergence speed, stability, and robustness across diverse NLP tasks, exploring low-resource and zero-shot learning techniques, exploring hardware acceleration techniques, improving explainability and interpretability of NLP models, addressing robustness and fairness issues, tailoring optimization strategies to specific domains or applications, exploring meta-learning and AutoML approaches, and exploring continual learning and lifelong learning techniques.

By exploring these future directions, researchers and practitioners can advance the contemporary field of NLP optimization addressing evolving challenges and unlocking new opportunities for refining the competency, effectiveness, and fairness of language processing systems.

6.8 Summary

In summary, optimization is crucial in NLP models, enhancing performance, scalability, and efficiency across various tasks. Techniques like parameter tuning, algorithmic optimization, feature engineering representation learning, and domain-specific strategies are used to improve model effectiveness, training efficiency, and the progress of robust, interpretable, and scalable language processing systems. The future of optimization in NLP holds great promise, with research focusing on efficient model architectures, multi-task learning, adaptive optimization algorithms, and domain-specific optimization. By embracing these future directions, NLP practitioners can unlock new capabilities and address complex challenges in language understanding, generation, and interaction. Optimization in NLP is not just a technical endeavour but a catalyst for innovation and transformation in human language interaction.

References

[1] Gillis, A. S., Lutkevich, B., & Burns, E. (2024 Feb 15). Natural Language Processing (NLP). Enterprise AI.

[2] Gokcuoglu, A. Y. (2023 Dec 25). 7 Real-life examples of Natural Language Processing.

[3] Gill, J. K. (2023 Nov 23).What are the Differences Between NLP, NLU, and NLG? XenonStack.

[4] 11 Real-Life Examples of NLP in Action. [Internet]. Available from: https://levity.ai/blog/11-nlp-real-life-examples.

[5] O'Neal, H. (2023 Nov 7). What is the difference between NLP and NLU: Business Use Cases. MetaDialog. Available from: https://www.metadialog.com/blog/difference-between-nlu-and-nlp/.

[6] Natural Language Processing Optimization | AI/ML Development Solutions. [Internet]. Available from: https://aimlprogramming.com/services/natural-language-processing-optimization/.

[7] Kouloumpis, E., Wilson, T., & Moore, J. (2011). Twitter sentiment analysis: The good the bad and the omg! *Proceedings of the International AAAI Conference on Web and Social Media*, *5*(1), 538–541.

[8] Thet, T. T., Na, J. C., & Khoo, C. S. (2010). Aspect-based sentiment analysis of movie reviews on discussion boards. *Journal of Information Science*, *36*(6), 823–848.

[9] Natural Language Processing (NLP) [A Complete Guide]. DeepLearning.AI. 2023 Jan 11. Available from: https://www.deeplearning.ai/resources/natural-language-processing/

[10] Radford, A., Narasimhan, K., Salimans, T., & Sutskever, I. (2018). Improving language understanding by generative pre-training.

[11] Wang, L., Feng, M., Zhou, B., Xiang, B., & Mahadevan, S. (2015). Efficient Hyper-parameter Optimization for NLP Applications (pp. 2112–2117). doi: 10.18653/v1/D15-1253.

[12] Wu, Y., Schuster, M., Chen, Z., Le, Q. V., Norouzi, M., Macherey, W. et al. (2016). Google's neural machine translation system: Bridging the gap between human and machine translation. arXiv preprint arXiv:1609.08144.

[13] Devlin, J., Chang, M. W., Lee, K., & Toutanova, K. B. (2018). Pre-training of deep bidirectional transformers for language understanding. arXiv preprint arXiv:1810.04805.

[14] Mikolov, T., Chen, K., Corrado, G., & Dean, J. (2013). Efficient estimation of word representations in vector space. arXiv preprint arXiv:1301.3781.

[15] Mikolov, T., Sutskever, I., Chen, K., Corrado, G. S., & Dean, J. (2013). Distributed Representations of Words and Phrases and their Compositionality. *Advances in Neural Information Processing Systems*, 26.

[16] Zhang, X., Zhao, J., & LeCun, Y. (2015). Character-level convolutional networks for text classification. *Advances in Neural Information Processing Systems*, 28.

[17] Liu, Y., Ott, M., Goyal, N., Du, J., Joshi, M., Chen, D. et al. (2019). Roberta: A robustly optimized Bert pretraining approach. arXiv preprint arXiv:1907.11692.

[18] Yang, Z., Dai, Z., Yang, Y., Carbonell, J., Salakhutdinov, R. R., & Le, Q. V. (2019). Xlnet: Generalized autoregressive pretraining for language understanding. *Advances in Neural Information Processing Systems*, 32.

[19] Kingma, D. P., & Ba, J. (2014). Adam: A method for stochastic optimization. arXiv preprint arXiv:1412.6980.

[20] Liu, L., Jiang, H., He, P., Chen, W., Liu, X., Gao, J. et al. (2019). On the Variance of the Adaptive Learning Rate and Beyond.

[21] Conrad, J., & Branting, L. (2018). Introduction to the Special Issue on Legal Text Analytics. *Artificial Intelligence and Law, 26*. doi: 10.1007/s10506-018-9227-z.

[22] Dai, Y., Yu, H., Jiang, Y., Tang, C., Li, Y., & Sun, J. A Survey on Dialog Management: Recent Advances and Challenges. ArXiv, 2020.abs/2005.02233.

[23] Lee, J., Yoon, W., Kim, S., Kim, D., Kim, S., So, C. H., & Kang, J. (2019). BioBERT: A pre-trained biomedical language representation model for biomedical text mining. *Bioinformatics, 36*(4), 1234–1240. doi: 10.1093/bioinformatics/btz682.

[24] Narayanan, D., Shoeybi, M., Casper, J., LeGresley, P., Patwary, M., Korthikanti, V. et al. (2021 Nov). Efficient large-scale language model training on gpu clusters using megatron-lm. *Proceedings of the International Conference for High Performance Computing, Networking, Storage and Analysis*, 1–15.

[25] Puri, R., Kirby, R., Yakovenko, N., & Catanzaro, B. (2018). Large Scale Language Modeling: Converging on 40GB of Text in Four Hours. *2018 30th International Symposium on Computer Architecture and High Performance Computing (SBAC-PAD)*, 290–297. doi: 10.1109/CAHPC.2018.8645935.

[26] Kumar,, Ananth, T., Arun Samuel, T. S., Dinesh Jackson Samuel, R., & Niranjanamurthy, M. eds. (2022). Privacy and security challenges in cloud computing: A holistic approach. CRC Press.

[27] Yang, Z., Yang, D., Dyer, C., He, X., Smola, A., & Hovy, E. (2016 Jun). Hierarchical attention networks for document classification. In Proceedings of the 2016 Conference of the North American Chapter of the Association for Computational Linguistics: Human Language Technologies (pp. 1480–1489).

[28] Chen, Y. (2015). Convolutional neural network for sentence classification (Master's thesis, University of Waterloo).

[29] Scalenut. Best NLP Optimization Tools in 2023. 2024 Feb 16. Available from: www.scalenut.com/blogs/best-nlp-optimization-tools-in-2023.

[30] Introduction – NLP.py: An Object-Oriented Environment for Large-Scale Optimization. [Internet]. Available from: https://pythonoptimizers.github.io/NLP.py/latest/introduction.htm

[31] Team, R. (2021 Mar 1). NLP Business Applications | NLP AI. Available from: https://www.repustate.com/blog/using-nlp-for-business-success/#why-is-nlp-so-important.

[32] Quan, N. (2024 Jan 26). Top 8 Applications of Natural Language Processing (NLP). Eastgate Software. Available from: https://eastgate-software.com/top-8-applications-of-natural-language-processing-nlp/.

[33] Jin, G. (2022). Application optimization of NLP system under deep learning technology in text semantics and text classification. In 2022 International Conference on Education, Network and Information Technology (ICENIT) (pp. 279–283). doi: 10.1109/ICENIT57306.2022.00068.

[34] Shah, K., Patel, H., Sanghvi, D. et al. (2020). A comparative analysis of logistic regression, random forest and KNN models for the text classification. *Augmented Human Research*, *5*, 12. doi: 10.1007/s41133-020-00032-0.

[35] Boyapati, S., Swarna, S. R., & Kumar, A. (2020). Quantum neural networks for dynamic route identification to avoid traffic. In 2020 Fourth International Conference on I-SMAC (Iot in Social, Mobile, Analytics and Cloud)(I-SMAC) (pp. 1018–1022). IEEE.

[36] Alba, P. R., Lynch, J. A., Gao, A. et al. (2020). Using natural language processing (NLP) tools to identify veterans with metastatic prostate cancer (mPCa). *Journal of Clinical Oncology*, *38*, 60-60.

[37] Gökgöl, M., & Orhan, Z. (2019). OP41 intercultural medical decision support system using natural language processing (NLP). *International Journal of Technology Assessment in Health Care*, *35*(S1), 10-10. doi: 10.1017/S0266462319001090.

[38] Dernoncourt, F., Lee, J. Y., & Szolovits, P. NeuroNER: An easy-to-use program for named-entity recognition based on neural networks. arXiv preprint arXiv:1705.05487. 2017.

[39] Zhang, Y., Wang, X., Hou, Z., & Li, J. (2018). Clinical named entity recognition from Chinese electronic health records via machine learning methods. *JMIR Medical Informatics*, *6*(4), e9965.

[40] Vaswani, A., Shazeer, N., Parmar, N., Uszkoreit, J., Jones, L., Gomez, A. N. et al. (2017). Attention is all you need. *Advances in Neural Information Processing Systems*, 30.

[41] Keskar, N. S., McCann, B., Varshney, L. R., Xiong, C., & Socher, R. (2019). Ctrl: A conditional transformer language model for controllable generation. arXiv preprint arXiv:1909.05858.

[42] Liu, Y., & Lapata, M. (2019). Text summarization with pretrained encoders. arXiv preprint arXiv:1908.08345.

[43] Rajpurkar, P., Zhang, J., Lopyrev, K., & Liang, P. (2016). Squad: 100,000+ questions for machine comprehension of text. arXiv preprint arXiv:1606.05250.

[44] Vinyals, O., & Le, Q. (2015). A neural conversational model. arXiv preprint arXiv:1506.05869.

[45] See, A., Liu, P. J., & Manning, C. D. (2017). Get to the point: Summarization with pointer-generator networks. arXiv preprint arXiv:1704.04368.

[46] Dhiman, H., Vyas, N., Pagrotra, A., & Dutt, V. (2023). Real-time traffic analysis and driver safety enhancement via IoT and computer vision with high-resolution project lamps for intelligent road illumination. [Internet]. In 2023 1st International Conference on Circuits, Power and Intelligent Systems (CCPIS). IEEE. Available from: http://dx.doi.org/10.1109/CCPIS59145.2023.10291872.

[47] Yekrangi, M., & Nikolov, N. S. (2023). Domain-specific sentiment analysis: an optimized deep learning approach for the financial markets. *IEEE Access*, *11*. 70248–70262. doi: 10.1109/ACCESS.2023.3293733.

[48] Sanh, V., Debut, L., Chaumond, J., & Wolf, T. (2019). DistilBERT, a distilled version of BERT: Smaller, faster, cheaper and lighter. arXiv preprint arXiv:1910.01108.

[49] Kaplan, J., McCandlish, S., Henighan, T., Brown, T. B., Chess, B., Child, R. et al. (2020). Scaling laws for neural language models. arXiv preprint arXiv:2001.08361.

[50] Baevski, A., & Auli, M. (2018). Adaptive input representations for neural language modeling. arXiv preprint arXiv:1809.10853.

Nancy Jasmine Goldena and Thangapriya

7 Unveiling the Intriguing Applications of Mathematical Optimization in Artificial Intelligence

Abstract: Optimization algorithms play a vital role in addressing large-scale, data-intensive problems inherent in artificial intelligence (AI) applications. As AI continues to evolve, researchers are confronted with a multitude of opportunities and challenges, prompting exploration into promising avenues for research and development. This chapter provides an in-depth examination of popular optimization algorithms and their relevance in the context of AI. Furthermore, it investigates the expansive landscape of AI applications, highlighting the evolving trends in optimization methodologies. By synthesizing insights from both fields, this chapter aims to illuminate the transformative potential of mathematical optimization (MO) in advancing the capabilities of AI. This chapter explores the interdependent relationship between MO techniques and AI, elucidating their profound impact on diverse domains.

Keywords: AutoML optimization, computer vision optimization, deep learning optimization, finance optimization, healthcare optimization, machine learning optimization, NLP optimization, reinforcement learning optimization, robotics optimization, supply chain management optimization

7.1 Introduction

The convergence of mathematical optimization (MO) and artificial intelligence (AI) heralds a new era of innovation and discovery, offering boundless opportunities to tackle complex challenges across diverse domains. In the vast landscape of computer science, a multitude of problems revolve around the fundamental task of optimizing objectives, whether it entails maximizing efficiency, minimizing errors, or finding optimal solutions amidst vast solution spaces. Within the realm of AI, this optimization imperative manifests prominently in the training of predictive models, where the minimization of loss functions such as mean squared error forms the cornerstone of algorithmic refinement. MO, as a discipline, encompasses a rich array of methodologies meticulously crafted to address a spectrum of problem types, ranging from well-

Acknowledgements: We express our gratitude to the God Almighty for providing guidance and inspiration throughout the process, allowing us to complete the work in a timely manner. We would like to express the blessings and love received, which contributed to the successful completion of the chapter.

https://doi.org/10.1515/9783111436180-007

structured convex problems to inherently combinatorial challenges. Its pervasive influence extends beyond the confines of computer science, permeating engineering, economics, mathematics, and various other scientific domains. Indeed, the quest for optimal solutions transcends disciplinary boundaries, driving researchers and practitioners alike to explore innovative techniques and approaches [1].

In the design arena, optimization emerges as a critical and transformative force, empowering designers to create solutions that are not only efficient but also economical in terms of time and resources [2]. However, the complexity inherent in many engineering optimization problems often surpasses the capabilities of traditional approaches, necessitating the development of sophisticated methodologies capable of navigating intricate solution spaces with finesse [3]. The combination of AI with MO represents a paradigm shift in problem-solving capabilities, endowing computational systems with enhanced "brainpower" to tackle multifaceted challenges with unparalleled efficacy. Whether optimizing logistical operations, refining supply chains, or orchestrating complex decision-making processes, this fusion of disciplines holds the promise of revolutionizing industries and reshaping the fabric of society. Through a comprehensive examination of optimization algorithms, AI applications, and emerging trends, this chapter aims to illuminate the profound impact of this convergence on the technological landscape and beyond.

7.2 Research Domains of MO in AI

AI research spans a wide array of domains where MO techniques can be applied to enhance efficiency, accuracy, and scalability. Here are some key AI research domains where MO techniques find extensive utility, as shown in Figure 7.1.

7.2.1 Machine Learning Model Optimization

MO techniques are essential in optimizing machine learning (ML) models, including parameter tuning, hyperparameter optimization, and model selection. Methods such as gradient-based optimization, evolutionary algorithms, and Bayesian optimization are commonly employed to enhance model performance and generalization. It involves fine-tuning various parameters and configurations to improve model performance, scalability, and efficiency. Different types of ML are shown in Figure 7.2.

Here are some recent advancements and techniques in ML model optimization:

Automated Hyperparameter Tuning: This is a crucial advancement in ML, enabling the development of more efficient and effective models. Techniques such as

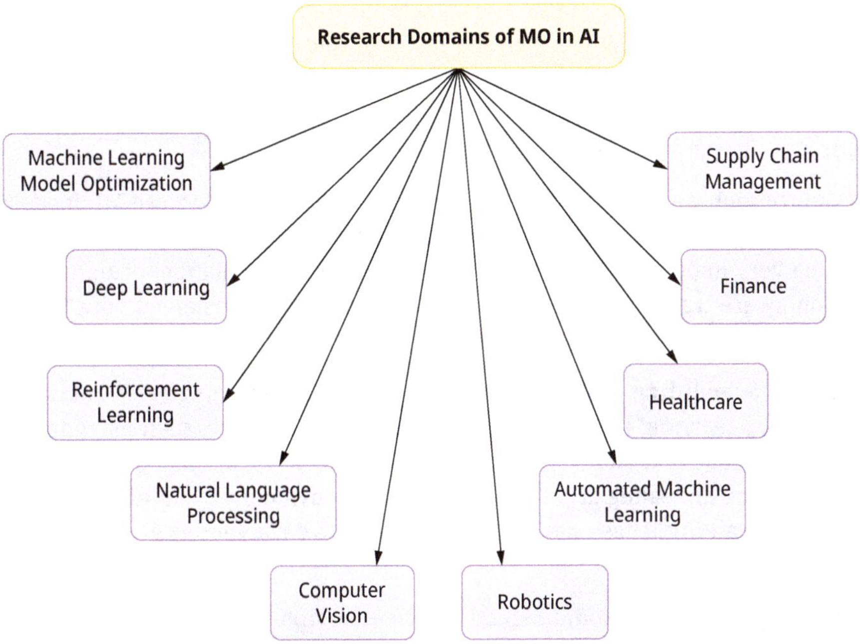

Figure 7.1: Research domains of MO in AI.

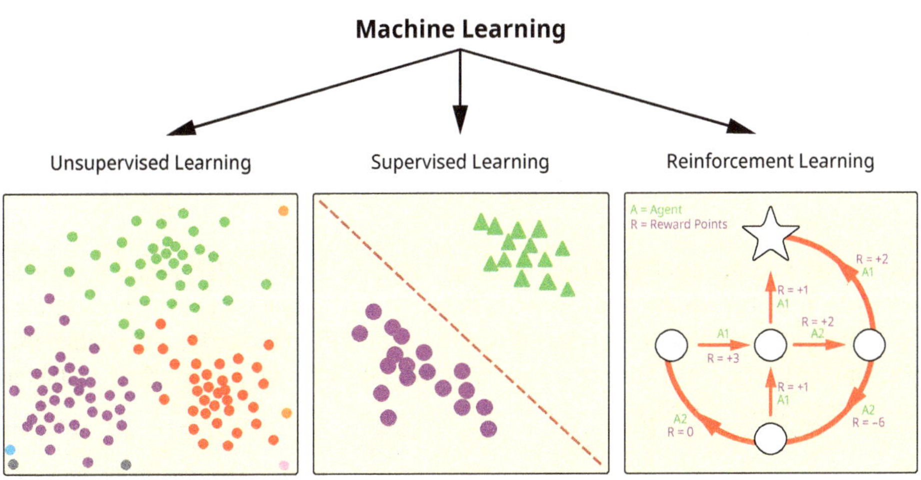

Figure 7.2: ML types.

Bayesian optimization, genetic algorithms, and reinforcement learning (RL)-based approaches have become increasingly popular. These methods efficiently explore the hyperparameter space to identify optimal configurations, resulting in enhanced model performance [10].

Neural Architecture Search (NAS): These methods automate the design of neural network architectures, making it easier and faster to find the best structures for specific tasks. Recent advancements in NAS algorithms, such as differentiable architecture search and evolutionary algorithms, have led to the discovery of novel architectures with superior performance and efficiency [11].

Model Compression and Quantization: With the increasing need to use ML models on devices with limited resources, techniques to compress and simplify these models have become very important. New methods like knowledge distillation, pruning, and quantization-aware training aim to make models smaller and faster without losing much performance [12].

Learning Rate Scheduling and Optimization: Optimizing learning rates is crucial for training deep learning models effectively. Recent techniques, such as adaptive learning rate methods (like Adam and RMSprop) and learning rate scheduling strategies (such as cosine annealing and warmup schedules), dynamically adjust the learning rate during training. These methods help to improve convergence and prevent overfitting [13].

Bayesian Optimization for Model Selection: Bayesian optimization methods are increasingly used for model selection and configuration optimization. These techniques efficiently explore the space of models and hyperparameters by leveraging probabilistic models of the objective function, leading to faster convergence and improved model performance.

Adversarial Robustness and Regularization: Adversarial attacks pose a major threat to the robustness of ML models. Recent research focuses on developing regularization techniques and adversarial training methods to enhance model robustness against adversarial perturbations, thereby improving generalization performance and security.

Distributed and Parallel Training: Training large-scale ML models requires distributed and parallel computing infrastructure. Recent advancements in distributed training frameworks, such as TensorFlow distributed and PyTorch distributed, enable efficient parallelization of training across multiple GPUs or devices, leading to faster convergence and scalability.

Overall, recent advancements in ML model optimization encompass a broad range of techniques and methodologies aimed at improving model performance, scalability, and efficiency across various applications and domains [15].

7.2.2 Deep Learning

Deep learning models often involve vast numbers of parameters, making optimization challenging. MO techniques, such as stochastic gradient descent (SGD) variants, adaptive learning rate methods, and network pruning algorithms, are instrumental in training deep neural networks efficiently and improving their performance. Deep learning MO is the process of minimizing a loss function by optimizing the parameters of deep neural networks. To successfully train deep learning models and achieve high performance on a variety of tasks, including image classification, object recognition, natural language processing, and RL, this optimization procedure is essential. Figure 7.3 shows the different types of neural networks.

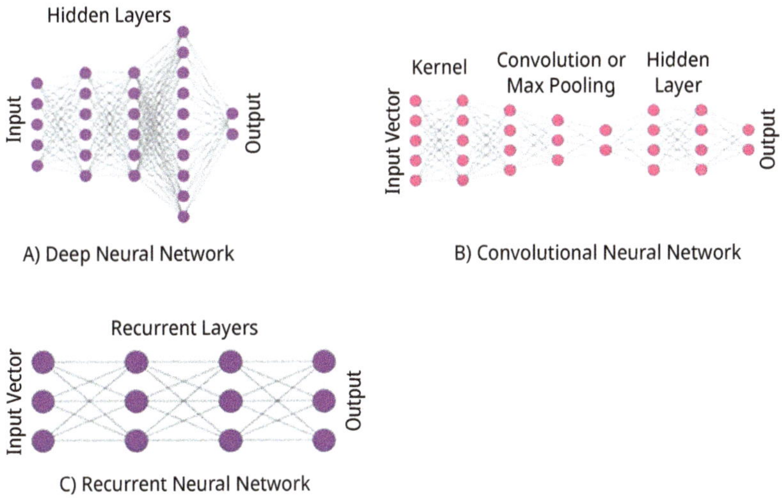

Figure 7.3: Different types of neural networks.

Here are some key aspects and techniques of deep learning MO:

Gradient-Based Optimization: The majority of deep learning optimization strategies are built on the gradient descent algorithm. It involves modifying the model parameters iteratively in the direction of the loss function's negative gradient relative to the parameters. There are several gradient descent variations, each with a distinct tradeoff between memory consumption and convergence speed, such as SGD, mini-batch gradient descent, and batch gradient descent.

Optimization Algorithms: To enhance the fundamental gradient descent technique, numerous sophisticated optimization algorithms have been created. These include Adam (adaptive moment estimation), which combines momentum with adaptive learning rates based on the first and second moments of the gradients, and momentum optimization, which speeds up convergence by building up a moving average of previous gradients.

Learning Rate Scheduling: To enhance convergence and avoid overfitting, learning rate scheduling algorithms dynamically modify the learning rate during training. A few popular scheduling strategies are cyclic learning rates, which alternate the learning rate over training epochs, step decay, and exponential decay.

Weight Initialization: Proper initialization of model weights is crucial for efficient optimization and for preventing vanishing or exploding gradients. Techniques such as Xavier initialization and orthogonal initialization help initialize weights in a way that facilitates convergence and improves training stability.

Batch Normalization: The process of normalizing each layer's activations across mini-batches during training is known as batch normalization. By decreasing internal covariate shift, it helps stabilize and accelerates the training process and permits larger learning rates without running the risk of divergence.

Regularization Techniques: Regularization techniques, including weight decay, dropout, and L1 and L2 regularization aid in preventing overfitting and enhancing model generalization. These techniques penalize large parameter values or randomly drop units during training to encourage simpler and more robust models.

Advanced Architectures: Architectural innovations such as residual connections, skip connections, and attention mechanisms have been introduced to facilitate optimization and improve gradient flow through deep neural networks. These architectural modifications enable the training of deeper and more complex models while mitigating the vanishing gradient problem.

Parallel and Distributed Training: To train large-scale deep learning models efficiently, parallel and distributed training techniques are employed. These techniques distribute the computation and memory across multiple devices or machines, enabling faster convergence and scalability to larger datasets and models [31].

In general, deep learning MO includes a broad range of approaches and strategies with the goal of effectively training deep neural networks and achieving innovative outcomes on different tasks [32–40]. Furthering the capabilities of deep learning models and expanding the frontiers of AI requires ongoing research and innovation in this area [32–40].

7.2.3 Reinforcement Learning

RL teaches agents how to maximize cumulative rewards by making consecutive decisions. MO techniques are utilized in policy optimization, value function approximation, and exploration-exploitation trade-offs. Optimization plays a crucial role in the learning of optimal policies by algorithms such as Q-learning, policy gradient approaches, and actor-critic systems [28]. RL MO involves optimizing the policies or value functions of an agent to maximize cumulative rewards in a sequential decision-making environment. It is a fundamental aspect of training RL agents to perform tasks efficiently and effectively. Here are some key components and techniques of RL MO:

Policy Optimization: Policy-based RL involves the agent directly learning a policy by mapping states to actions. The policy parameters are updated to maximize predicted rewards using policy optimization techniques such as policy gradient methods, like REINFORCE, and actor-critic algorithms, such as A2C (advantage actor-critic) and PPO (proximal policy optimization) [29].

Value Function Approximation: Value-based RL methods estimate the value function, which represents the expected cumulative rewards of following a particular policy. Value function approximation techniques, such as Q-learning, deep Q-networks (DQN), and double DQN, optimize the parameters of a value function approximator to accurately estimate state-action values [30].

Exploration-Exploitation Trade-Offs: Effective RL requires striking a balance between exploitation, which uses existing policies to maximize rewards, and exploration, which tries new behaviours to find possibly better policies. Exploration strategies, including ε-greedy, Boltzmann exploration, and Thompson sampling, optimize the agent's exploration policy to efficiently explore the environment.

Reward Shaping and Incentive Design: Techniques for shaping rewards modify the reward signal to direct the agent towards desired behaviour and improve learning. Incentive design involves carefully designing the reward function to align with the task objectives and provide informative feedback to the agent, facilitating efficient optimization.

Temporal Difference (TD) Learning: These techniques, including the SARSA (state-action-reward-state-action) and TD(λ) algorithms, improve the agent's estimates based on TDs between successive time steps to optimize its policy or value function. These techniques enhance convergence and facilitate online learning [25].

Off-Policy Learning: Off-policy RL methods decouple the exploration policy from the target policy, allowing the agent to learn from historical data collected using a different policy. Techniques such as Q-learning with experience replay and deep deterministic policy gradient optimize the agent's policy or value function using off-policy data samples.

Model-Based RL: This technique simulates potential paths for policy evaluation and planning by learning a model of the dynamics of the environment. Model-based optimization algorithms, such as model predictive control (MPC) and Dyna-Q, optimize the agent's policy or value function using the learned environment model.

Multi-agent RL: This type of RL involves the simultaneous interaction of several agents with the environment. Optimization techniques, such as multi-agent actor-critic and decentralized policy optimization, optimize the policies of multiple agents to achieve coordination and cooperation.

Overall, RL MO encompasses a diverse range of techniques and methodologies aimed at training RL agents to make optimal decisions in complex and uncertain environments. Continuous research and innovation in this field are essential for advancing the capabilities of RL and enabling the development of intelligent autonomous systems [22–24].

7.2.4 Natural Language Processing (NLP)

Optimization strategies are useful for NLP applications, including analysing sentiment, text summarization, and automated translation. MO methods are used to train neural language models, optimize attention mechanisms, and fine-tune pre-trained language representations. NLP applications are shown in Figure 7.4.

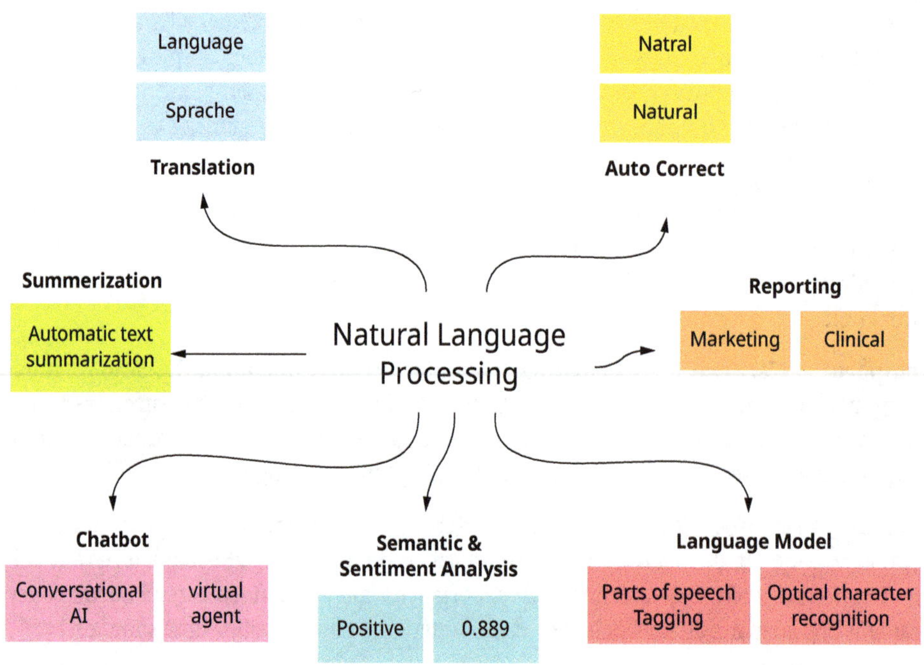

Figure 7.4: NLP applications.

NLP MO is the process of optimizing different parts and models that are used in NLP operations in order to increase scalability, efficiency, and performance. Here are some key aspects and techniques of NLP MO:

Sequence-to-Sequence Model Optimization: Recurrent neural networks (RNNs) and transformer architectures (e.g., BERT and GPT) are examples of sequence-to-sequence (Seq2Seq) models that are optimized to produce text that is logical and useful to the context. Optimization techniques, including gradient-based methods and advanced optimization algorithms like Adam, are used to train these models on large text corpora efficiently [20].

Attention Mechanisms' Optimization: To achieve state-of-the-art performance in NLP tasks, attention mechanisms, which allow models to concentrate on essential portions of input sequences, are essential. Model interpretability and computational efficiency are enhanced through the optimization of techniques like multi-head attention and self-attention [21].

Language Model Fine-Tuning: To adapt pre-trained language models, like BERT and GPT, to particular NLP tasks, such as text classification, named entity identification, and machine translation, they are refined on task-specific datasets. Optimization methods, including transfer learning and differential learning rate scheduling, are used to fine-tune these models effectively [14].

Embedding Optimization: Word embeddings and contextual embeddings play a crucial role in representing words and sentences in NLP tasks. Optimization techniques, such as skip-gram models, continuous bag-of-words models, and contextual embedding methods like ELMo (embeddings from language models) and contextualized word embeddings, optimize the learning of embeddings to capture semantic and syntactic information effectively.

Text Generation Optimization: Text generation models, including language models and sequence-to-sequence models with attention mechanisms, are optimized to generate coherent and diverse text samples. The creation process is controlled, while techniques including temperature scaling, nucleus sampling, and beam search optimization are applied to enhance the quality of the created text.

Named Entity Recognition (NER) Optimization: NER models identify and classify named entities (e.g., persons, organizations, locations) in text. Optimization techniques, including conditional random fields (CRFs) and neural network-based approaches like BiLSTM-CRF (bidirectional LSTM with CRF), are used to optimize NER models for accurate entity recognition.

Sentiment Analysis Optimization: Sentiment analysis models classify the sentiment or emotion expressed in text, such as positive, negative, or neutral. Optimization techniques, including feature engineering, ensemble methods, and deep learning architectures like convolutional neural networks (CNNs) and RNNs, are used to optimize sentiment analysis models for improved classification accuracy.

Machine Translation Optimization: Machine translation models translate text from one language to another. Optimization techniques, including attention mechanisms, transformer architectures, and RL-based approaches, are used to optimize machine translation models for accurate and fluent translations.

Overall, NLP MO encompasses a broad range of techniques and methodologies aimed at optimizing various components and models used in NLP tasks to advance the capabilities of natural language understanding and generation systems. Continuous research and innovation in this field are essential for pushing the boundaries of NLP and enabling the development of intelligent language processing applications [16–19].

7.2.5 Computer Vision

In computer vision tasks such as object identification, image segmentation, and recognition of images, optimization is essential. MO approaches are used to optimize CNNs to increase classification accuracy and extract useful features. In order to increase productivity, accuracy, and efficiency, computer vision MO entails optimizing models and algorithms utilized in computer vision activities. A simple example of computer vision is shown in Figure 7.5.

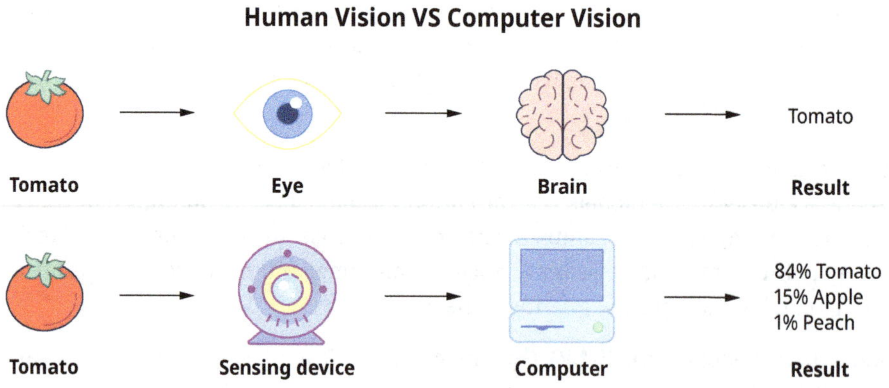

Figure 7.5: Simple example of computer vision.

Here are some key aspects and techniques of computer vision:

CNN Optimization: The foundation for recent computer vision systems, CNNs are utilized for tasks including object identification, image segmentation, and image classification. Optimization techniques, including gradient-based methods like SGD and advanced optimization algorithms like Adam, are used to train CNNs on large-scale image datasets efficiently.

Transfer Learning and Fine-Tuning: Using pre-trained CNN models (such as ResNet, VGG, and Inception) as feature extractors and enhancing them on datasets relevant to a certain task is known as transfer learning. Optimization methods, including differential learning rate scheduling and batch normalization, are used to adapt pre-trained models to new tasks and improve performance.

Object Detection Optimization: Object detection models localize and classify objects within images. Optimization techniques, including region proposal networks, anchor box optimization, and non-maximum suppression, are used to optimize object detection models for accurate and efficient object localization and recognition.

Semantic Segmentation Optimization: Semantic segmentation models assign semantic labels to each pixel in an image, delineating object boundaries and regions. Optimization methods, including dilated convolutions, skip connections, and CRFs, are used to optimize semantic segmentation models for accurate pixel-wise classification.

Instance Segmentation Optimization: Instance segmentation models not only segment objects in an image but also distinguish between individual object instances. To optimize instance segmentation models for accurate object delineation and instance recognition, methods such as Mask R-CNN (region-based CNN) and feature pyramid networks can be used.

Image Generation Optimization: Image generation models, such as generative adversarial networks and variational auto encoders, generate realistic images from noise or latent representations. Optimization techniques, including adversarial training, perceptual loss functions, and regularization methods [41], are used to optimize image generation models for generating high-quality and diverse images.

Feature Extraction Optimization: By obtaining meaningful representations from raw image data, feature extraction is essential to computer vision problems. Optimization techniques, including convolutional layer design, pooling operations, and feature normalization, are used to optimize feature extraction pipelines for capturing relevant visual patterns and semantics.

Image Registration and Alignment Optimization: Image registration and alignment techniques align multiple images to a common reference frame for tasks such as image stitching and medical image analysis. Optimization methods, including iterative closest point algorithms, optical flow estimation, and geometric transformations, are used to optimize image registration and alignment for accurate and robust alignment of image data.

Overall, computer vision MO encompasses a wide range of techniques and methodologies aimed at optimizing models and algorithms for various computer vision tasks. Continuous research and innovation in this field are essential for advancing the capabilities of computer vision systems and enabling the development of intelligent visual perception applications [46–42].

7.2.6 Optimization in Robotics

Optimization is fundamental in robotics research for motion planning, trajectory optimization, and robot control. MO algorithms are employed to solve inverse kinematics problems, optimize robot trajectories in dynamic environments, and learn optimal control policies for robotic manipulation tasks. In robotics, MO plays a crucial role in various aspects of robot design, control, planning, and perception. Here are some key aspects and techniques of robotics MO:

Motion Planning Optimization: Motion planning is the process of creating paths for robots to move in their surroundings that avoid collisions while maximizing factors like path length, smoothness, and energy efficiency. Optimization techniques, including rapidly exploring random trees, sampling-based planners, and trajectory optimization algorithms like sequential quadratic programming and direct collocation, are used to find optimal robot trajectories efficiently.

Inverse Kinematic Optimization: Inverse kinematics problems involve determining the joint configurations of a robot given desired end-effector positions and orientations. Optimization techniques, including gradient-based methods and numerical solvers, are used to solve inverse kinematics problems efficiently and accurately, considering constraints such as joint limits and collision avoidance.

Robot Control Optimization: Robot control involves generating control signals to actuate robot joints and achieve desired motions. For tasks including motion tracking, energy management, and impedance management, control policies are optimized using techniques like RL-based methods, such as linear quadratic regulators and MPC.

Sensor Fusion and Localization Optimization: Sensor fusion combines data from multiple sensors (e.g., cameras, lidars, and inertial sensors) to estimate the robot's state and environment accurately. Optimization techniques, including Kalman filters, particle filters, and graph-based optimization methods like pose graph optimization, are used to fuse sensor data and perform localization and mapping tasks efficiently.

Path Planning and Optimization in Manipulation: Path planning and optimization techniques are used to plan collision-free paths for robot manipulators to grasp and manipulate objects in cluttered environments. Optimization methods, including grasp optimization, task-space control, and optimization-based motion planning algorithms like CHOMP (covariant Hamiltonian optimization for motion planning), are used to optimize manipulation trajectories for dexterous and efficient object manipulation.

Optimal Control for Dynamic Systems: Optimal control theory is used to find control policies that optimize a performance criterion over a finite or infinite time horizon. In robotics, optimal control techniques, including dynamic programming, Pontryagin's maximum principle, and direct collocation methods, are used to optimize control inputs for dynamic systems such as robotic manipulators and mobile robots, considering constraints such as dynamics, actuator limits, and environmental uncertainties.

Task Allocation and Multi-robot Coordination Optimization: In multi-robot systems, task allocation and coordination optimization techniques are used to allocate tasks among robots and optimize their coordination to achieve collective objectives efficiently. Optimization methods, including combinatorial optimization, distributed optimization algorithms, and game-theoretic approaches, are used to optimize task allocation, role assignment, and cooperative behaviours in multi-robot systems.

Overall, robotics MO encompasses a wide range of techniques and methodologies aimed at optimizing various aspects of robot design, control, planning, and perception. Continuous research and innovation in this field are essential for advancing the capabilities of robots and enabling them to perform complex tasks autonomously and efficiently in real-world environments [47–50].

7.2.7 Automated Machine Learning (AutoML)

Model selection, hyperparameter tuning, and feature engineering are all automated by AutoML systems. To find high-performing ML pipelines, model space is effectively searched using MO approaches such as evolutionary strategies, genetic algorithms, and Bayesian optimization. AutoML MO involves optimizing the selection and configuration of ML pipelines to achieve optimal performance on a given task. Here are some key aspects and techniques of AutoML MO:

Hyperparameter Optimization: Hyperparameters, which include network design, regularization strength, and learning rate, control how ML models learn. In order to quickly search the hyperparameter space and identify suitable configurations for model training, hyperparameter optimization techniques such as grid search, random search, Bayesian optimization, and genetic algorithms are applied.

Model Selection and Architecture Search: Selecting the right ML architecture or algorithm for a job is known as model selection. Automated model selection techniques, such as meta-learning, NAS, and ensemble methods, optimize the selection of models to achieve optimal performance across diverse datasets and tasks.

Feature Engineering and Selection: To enhance model performance, feature engineering involves producing insightful features from raw data. Automated feature engineering techniques, such as feature extraction, transformation, and selection algorithms, optimize the feature space to capture relevant patterns and reduce dimensionality effectively.

Pipeline Configuration Optimization: ML pipelines consist of multiple preprocessing, feature extraction, and model training steps. Automated pipeline configuration techniques, such as pipeline optimization algorithms and Bayesian optimization, optimize the configuration of ML pipelines to achieve optimal performance and efficiency for a given task.

Algorithm Selection and Tuning: Automated algorithm selection techniques determine the most suitable ML algorithm or ensemble of algorithms for a given task. Techniques such as meta-learning, algorithm selection frameworks, and multi-armed bandit algorithms optimize the selection and tuning of algorithms to maximize predictive performance and efficiency.

Ensemble Learning and Stacking Optimization: To increase overall performance and robustness, ensemble learning integrates predictions from several ML models. Stacking, bagging, and boosting algorithms are examples of automated ensemble learning strategies that maximize the mix and composition of individual models to achieve better results than single models.

Transfer Learning and Domain Adaptation: To enhance performance on a target task with a limited amount of labelled data, transfer learning techniques make use of knowledge from previously trained models or related tasks. To improve model generalization and adaptability, automated transfer learning techniques like feature extraction, domain adaptation algorithms, and fine-tuning maximize the transfer of knowledge from source domains to target domains.

Overall, AutoML MO encompasses a wide range of techniques and methodologies aimed at automating and optimizing the ML pipeline selection, configuration, and tuning process. Continuous research and innovation in this field are essential for democratizing ML and enabling non-experts to build high-performing models efficiently for various applications and domains [51–55].

7.2.8 Healthcare

Optimization is applied to medical image processing, disease diagnosis, treatment planning, and drug discovery in the healthcare industry. Personalized treatment strategies, successful clinical trials, and the best drug combinations for better patient outcomes are all made possible with the use of MO methods. MO is essential to the execution of healthcare in a number of ways, including the distribution of resources, organizing, planning for treatment, and decision-making. The following are some essential elements and methods of healthcare MO:

Resource Allocation Optimization: Healthcare resources must be distributed effectively to fulfil patient demand and improve the quality of healthcare. These resources include hospital beds, medical personnel, equipment, and medication. Hospitals and other healthcare institutions use MO approaches, such as linear programming, integer programming, and queuing theory, to efficiently allocate resources, minimize wait times, and maximize resource utilization.

Patient Scheduling and Appointment Optimization: The goal of patient scheduling is to minimize wait times for patients, maximize the productivity of healthcare providers, and minimize the number of absences by scheduling appointments for medical consultations, diagnostic tests, operations, and treatments as efficiently as possible. In order to maximize patient scheduling while improving healthcare accessibility and effectiveness, optimization approaches such as scheduling algorithms, queue management systems, and appointment reminder systems are applied.

Treatment Planning and Optimization: To maximize therapy efficacy, reduce side effects, and enhance patient outcomes, MO techniques – such as mathematical modelling, simulation, and optimization algorithms – are used to optimize treatment plans, dose scheduling, and medication combinations.

Medical Imaging Optimization: Large amounts of data are produced by medical imaging methods like MRI, CT, PET, and X-ray imaging, which require effective processing, analysis, and interpretation. Medical image acquisition, reconstruction, segmentation, and analysis are optimized using optimization techniques, such as ML-based methodologies, image processing techniques, and restoration algorithms, to enable better evaluation, planning of treatment, and monitoring of diseases.

Healthcare Supply Chain Optimization: Healthcare supply chains encompass the acquisition, allocation, and supervision of medical supplies, medications, and equipment to healthcare establishments. In order to maximize supply chain efficiency, reduce shortfalls and deficiencies, and guarantee timely shipment of necessary healthcare services, supply chain modelling, demand forecasting, and inventory optimization are some of the MO approaches that are utilized.

Clinical Trial Optimization: To assess the safety and effectiveness of novel medications, therapies, and medical devices, clinical trials are important. Clinical trial procedures are maximized, expenses minimized, trial durations minimized, and the discovery of innovative healthcare interventions is accelerated through the use of optimization techniques such as trial design optimization, patient recruitment optimization, and adaptive trial design methodologies.

Healthcare Policy and Decision-Making Optimization: Healthcare policy issues such as the distribution of resources, finance, reimbursement rules, and treatment recommendations can all have an enormous effect on healthcare outcomes, costs, and inequalities. To optimize public health and well-being, healthcare policy decisions, healthcare intervention prioritization, and resource allocation are informed by optimization approaches such as decision analysis, cost-effectiveness analysis, and health economic modelling.

To improve patient outcomes, increase access to healthcare, and maximize resource use in healthcare systems, a broad range of approaches and methodologies are collectively referred to as healthcare MO. These methodologies and techniques are intended to optimize different areas of healthcare delivery, operations, and decision-making. To solve the complex issues facing the healthcare sector and improve the efficiency, performance, and quality of healthcare services provided globally, continuous research and innovation in this area are essential [56–59].

7.2.9 Finance

In financial modelling and algorithmic trading, optimization is employed for portfolio optimization, risk management, and trading strategy optimization. MO techniques enable investors to allocate assets effectively, minimize risk exposure, and maximize returns in dynamic market conditions. In finance, MO plays a crucial role in various areas, including portfolio management, risk assessment, trading strategies, and asset allocation. Here are some key aspects and techniques of finance MO:

Portfolio Optimization: This involves constructing investment portfolios that maximize returns while minimizing risk. MO techniques, such as mean-variance optimization, quadratic programming, and stochastic programming, are used to allocate assets across different investment options to achieve optimal risk-return trade-offs based on investor preferences and constraints [27].

Asset Allocation Optimization: Determining the best combination of asset types, such as stocks, bonds, real estate, and commodities, in a portfolio to meet financial goals is known as asset allocation. Optimization techniques, including dynamic asset allocation models, risk parity strategies, and tactical asset allocation approaches, are used to optimize asset allocation decisions over time based on market conditions, economic forecasts, and risk factors.

Risk Management Optimization: Risk management includes recognizing, evaluating, and reducing financial risks, including operational, liquidity, credit, and market risks. Value-at-risk modelling, conditional value-at-risk optimization, and stress testing procedures are a few of the MO approaches that are used to optimize risk management strategies, protect portfolios from unfavourable events, and ensure regulations are followed.

Portfolio Rebalancing Optimization: Rebalancing a portfolio requires periodically modifying its asset allocation to preserve target asset class weights and risk-return characteristics. Convex optimization and linear programming are two optimization approaches that are used to efficiently rebalance portfolios while minimizing tracking error and transaction costs in comparison to benchmark indices.

Trading Strategy Optimization: Trading strategies involve making buy or sell decisions in financial markets to exploit market inefficiencies and generate profits. Various techniques, including algorithmic trading algorithms, market microstructure models, and high-frequency trading strategies, are used to optimize trading execution, minimize trading costs, and maximize trading profits based on market data and order flow dynamics.

Option Pricing and Hedging Optimization: Financial derivatives are valued, and their fair market prices are evaluated using option pricing models, such as the Black-Scholes-Merton and binomial option pricing models. Option pricing models are solved effectively and precisely by optimization approaches such as numerical methods, finite difference methods, and Monte Carlo simulation. Additionally, hedging strategies, such as delta hedging and dynamic hedging, are optimized to manage the risk exposure of options and derivative positions effectively.

Capital Allocation Optimization: Capital allocation involves distributing financial resources across different business units, projects, or investment opportunities to maximize shareholder value and return on investment. MO techniques, including capital budgeting models, project selection algorithms, and capital allocation frameworks like the capital asset pricing model, are used to optimize capital allocation decisions and allocate resources to projects with the highest expected returns and risk-adjusted profitability.

Overall, finance MO encompasses a wide range of techniques and methodologies aimed at optimizing various aspects of financial decision-making, risk management, and portfolio management to achieve investment objectives, enhance financial performance, and mitigate risks in financial markets. Continuous research and innovation in this field are essential for developing advanced optimization models and algorithms to address the evolving challenges and complexities of the global financial system [60–62].

7.2.10 Supply Chain Management

The optimization of supply chain activities, such as demand forecasting, logistics planning, and inventory management, is dependent upon optimization. Businesses can increase supply chain efficiency overall, optimize manufacturing schedules, and save transportation expenses by utilizing MO strategies. The aim of supply chain management optimization is to optimize the supply chain in various aspects, including inventory control, production scheduling, distribution logistics, and supply chain network design. Here are some key aspects and techniques of supply chain management MO:

Inventory Optimization: The goal of inventory management is to maximize stock levels in order to strike a balance between carrying and stockout expenditures. Using MO techniques, such as the economic order quantity model, the economic production quantity model, and stochastic inventory models, the optimal ordering policies, safety stock levels, and reorder points are established to minimize inventory costs while ensuring sufficient product availability.

Production Planning and Scheduling Optimization: To effectively fulfil customer demand, production planning requires optimizing production schedules, capacity utilization, and allocation of resources. When maximizing production speed, minimizing production costs, and optimizing production schedules, MO techniques – such as linear, integer, and mixed-integer linear programming – take capacity utilization boundaries, lead times for production, and demand variability into consideration.

Distribution Logistics Optimization: The field of distribution logistics concerns the efficient flow of commodities via distribution centres, transportation systems, and distribution channels from vendors to consumers. Distribution routes, warehouse locations, and modes of transportation are optimized using MO approaches, such as vehicle routing optimization, facility location models, and network flow optimization, to save transportation costs, shorten delivery times, and enhance service quality.

Supply Chain Network Design Optimization: The process of optimizing supply chain networks' configuration, which includes determining where and how many suppliers, manufacturing sites, warehouses, and distribution centres are located. Supply chain networks are designed using MO approaches, such as network optimization models, location-based facility models, and designing network algorithms, to minimize overall supply chain costs, minimize transportation distances, and boost supply chain flexibility and robustness.

Demand Forecasting and Demand Planning Optimization: The goal of demand forecasting and demand planning is to anticipate future product demand and adjust production and inventory levels appropriately. MO techniques are used to accurately predict demand and maximize production and inventory schedules to meet the requirements with the least amount of leftover stock and stockouts.

Supplier Selection and Procurement Optimization: The procurement and supplier selection process includes locating and selecting suppliers based on criteria like cost, reliability, quality, and delivery schedule. To save procurement costs, minimize supply chain risks, and enhance supplier performance, supplier selection decisions and procurement strategies are optimized through the use of MO techniques, including supplier assessment models, multi-criteria decision analysis, and supplier portfolio optimization.

Collaborative Planning, Forecasting, and Replenishment (CPFR) Optimization: Supply chain efficiency is increased and operations are synchronized by means of collaborative planning and forecasting between trading partners. MO approaches, such as collaborative planning algorithms, demand forecasting models, and replenishment optimization algorithms, are used to optimize CPFR processes, increase demand visibility, and minimize supply chain unpredictability and uncertainty.

Overall, supply chain management MO encompasses a variety of approaches and strategies targeted at streamlining the supply chain in order to increase productivity, cut expenses, and boost consumer satisfaction. The development of sophisticated optimization models and algorithms to handle the complexity and unpredictability of contemporary supply chains and achieve sustainable supply chain performance depends on continual research and innovation in this area. Finally, MO techniques serve as powerful tools across various AI research domains, empowering researchers and practitioners to tackle complex problems and drive innovation in diverse applications [63–65].

7.3 Popular MO Techniques Used in AI

Several MO techniques are widely used in AI to solve various optimization problems efficiently, as shown in Figure 7.6.

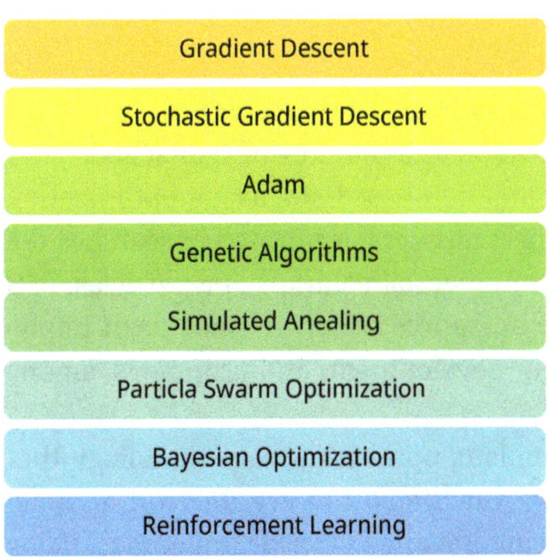

Figure 7.6: Popular mathematical optimization techniques used in artificial intelligence.

Here are some popular MO techniques used in AI:

- **Gradient Descent**: This is a first-order optimization process that moves iteratively in the direction of the function's negative gradient to minimize a function. It is frequently used to update model parameters and minimize loss functions in ML training, such as neural networks.
- **Stochastic Gradient Descent (SGD)**: SGD is a gradient descent variation that uses a random subset of training instances (mini-batches) at each iteration to optimize the objective function. Because of its high computing efficiency and capacity to handle massive databases, it is frequently utilized in large-scale learning tasks.
- **Adam**: The advantageous features of momentum optimization and RMSprop are combined in the adaptive learning rate optimization approach known as Adam. By adjusting the learning rate for every parameter according to the gradient's first and second moments, deep neural network training becomes more effective and converges more quickly.
- **Genetic Algorithms**: Inspired by the theory of natural selection and the evolution of genes, genetic algorithms are a category of evolutionary algorithms. They generate a population of potential solutions to an optimization issue over several generations by employing strategies like the crossover phenomenon, mutation, and filtering. When dealing with optimization problems that conventional gradient-based techniques may find difficult, such as combinatorial optimization and function optimization with non-smooth, irregular, or multivariate objective functions, genetic algorithms can be used.
- **Simulated Annealing**: A probabilistic optimization technique called "simulated annealing" has been inspired by the metallurgical annealing procedure. To manage the acceptance of inadequate solutions and enable the algorithm to stretch out of local optima and more deeply explore the solution space, the temperature parameter is gradually decreased. In particular, optimisation problems with a harsh, multi-modal search space benefit greatly from simulated annealing.
- **Particle Swarm Optimization (PSO)**: The social dynamics of fish schools and bird flocks provide inspiration for the population-based stochastic optimization technique known as PSO. To identify the best answers, it iteratively searches the search space and maintains a population of potential solutions, or particles. PSO is frequently employed in continuous optimization issues and has been implemented in selecting features, neural system development, and groupings, among other AI applications.
- **Bayesian Optimization**: This is a probabilistic optimization methodology that models and optimizes undefined objective functions that are difficult or time-consuming to assess. It achieves this by using Bayesian statistical methods. Using a probabilistic surrogate model, it successively assesses the objective function at promising regions in the search space and updates the model to concentrate the search on areas most likely to have optimal solutions. In hyperparameter tuning

for ML models and other black-box optimization issues, Bayesian optimization is frequently utilized.
- **Reinforcement Learning**: An optimization technique called "reinforcement learning" (RL) allows an agent to make successive decisions by interacting with its surroundings to maximize cumulative rewards. Robotics, gaming, and autonomous systems are just a few of the domains where RL techniques, such as Q-learning, policy gradients, and actor-critic methods, maximize the agent's policy or value function to learn the best possible decision-making strategies [26].

These are just a few examples of popular MO techniques used in AI [6–8]. Depending on the problem domain and specific requirements, different optimization algorithms may be more suitable for solving different types of optimization problems efficiently.

7.4 Popular AI Tools for MO

Several AI tools and libraries are commonly used for MO in various domains, as shown in Figure 7.7.

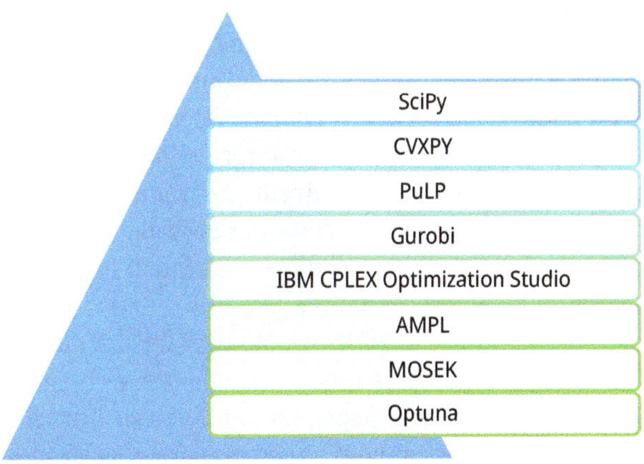

Figure 7.7: Popular artificial intelligence tools and libraries for mathematical optimization.

Here are some popular ones:
- **SciPy**: This is a Python library that is available for scientific computing and is open-source. It has modules for linear algebra, integration, interpolation, optimization, and other mathematical operations. The **scipy.optimize** module provides various optimization algorithms, including nonlinear optimization, constrained optimization, and global optimization methods.

- **CVXPY**: This modelling language for convex optimization problems is incorporated in Python. It lets users write optimization problems using a natural mathematical syntax, and it automatically converts such issues into standard form so that a variety of convex optimization algorithms can resolve them.
- **PuLP**: This is a modelling package that is available for free. For the purpose of defining and resolving linear and integer programming, mixed-integer programming, and network flow optimization, it offers a simple user interface.
- **Gurobi**: It is a commercial optimization solver that can handle many different types of MO problems, including convex optimization, mixed-integer programming, quadratic programming, and linear programming (LP, QP, and MIP). It provides high-performance optimization tools and APIs for Python, MATLAB, and C++, among other computer languages.
- **IBM CPLEX Optimization Studio**: A full spectrum of optimization tools and solutions for combinatorial, mathematical, and constraint programming is available in the commercial optimization suite IBM CPLEX Optimization Studio. It has sophisticated optimization methods and modelling capabilities, and it has APIs for a number of computer languages, including Python, Java, and C++.
- **AMPL**: A mathematical programming language, or AMPL, is a modelling language and environment used to create and resolve MO issues. Numerous forms of optimization problems are supported by it, such as constraint programming, mixed-integer programming, linear programming, and nonlinear programming. In addition to integrating with well-known optimization solvers like Gurobi, CPLEX, and GLPK, AMPL offers solvers for a wide range of optimization problems.
- **MOSEK**: This is a commercial optimization solution designed for mathematical programming, encompassing mixed-integer, conical, and linear programming. High-performance algorithms and APIs are available for multiple programming languages, such as Python, MATLAB, and C++. It finds extensive application in diverse areas, such as operations research, engineering, and finance.
- **Optuna**: An open-source hyperparameter optimization framework for ML models is called Optuna. It offers algorithms for grid-based search, random search, probabilistic optimization, and other optimization approaches to effectively find the best possible hyperparameters for ML algorithms. With its user-friendly design, Optuna smoothly interfaces with well-known ML libraries like TensorFlow, PyTorch, and scikit-learn.

These are just a few examples of popular AI tools and libraries for MO [4, 5]. Depending on the specific requirements of the optimization problem and the programming language, people can choose different tools and libraries that best suit their needs.

7.5 Research Directions Based on MO in AI

Here are some trending research directions focusing on MO, as discussed in Table 7.1.

Table 7.1: Research directions based on mathematical optimization in artificial intelligence.

S. no.	Topics	Description
1	Optimizing inventory management and pricing strategies	Applying MO techniques to optimize inventory levels and pricing strategies in retail and supply chain management, leading to improved profitability and customer satisfaction.
2	Short-term load forecasting using AI	Utilizing MO models and ML algorithms to forecast short-term load demand in energy systems, enabling efficient resource allocation and grid management.
3	Prediction of stock market using AI	Employing MO algorithms and ML techniques to predict stock market trends and make informed investment decisions, optimizing portfolio management and risk mitigation strategies.
4	Confluence of AI and blockchain-powered smart contracts in the finance system	Integrating MO techniques with blockchain technology and smart contracts to automate financial transactions and optimize contract execution in decentralized finance systems.
5	Analysing data patterns to predict financial risks	Leveraging MO models and AI algorithms to analyse data patterns and predict financial risks, enabling proactive risk management and mitigation strategies in banking and financial institutions.
6	Forecasting evolving market circumstances	Businesses are able to modify their strategies and maintain their competitiveness in dynamic market situations by using ML algorithms and MO techniques to analyse market data and forecast changes in the marketplace.
7	Monitoring network behaviour and surveillances	Employing MO methods and AI algorithms to quickly identify and address security threats by using AI algorithms and MO techniques to monitor network activity, identify abnormalities, and respond quickly to security risks, strengthening cybersecurity protocols and safeguarding private data.

These topics showcase the research directions of MO techniques in various domains such as finance, energy, cybersecurity, and supply chain management, highlighting their importance in addressing complex optimization problems and driving innovation in AI-powered solutions [66–70].

7.6 Conclusion

In conclusion, the exploration of MO reveals it to be the silent engine driving the ML algorithms that power the majority of modern intelligence within computers. The importance of optimization keeps growing as the field of AI develops. From driverless cars to natural language processing, it serves as a link between theoretical mathematical concepts and real-world applications. Having a solid understanding of optimization's mathematical foundations allows us to innovate and enhance existing approaches. AI continues to shape our world, offering boundless opportunities for research, innovation, and societal transformation. By delving into these diverse AI topics, researchers contribute to the advancement of AI and its positive impact on humanity. As we navigate the ever-expanding landscape of AI, we remain committed to responsible innovation, ethical considerations, and the pursuit of knowledge for the betterment of society. Through the lens of various AI topics, we have explored the profound impact of MO techniques and AI algorithms on solving complex problems and driving innovation. AI's impact extends beyond individual domains, influencing healthcare, social good, autonomous vehicles, ethics, and education. AI revolutionizes medical diagnosis, drug discovery, and personalized treatment, contributing to advancements in healthcare. It addresses global challenges through environmental monitoring, disaster response, and poverty alleviation, harnessing technology for social good. Autonomous vehicles redefine transportation with advancements in sensor fusion, path planning, and safety protocols, paving the way for safer and more efficient mobility solutions. Ethical considerations in AI ethics and bias drive research efforts towards fairness, bias detection, and ethical decision-making in AI systems.

Conflicts of Interest: The authors have no conflicts of interest to declare.

References

[1] White, S. K. (13 Apr 2024). 12 most popular AI use cases in the enterprise today. *CIO*, www.cio.com/article/652775/12-most-popular-ai-use-cases-in-the-enterprise-today.html.

[2] Aiforgoodstg. (25 Feb 2022). 4 Current and Future Trends in AI. AI For Good. http://aiforgood.itu.int/4-current-and-future-trends-in-ai.

[3] Craig, L. (4 Jan 2024). 10 top AI and machine learning trends for 2024. *Enterprise AI*, www.techtarget.com/searchenterpriseai/tip/9-top-AI-and-machine-learning-trends.

[4] Lukan, E. (13 Mar 2024). 168 Best AI Tools of 2024. www.synthesia.io/post/ai-tools.

[5] AI Tools | Runway. http://runwayml.com/ai-tools.

[6] Thangamuthu, K. (6 Dec 2023). Unlocking the Future: 7 Revolutionary AI Techniques. www.linkedin.com/pulse/unlocking-future-7-revolutionary-ai-techniques-kumar-thangamuthu-kvm5e.

[7] Kothari, S. (26 July 2023). Top Artificial Intelligence Techniques: Cracking the Code. *Simplilearn.com*, www.simplilearn.com/tutorials/artificial-intelligence-tutorial/top-ai-techniques.

[8] Michael. (8 Mar 2024). Exploring the Top Algorithms for Semantic Segmentation. *Keymakr's Blog Features the Latest News and Updates*, http://keymakr.com/blog/exploring-the-top-algorithms-for-

semantic-segmentation/#:~:text=Semantic%20segmentation%20is%20a%20process%20in%20computer%20vision%20that%20involves,the%20objects%20in%20the%20image.
[9] Kumari, K. (9 Apr 2024). A basic introduction to object detection. *Analytics Vidhya*, www.analyticsvidhya.com/blog/2022/03/a-basic-introduction-to-object-detection.
[10] GeeksforGeeks. (18 Feb 2024). Transfer Learning With Fine-tuning. *GeeksforGeeks*. www.geeksforgeeks.org/transfer-learning-with-fine-tuning-in-nlp.
[11] Habib, G., & Qureshi, S. (1 July 2022). Optimization and Acceleration of Convolutional Neural Networks: A Survey. *Journal of King Saud University. Computer and Information Sciences/Mağalaẗ Ğam'aẗ Al-malīk Saud : Ùlm Al-ḥasib Wa Al-ma'lumat, 4244–4268*(7), https://doi.org/10.1016/j.jksuci.2020.10.004.
[12] Computer Vision. Wikipedia, 7 Apr. 2024, http://en.wikipedia.org/wiki/Computer_vision.
[13] GeeksforGeeks. (11 Jan 2024). Machine Translation in AI. *GeeksforGeeks*. www.geeksforgeeks.org/machine-translation-of-languages-in-artificial-intelligence.
[14] Jim, J. R. et al. (1 Feb 2024). Recent advancements and challenges of NLP-based sentiment analysis: A state-of-the-art review. *Natural Language Processing Journal, 100059*, https://doi.org/10.1016/j.nlp.2024.100059.
[15] GeeksforGeeks. (3 Jan 2024). Named Entity Recognition. *GeeksforGeeks*. www.geeksforgeeks.org/named-entity-recognition.
[16] Natural Language Processing: Text Generation With Python. www.turing.com/kb/natural-language-processing-understanding-analyzing-generating-text-with-python.
[17] Turing. (10 Feb 2022). A Guide on Word Embeddings in NLP. www.turing.com/kb/guide-on-word-embeddings-in-nlp.
[18] Banjara, B. (25 Apr 2024). A Comprehensive Guide to Fine-Tuning Large Language Models. *Analytics Vidhya*, www.analyticsvidhya.com/blog/2023/08/fine-tuning-large-language-models.
[19] Van Otten, N. (1 Nov 2023). Top 6 Most Useful Attention Mechanism in NLP Explained and When to Use Them. *Spot Intelligence*, http://spotintelligence.com/2023/01/12/attention-mechanism-in-nlp/#:~:text=The%20attention%20mechanism%20allows%20neural,importance%20of%20various%20input%20components.
[20] GeeksforGeeks. (3 Jan 2024). Seq2seq Model in Machine Learning. *GeeksforGeeks*. www.geeksforgeeks.org/seq2seq-model-in-machine-learning.
[21] What Is Natural Language Processing? | IBM. www.ibm.com/topics/natural-language-processing.
[22] Multi-agent Reinforcement Learning. Wikipedia, 14 Feb. 2024, http://en.wikipedia.org/wiki/Multi-agent_reinforcement_learning.
[23] Integrate.Ai. (29 Oct. 2018). What Is Model-Based Reinforcement Learning? – The integrate.ai Blog – Medium. Medium, http://medium.com/the-official-integrate-ai-blog/understanding-reinforcement-learning-93d4e34e5698.
[24] Suran, A. (30 Mar. 2022). On-Policy V/S Off-Policy learning – towards data science. *Medium*, http://towardsdatascience.com/on-policy-v-s-off-policy-learning-75089916bc2f.
[25] Furqatovich, I. U., Mukhriddin Musoqulo'g'li, R., Karshievna, U. D., Ananth, C., & Kumar, T. A. (2024 Apr 4). Artificial intelligence powered strategies to enhance the competitiveness of healthcare institutions in Uzbekistan. In 2024 Ninth International Conference on Science Technology Engineering and Mathematics (ICONSTEM) (pp. 1–6). IEEE.
[26] Reward Shaping – Introduction to Reinforcement Learning. http://gibberblot.github.io/rl-notes/single-agent/reward-shaping.html.
[27] Rocca, J. (7 Jan 2022). The Exploration-exploitation Trade-off: Intuitions and Strategies | By Joseph Rocca | towards data science. *Medium*, http://towardsdatascience.com/the-exploration-exploitation-dilemma-f5622fbe1e82.

[28] Salloum, Z. (12 Dec 2021). Function Approximation in Reinforcement Learning – Towards Data Science. *Medium*, http://towardsdatascience.com/function-approximation-in-reinforcement-learning-85a4864d566.

[29] Part 3: Intro to Policy Optimization – Spinning up Documentation. http://spinningup.openai.com/en/latest/spinningup/rl_intro3.html.

[30] What Is Reinforcement Learning? – Reinforcement Learning Explained – AWS. http://Amazon Web Services, Inc., aws.amazon.com/what-is/reinforcement-learning/#:~:text=Reinforcement%20learning%20(RL)%20is%20a,use%20to%20achieve%20their%20goals.

[31] Verma, Y. (18 Mar. 2024). A Guide to Parallel and Distributed Deep Learning for Beginners. *Analytics India Magazine*, http://analyticsindiamag.com/a-guide-to-parallel-and-distributed-deep-learning-for-beginners.

[32] Shaikh, J. (15 June 2023). 10 advanced deep learning architectures data scientists should Know!. *Analytics Vidhya*, www.analyticsvidhya.com/blog/2017/08/10-advanced-deep-learning-architectures-data-scientists.

[33] Jain, S. J. (8 Feb. 2024). An overview of regularization in deep learning (With Python Code). *Analytics Vidhya*, www.analyticsvidhya.com/blog/2018/04/fundamentals-deep-learning-regularization-techniques.

[34] DeepAI. (25 June 2020). Batch Normalization. *DeepAI*. http://deepai.org/machine-learning-glossary-and-terms/batch-normalization#:~:text=Batch%20Normalization%20is%20a%20technique,a%20standard%20deviation%20of%20one.

[35] Brownlee, J. (7 Feb. 2021). Weight Initialization for Deep Learning Neural Networks. *MachineLearningMastery.com*, http://machinelearningmastery.com/weight-initialization-for-deep-learning-neural-networks.

[36] Lau, S. (20 June 2018). Learning Rate Schedules and Adaptive Learning Rate Methods for Deep Learning. *Medium*, http://towardsdatascience.com/learning-rate-schedules-and-adaptive-learning-rate-methods-for-deep-learning-2c8f433990d1#:~:text=Learning%20rate%20schedules%20seek%20to,step%20decay%20and%20exponential%20decay.

[37] Secherla, S. (6 Sept. 2022). Understanding Optimization Algorithms in Machine Learning. *Medium*, http://towardsdatascience.com/understanding-optimization-algorithms-in-machine-learning-edfdb4df766b.

[38] Goyal, C. (14 June 2021). Complete Guide to Gradient-Based Optimizers in Deep Learning. *Analytics Vidhya*, www.analyticsvidhya.com/blog/2021/06/complete-guide-to-gradient-based-optimizers.

[39] What Is Deep Learning? | IBM. www.ibm.com/topics/deep-learning#:~:text=Deep%20learning%20is%20a%20subset,AI)%20in%20our%20lives%20today.

[40] Distributed and Parallel Training Tutorials – PyTorch Tutorials 2.3.0+cu121 Documentation. http://pytorch.org/tutorials/distributed/home.html.

[41] Adversarial Robustness via Fisher-Rao Regularization. *IEEE Journals & Magazine | IEEE Xplore*. 1 Mar. 2023. http://ieeexplore.ieee.org/document/9773978.

[42] Moving Towards Automating Model Selection Using Bayesian Optimization – MATLAB and Simulink. www.mathworks.com/help/stats/towards-automating-model-selection.html.

[43] Few-Shot Learning and Meta-Learning | Tutorial – Borealis AI. *Borealis AI*. 28 July 2023. www.borealisai.com/research-blogs/tutorial-2-few-shot-learning-and-meta-learning-i.

[44] Thailappan, D. (8 Mar. 2024). An Introduction to Few-Shot Learning. Analytics Vidhya, www.analyticsvidhya.com/blog/2021/05/an-introduction-to-few-shot-learning.

[45] 12.11. Learning Rate Scheduling – Dive Into Deep Learning 1.0.3 Documentation. http://d2l.ai/chapter_optimization/lr-scheduler.html.

[46] Pokhrel, S. 4 Popular Model Compression Techniques Explained. http://xailient.com/blog/4-popular-model-compression-techniques-explained.

[47] Ricaurte-Quijano, P., & Carli-Álvarez, A. (Oxbridgepublishinghouse Oct. 2016). The Wiki Learning Project: Wikipedia as an Open Learning Environment. *Comunicar, 24*(49), 61–69. Crossref. https://doi.org/10.3916/c49-2016-06.
[48] Zebin, Y., & Aijun, Z. (May 2020). Hyperparameter tuning methods in automated machine learning. In SCIENTIA SINICA Mathematica (p. 695, Vol. 50, no. 5). Science China Press., Co. Ltd. Crossref. https://doi.org/10.1360/n012019-00092.
[49] IEEExplore Digital Library. (July 2010). *Choice Reviews Online*, *47*(11), 47–6268. American Library Association. Crossref. https://doi.org/10.5860/choice.47-6268.
[50] Trung, L. H. (2019). *EurasianUnionScientists*, *4*(65), Logic + LLC, Crossref. https://euroasia-science.ru/pdf-arxiv/the-controllability-function-of-polynomial-for-descriptor-systems-23-31/, https://doi.org/10.31618/esu.2413-9335.2019.4.65.275.
[51] Naresh, V. S. et al. (Apr. 2024). Predictive machine learning in optimizing the performance of electric vehicle batteries: Techniques, challenges, and solutions. *WIREs Data Mining and Knowledge Discovery*, Wile, Crossref. https://doi.org/10.1002/widm.1539.
[52] Azevedo, B. F. et al. (Jan. 2024). Hybrid approaches to optimization and machine learning methods: A systematic literature review. In Machine Learning. Springer Science and Business, Media LLC. Crossref. https://doi.org/10.1007/s10994-023-06467-x.
[53] De Harder, H. (13 July 2023). Five ways to combine mathematical optimization and machine learning. *Medium*, http://towardsdatascience.com/four-ways-to-combine-mathematical-optimization-and-machine-learning-8cb874276254.
[54] Shepherd, T. A.I. and Mathematical Optimization: A Striking Synergy. www.twoday.lt/blog/a.i.-and-mathematical-optimization-a-striking-synergy.
[55] Shepherd, T. GL: A.I. and Mathematical Optimization: What&Rsquo;S the Difference? www.twoday.com/blog/ai-and-mathematical-optimization-whats-the-difference.
[56] Islam, J. et al. (1 Mar. 2020). A Holistic Review on Artificial Intelligence Techniques for Well Placement Optimization Problem. *Advances in Engineering Software*, *102767*. https://doi.org/10.1016/j.advengsoft.2019.102767.
[57] Artificial Intelligence Approaches for UAV Navigation: Recent Advances and Future Challenges. *IEEE Journals & Magazine | IEEE Xplore*. 2022. http://ieeexplore.ieee.org/abstract/document/9729807.
[58] Mohamadou, Y. et al. (6 July 2020). A review of mathematical modeling, artificial intelligence and datasets used in the study, prediction and management of COVID-19. *Applied Intelligence*, *3913–3925* (11), https://doi.org/10.1007/s10489-020-01770-9.
[59] Li, Y. et al. (1 May 2023). Role of artificial intelligence on green economic development: joint determinates of natural resources and green total factor productivity. *Resources Policy*, *103508*, https://doi.org/10.1016/j.resourpol.2023.103508.
[60] Yao, Z. et al. (1 July 2023). combustion optimization of a coal-fired power plant boiler using artificial intelligence neural networks. *Fuel*, *128145*, https://doi.org/10.1016/j.fuel.2023.128145.
[61] He, C. et al. (19 Jan. 2023). A review on artificial intelligence enabled design, synthesis, and process optimization of chemical products for industry 4.0. *Processes*, *330*(2), https://doi.org/10.3390/pr11020330.
[62] Sohail, A. (30 Aug. 2021). Genetic algorithms in the fields of artificial intelligence and data sciences. *Annals of Data Science*, *1007–1018*(4), https://doi.org/10.1007/s40745-021-00354-9.
[63] Nutakki, M., & Mandava, S. (1 Mar. 2023). Review on optimization techniques and role of artificial intelligence in home energy management systems. *Engineering Applications of Artificial Intelligence*, *105721*, https://doi.org/10.1016/j.engappai.2022.105721.
[64] Neira, D. E. B. et al. (1 Feb. 2024). Utilizing modern computer architectures to solve mathematical optimization problems: A survey. *Computers & Chemical Engineering*, *108627*, https://doi.org/10.1016/j.compchemeng.2024.108627.

[65] Carrizosa, E. et al. (1 Mar. 2024). Generating collective counterfactual explanations in score-based classification via mathematical optimization. *Expert Systems With Applications, 121954*, https://doi.org/10.1016/j.eswa.2023.121954.

[66] Nishant, R. et al. (29 May 2023). The formal rationality of artificial intelligence-based algorithms and the problem of bias. *JIT. Journal of Information Technology/Journal of Information Technology*, https://doi.org/10.1177/02683962231176842.

[67] Zhao, T. (26 Mar. 2024). Artificial intelligence in mathematical modeling of complex systems. *ICST Transactions on E-education and E-learning, 10*, https://doi.org/10.4108/eetel.5256.

[68] Brownlee, J. (11 Oct. 2021). How to Choose an Optimization Algorithm. *MachineLearningMastery.com*, http://machinelearningmastery.com/tour-of-optimization-algorithms.

[69] Mohammadi, A., & Sheikholeslam, F. (1 Nov. 2023). Intelligent optimization: Literature review and state-of-the-art algorithms (1965–2022). *Engineering Applications of Artificial Intelligence, 106959*, https://doi.org/10.1016/j.engappai.2023.106959.

[70] Secherla, S. (6 Sept. 2022). Understanding optimization algorithms in machine learning. *Medium*, http://towardsdatascience.com/understanding-optimization-algorithms-in-machine-learning-edfdb4df766b.

Yogesh Kumar Sharma, Uppuluri Lakshmi Soundharya, Ravi Rastogi, and Anantha Raman G. R.

8 Unleashing the Power of Evolutionary Algorithms: Advanced Optimization Techniques in Artificial Intelligence

Abstract: The exploration of practical optimization approaches is constant within artificial intelligence (AI). Among several methods, evolutionary algorithms (EAs) stand out as powerful tools, deriving influence from the selection processes by nature and the evolution of genes. This chapter explores the complexities of EAs, explaining how they work, their many uses in AI, the obstacles they present, and the directions of future improvements. This study thoroughly examines modern optimization algorithms, discussing difficulties and suggesting possible directions for further investigation. The examination covers various optimization methods in several fields, including mathematical programming, machine learning, and AI. The text examines the fundamental difficulties encountered by current algorithms, such as problems with scalability, speed of convergence, and capacity to adapt to different problem domains. In addition, the study thoroughly analyzes the influence of optimization algorithms on practical applications, considering their efficacy and constraints. The study highlights developing trends, including hybrid techniques and metaheuristic methodologies, which show promise in addressing contemporary difficulties. This chapter offers a comprehensive overview of the most advanced optimization algorithms. It serves as a helpful reference for academics, practitioners, and decision-makers, helping them overcome current restrictions and explore new possibilities in the field of optimization research.

Keywords: Evolutionary algorithm, optimization, artificial intelligence, hybrid techniques

8.1 Introduction

The development of web and multilingual technologies has presented significant hurdles in investigating practical situations and uses over recent years. The issues are collecting statistics from a vast information set, evaluating information across various angles, categorizing data, and summarizing the data connection. To tackle these issues, data mining is a preferred approach. Techniques for data extraction may be broadly classified into two groups: direct as well as indirect data extraction [1].

Techniques for data analysis may be categorized into three main groups: classification, estimation, and forecasting. Classification is the systematic process of classify-

https://doi.org/10.1515/9783111436180-008

ing a dataset into distinct classes, either based on previous understanding of established classes or according to the inherent structure of the dataset. It is often termed supervised learning. Within supervised learning, the total number of classes in a particular dataset is predetermined and used to assign class labels to unclassified instances. Indirect methods for data analysis may be classified into three distinct categories: clustering, prospecting, and visualization. Clusters are extensively used in the field of indirect data analysis. Clustering refers to grouping things in a cluster based on their high similarity to each other and distinctiveness from other clusters [2].

Identifying the relevant entities without any previous information is an unsupervised learning method. This has been employed in several technical domains, including biological informatics, data gathering, projections, and the segmentation of images. There are two main categories of clustering approaches: classical versus metaheuristic techniques. The five primary groups of classical clustering techniques include partitioned, grid-based, density-based, hierarchical, and model-based clustering. The famous traditional partitioned clustering method, K-means, is utilized because of its simplicity and efficacy [3].

Nevertheless, early convergence is a significant drawback. Meta-heuristic approaches are employed in clustering to identify the optimal global partition and address this issue. Clustering algorithms that rely on meta-heuristics may be classified into two distinct types. These clustering algorithms depend on individual and multi-objective criteria [4]. This study primarily focuses on a single-objective hybrid method clustering strategy based on partitions. This technique is employed to tackle two primary clustering challenges: first, calculating the optimal number of clusters before the clustering procedure, and second, identifying the most effective partitions for the clusters. To address these issues, most partitioned single-objective clustering approaches are iteratively used with varying numbers of classes as input. The resulting data partitioning that yields the most optimal cluster validation index is then chosen [5].

The accuracy of the index of a cluster is unable to determine the proper number of clusters accurately. Hence, it is necessary to construct a novel function of fitness that includes multiple cluster validation indexes to preserve both the inter-cluster and intra-cluster characteristics. A proficient feature selection method is necessary to enhance its efficacy and reduce duplication in unmanaged data sets. Selecting features in unsupervised learning poses a challenging difficulty due to the inability to use class labels to guide the search for pertinent information. To achieve efficient clustering, a novel single-objective hybrid strategy using meta-heuristic techniques is necessary [6, 7].

This chapter is organized as follows. Section 8.2 covers the EA algorithm's basic overview and types. Section 8.3 covers the applications of EA, Section 8.4 covers the challenges, limitations, and future directions of EAs, and Section 8.5 covers the conclusion of the chapter.

8.2 Evolutionary Algorithms (EA)

Evolutionary algorithms (EAs) are very effective heuristic search techniques that rely on the principles of Darwinian evolution. They possess remarkable qualities of resilience and adaptability, enabling them to identify optimal solutions for complex optimization issues efficiently. Using EAs significantly increases the likelihood of discovering a nearly optimal solution during the first phase of the optimization process. EAs do not rely on any fitness gradient information, can be quickly processed in parallel, and can overcome local minima where deterministic optimization approaches may be unsuccessful or infeasible [8]. The convergence of EAs is independent of the continuity or differentiability of the optimized function. EAs may be used not only as a substitute for classic optimization methods but also to enhance and expand their capabilities. In general, hybrid approaches that combine EAs with other techniques achieve superior performance. Figure 8.1 presents the key components of the EA algorithm.

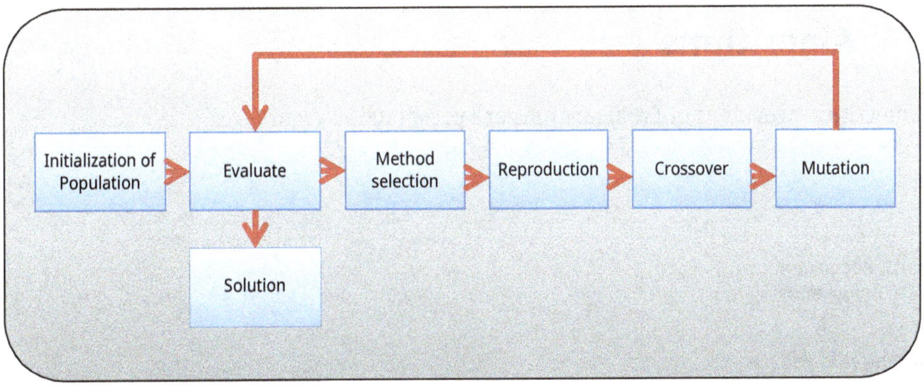

Figure 8.1: Key components of the EA algorithm.

EA algorithms can efficiently solve single and multi-objective optimization problems, which often need the integration of many disciplines [9]. Multi-objective optimization algorithms offer significant advantages compared to traditional methods. They can handle integer, discontinuous, or discrete design variables simultaneously. They are compatible with various game strategies and are unaffected by different Pareto front shapes. Additionally, they are capable of finding solutions in non-convex or discontinuous regions. EAs are practical heuristic search techniques that rely on Darwinian evolution. They possess vital attributes of resilience and adaptability, enabling them to identify optimal solutions for intricate optimization issues on a global scale [10]. Using EAs significantly increases the likelihood of discovering a nearly optimal solution during the first phase of the optimization process. EAs do not rely on any fitness gradient information, can be quickly processed in parallel, and can overcome local minima where deterministic optimization approaches may be unsuccessful or inapplicable. The convergence of EAs is unaffected by the continuity or differentiability of the optimized function [11].

EAs may serve as a substitute for conventional optimization methods and also have the potential to enhance and expand upon them. In general, hybrid approaches combining EAs with other techniques achieve superior performance. Enterprise applications can efficiently manage single and multi-objective optimization tasks, which often need the integration of many disciplines [12]. Multi-objective optimization algorithms offer significant advantages compared to traditional methods. They can handle integer, discontinuous, or discrete design variables simultaneously. These algorithms are compatible with various game strategies and are unaffected by different Pareto front shapes. Additionally, they are capable of finding solutions in non-convex or discontinuous regions.

8.3 Working and Critical Components of EA Algorithms

The critical workings of EA algorithms are presented in Figure 8.2.

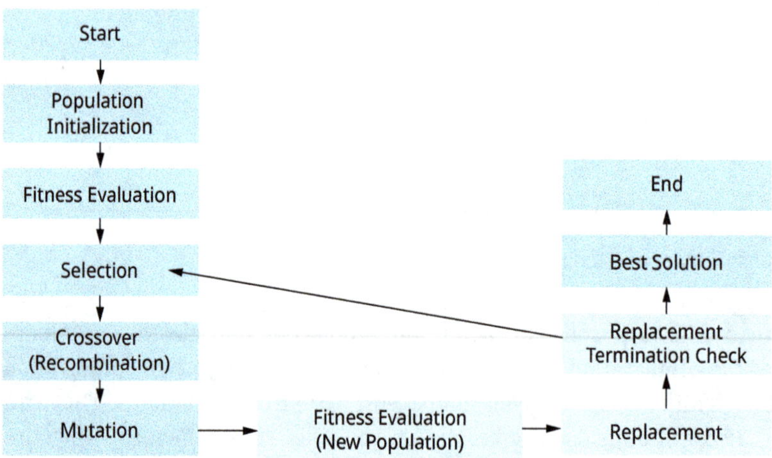

Figure 8.2: Working of the EA algorithm.

- **Initialization of Population:** The most straightforward and often-used approach for initializing the population in an EA is randomly creating solutions. Each gene in every solution is randomly allocated a value throughout its feasible range based on the method used to encode the issue [13].

- **Method Selection:** Increasing the selection pressure accelerates the convergence of a population toward a particular solution, perhaps limiting the exploration of the search space.
- **Reproduction:** Selection strategies are used to ascertain the genes that will undergo reproduction. Reproduction entails the selection of certain individuals who then exchange information via a crossover procedure. Each individual's genes undergo random mutations during the transition from one generation to the next [14].
- **Crossover**, also known as recombination, represents a genetic operator utilized in evolutionary computing to merge the genetic material of two parents and produce an additional child [15].
- **Mutation:** This catalyzes evolution by producing the genetic diversity essential for the progression of the evolutionary process. To comprehend the process of evolution, it is essential to analyze and describe the patterns and rates of mutation [16].
- **Replacement of Population:** The predominant technique for a substitute is generational replacement, in which the preceding generation is replaced with the newly produced offspring population.

8.4 Critical Types of Evolutionary Algorithms

The popular EAs are as follows.

8.4.1 Genetic Algorithms (GAs)

The genetic algorithm (GA) [17] is a well-established and well-recognized optimization approach that draws inspiration from natural processes. The GA emulates the natural processes in the environment, including the principles of Darwinian species development. GAs use a population of individuals, where each individual, referred to by its chromosome, indicates a possible solution to the issue at hand. The aim or function determines the issue being addressed. The value representing an individual's quality is ascribed to it based on how well it fits the objective function [18].

The value is often known as the individual's fitness and serves as a primary criterion for evaluation. Highly esteemed individuals are more likely to be chosen for the subsequent generation of individuals. GAs employ three general operators: "selection, crossover, and mutation." A selection process involves creating an entirely novel population using the fitness scores of members of the previous generation. Crossover typically involves exchanging parts of individuals between two selected individuals. Mutation involves randomly changing the values of specific genes [19, 20]:

Algorithm 1: *GA implementation*

```
# Define Genetic Algorithm parameters
population_size,num_generations,mutation_rate,crossover_rate
# Define an objective function
  # Evaluate the fitness of individual
# Generate an initial population
  # Generate random individuals
# Select parents based on fitness
  # Perform selection based on fitness
  # Crossover operation
  # Perform crossover to create offspring
# Mutation operation
def mutate (individual):
  # Perform mutation on individual
  pass
# Genetic Algorithm function
def genetic_algorithm ():
  # Generate an initial population
  # Iterate through generations
    # Evaluate the fitness of the population.
    # Select parents for crossover
    # Create offspring through crossover
    # Perform mutation on offspring
    # Replace old population with offspring
# Return the best individual
  # Run Genetic Algorithm
print best_solution
```

8.4.2 Genetic Programming (GP)

Genetic programming (GP) refers to using GAs to evolve computer programs. GP is a recently developed technique, a specialized version of a GA. GP focuses on solving certain sorts of problems by using customized genetic operators. Koza created GP to automatically generate program codes when the assessment criterion for their correct functioning is known. As the answer sought is a program, the developed alternative solutions are represented as trees rather than the more often used linear chromosomes (composed of bits or integers) in GAs [21]. The genetic programs are specifically designed to operate on trees. For example, crossover involves trading sub-trees, whereas mutation involves modifying a node's structure or leaf.

8.4.3 Evolution Strategies

The development of evolutionary strategies (ESs) varies from GAs, particularly in their selection mechanism. In a GA, a subsequent generation is formed by selecting members of the original population based on their potential for success value while maintaining a consistent population size. During the ES, a transient population is formed, which has a distinct size compared to the original population. The assumed parameters and other factors determine the number of individuals in this population. The fitness ratings are irrelevant at this stage; the members of the transient population experience genetic recombination and genetic variations [22, 23].

8.4.4 Evolutionary Programming

Evolutionary programming (EP) is a computational approach using principles inspired by biological evolution to solve complex problems. EP was created as a means of uncovering the grammatical structure of an unfamiliar language. EP gained popularity when it was introduced as a numerical optimization tool [24]. The EP is analogous to the ES, although it has one fundamental distinction [25]. In EP, a new population, is generated by applying mutations to each individual from the original population.

On the other hand, in ESs, every person has an equal chance of being chosen for the temporary population where genetic operations are carried out. Within the EP framework, mutations occur via an arbitrary perturbation of specific gene values in the mutated individual. The newly formed population and the population of parents have identical sizes. Ultimately, the next generation of individuals is formed by selecting the people with the highest rankings from parental and mutant populations.

Table 8.1: Comparative analysis of EA methods [26].

Aspect	Genetic algorithms (GA)	Particle swarm optimization (PSO)	Simulated annealing (SA)
Definition	Optimization technique inspired by natural selection and genetic evolution	Population-based optimization technique inspired by the social behavior of swarms	Probabilistic optimization technique inspired by the metallurgical annealing process
Representation	Solutions are represented as chromosomes or strings of genes	Solutions are represented as particles in a multidimensional search space	Solutions are represented as candidate solutions or states in the optimization problem

Table 8.1 (continued)

Aspect	Genetic algorithms (GA)	Particle swarm optimization (PSO)	Simulated annealing (SA)
Iterative process	Iteratively evaluates and improves candidate solutions through selection, crossover, and mutation operations	Iteratively updates particle positions based on personal and global best positions	Iteratively explores the solution space by accepting probabilistic changes
Exploration versus exploitation	Balances the exploration of new solutions with the exploitation of known promising solutions	Balances exploration of new regions with the exploitation of best-known solutions	Balances exploration to escape local optima with exploitation to converge on the optimal solution
Convergence behavior	Converges toward optimal or near-optimal solutions over iterations	Converges toward optimal solutions quickly, often requiring fewer iterations	It gradually converges toward optimal solutions, with the convergence rate influenced by the temperature schedule
Adaptability	May struggle to adapt to changing problem landscapes or dynamic environments	Generally adaptable to changes in problem dynamics, but may face challenges in noisy environments	Adapts to changes in the optimization landscape through temperature schedule adjustments
applications	Widely applied in various domains, including functional, neural network, and combinatorial optimization	Commonly used in the optimization of continuous functions, engineering design, and data clustering	Applied in tasks such as parameter optimization, combinatorial optimization, and scheduling problems
Strengths	Robust optimization capabilities for complex, high-dimensional problems	Fast convergence, simplicity of implementation, and effectiveness in continuous optimization	Balance between exploration and exploitation, effectively escaping local optima
Limitations	Computational complexity, balancing exploration and exploitation, and adaptability	Premature convergence, lack of robustness to noise, and difficulty in discrete optimization	Sensitivity to initial parameters, slow convergence, and temperature schedule tuning

8.5 Applications of EAs

One key reason for the increasing popularity of swarm intelligence algorithms is the rising need for intelligent optimization approaches in many commercial and technical tasks. EAs are well-suited for addressing optimization, scheduling, development, planning, and administration challenges [27].

8.5.1 Role of GAs in Real Life

GAs are a versatile method for optimization that may be used for a wide range of problems [28]. GAs effectively solve a wide range of optimization problems, including those with constraints, multiple modes, continuous variables, combinatorial elements, and multiple objectives. Consequently, GAs have a diverse array of practical uses in several fields. In this brief segment, we provide just a limited selection of them. The chapter addresses optimizing the arrangement of solar arrays, which is addressed using a GA. The numerical technique given is centered on a rotating representation of a stratosphere airship to optimize the arrangement of the solar array. The findings indicate that the suggested approach is beneficial during the first phase of erecting expansive, flexible solar arrays. Furthermore, it has been shown that by optimizing the solar array, the power production of the solar panel is greatly enhanced.

The research [29] presents an optimization model that utilizes the GA to minimize the total energy utilization of the high-sulfur oil and natural gas purifying process.

The GA-inspired technique was studied and implemented in the serial evolutionary filter [30]. As a result of this approach, the need for resampling, which is often utilized in many particle filters, is eliminated, and the variety of particles may be preserved. The results of the experiments demonstrate that the suggested sequential evolutionary filter outperforms three previous comparable filters in terms of state estimation. The chapter analyzed the advantages and disadvantages of combining a GA with local search.

The analysis was based on a challenging and current topic, particularly the route selection and allocation of the spectrum of multi-cast fluxes (RSA/M) within extensible optical fiber networks. They suggested a proficient optimization technique for resolving the RSA/M issue in eons. The suggested approach surpassed all other competing methods. In addition, the incorporation of Baldwin's effects has contributed to maintaining population variety in GAs.

The presentation of the local inventories routing framework for perishable items may be found in [31]. The suggested model incorporates the three layers of decision-making in the supply chain, including determining the optimal number and placement of warehouses, managing inventory levels at each shop, and optimizing the routes taken by each vehicle. The research demonstrates that the model proposed is NP-hard. Consequently, the authors devise a strategy based on GAs to address this challenge effectively. The provided technique has been shown to generate a near-optimal, high-quality solution within a realistic time frame.

The study [32] introduced a novel container loading technique that uses load distribution diagrams to incorporate load balance, weight limit, and stability restrictions. This method uses a multi-population biased random-key GA, including a novel fitness function that considers static and load balancing. The suggested technique is very successful due to its incorporation of weight balancing goals, stability ensured by complete base support, and adherence to mechanical equilibrium criteria.

A framework is presented in [33] to identify the investment strategy for enhancing the resilience of a train network to seismic hazards. This framework has four components. The third section presents an investment optimization model formulation, while the fourth section focuses on solving this model using a GA. The suggested methodology has been implemented in the actual Chinese railroad system. The findings demonstrate that the given architecture is more sensitive to the effects of earthquakes on the rail network than topology-based methodologies.

This study showcases the multi-objective optimization of operating expenses for space cooling and thermal comfort to achieve superior building energy efficiency. The primary goal of the proposed GA is to optimize the set point temperatures on an hourly basis, using a prediction of the environment along with occupancy patterns for the future day. The offered technique achieves cost savings of as much as 54% compared to the typical control approach [34].

A time-optimal train operating reference curve is created in [35] to minimize time while maximizing energy consumption. Multi-point to achieve energy savings: A multi-population GA addresses this challenge of optimizing several points combinatorically. The simulation results, which consider the actual conditions of the Shanghai line and the characteristics of the train, clearly illustrate the progress achieved with the suggested technique of multi-point coasting control.

The implementation of a GA to minimize the typical delay for a metropolitan signalized junction under oversaturated conditions is given in [34]. Alleviating city congestion is a pressing need for traffic management. An essential approach for decreasing congestion is the efficiency of traffic signalization. The existing traffic signal management system is not sufficiently optimized to manage the oversaturated state efficiently. The simulation results demonstrate that the GA effectively manages the traffic signals to minimize the mean delay to 56 s per automobile [36].

Utilize the versatile GA for solving node placement challenges in the context of [37]. Node placement issues arise in many technical domains, such as implementing radio-frequency tracking devices and mobile sensor networks. The flexible GA utilizes different lengths of encoded data, subarea-swap crossovers, and stochastic mutation and can automatically modify the total number of nodes and the attributes that correlate to them. The experimental findings demonstrate that the adaptable GA outperforms current methods in addressing node placement difficulties.

8.5.2 GP Is Applied to Real-Life Scenarios

The general practitioner has several practical uses. A handwritten character recognition method for pens connected with inertial sensors is given in [38]. The GP technique determines the characteristic function for every character within this framework. The testing findings demonstrate that the accuracy of the suggested technique surpasses the results obtained by the state-of-the-art research conducted in recognizing Persian/Arabic handwritten characters.

In a study conducted by WR [39], a new approach was introduced for predicting flood routing in natural channels. This approach used the "genetic expression programming" (GEP) method, which can be considered a variant of the GP method. The GEP approach utilizes a limited number of hydrologic factors, including input, outflow, and time. The efficacy of the suggested models is assessed using two metrics of goodness-of-fit. The efficacy of the suggested GEP systems is evaluated using three datasets obtained from existing literature. Empirical evidence has shown that the GEP models perform better than other solution strategies based on the Muskingum model. The report [40] analyzes EF's innovative fuel cell (FC) battery management approach. The primary goal of the suggested technique is to enhance the performance of FCs at various locations of operation without using DC/DC connecting converters.

8.5.3 Evolution Strategies (ES) in Real-Life Scenarios

The practicalities of ESs against gas are as follows. In research [41], an ES-integrated parallel optimization was proposed to reduce the weight of test steel frame members. Comparative cost analyses investigate steel frames with different beam-column connections and bracing. Numerical verification of the suggested technique uses three multi-story structures (10, 20, and 30-story). We use the data to suggest economically viable frames for multi-story steel constructions.

Research [42] proposed an electric battery to reduce its EF. In energy distribution networks, unpredictability is the leading cause of power outages. The authors suggest an intelligent battery that maximizes lifespan and meets all electric demand peaks. A customized steady-state ES allows the battery to modify its recharging method to various conditions. Experimental findings on synthetic and actual data show that the technique works. Research [43] proposed an improved non-dominated sorting ES algorithm that employs a specialized operator to steer the algorithm toward known SSO sites. The suggested strategy aims to maximize SSO reduction and minimize rehabilitation expenses. The eastern San Antonio Water System network tested the suggested approach. The provided technique enhances convergence by 70% over the examined alternatives [44].

In research [45], a wireless sensor malfunction diagnosis via fusion data analysis was proposed. Our hierarchical belief rule-based fault diagnostic model uses the

CMA-ES algorithm to optimize its starting parameters. Validating the technique requires a case study using Intel laboratory sensor data. The studies show that the suggested strategy outperforms the fuzzy system for experts and the back-propagating neural network model. Other than this application, some other applications of EA are as follows. Table 8.2 presents a comparative analysis of the applications of EAs in various fields.

Table 8.2: Comparative analysis of the applications of EAs in various fields.

Application	Description
Optimization of neural networks	Evolutionary algorithms optimize the architecture, hyperparameters, and weights of neural networks to improve performance in classification, regression, and image recognition tasks.
Function optimization	Evolutionary algorithms excel in optimizing complex, nonlinear, and multimodal functions, finding global optima where traditional gradient-based methods may struggle.
Multi-objective optimization	Evolutionary multi-objective optimization algorithms efficiently generate a set of Pareto-optimal solutions, enabling decision-makers to explore trade-offs between conflicting objectives in diverse applications.
Feature selection	Evolutionary algorithms identify subsets of features that maximize predictive performance while minimizing dimensionality, enhancing model interpretability and efficiency in tasks such as classification and regression.
Combinatorial optimization	Evolutionary algorithms efficiently solve combinatorial optimization problems like the traveling salesman problem, job scheduling, and vehicle routing by exploring large solution spaces to find optimal or near-optimal solutions.

8.6 Challenges and Future Directions

The key challenges and future directions of EAs are as follows.

8.6.1 Challenges

Electronic arts are a very intriguing field of study. Several open research problems include managing the trade-off between exploration and exploitation properties, implementing self-adaptive control of steering parameters, minimizing the number of parameters in the CMA-ES algorithm, developing new selection schemes, and enhancing their efficacy. The latter is very significant, particularly in the domain of design that evolves and evolvable hardware. Furthermore, there is a need for novel and more effective approaches to handle constraints [46].

Furthermore, further research is needed to explore the use of EAs in dynamic optimization issues, optimization in noisy and non-stationary settings, and multi-objective optimization problems, particularly those with a high number of choice factors. Additionally, additional investigation is required to understand how population size adapts in various optimization settings. To address costly issues more effectively, it is necessary to devise innovative approaches that may enhance competitiveness. One of the issues that must be addressed is how to handle constraints in EAs [47] when applied to engineering optimization problems. The constraint handling methods can be categorized into six main groups: penalty methods, methods operating within the feasible region, methods employing parallel approaches as the population gets closer, methods assuming the superiority of feasible individuals, techniques utilizing multi-objective optimization techniques, and hybrid methods. Naturally, each of these categories may be further subdivided into several subcategories.

The categorization of the methods used to handle constraints using EAs may be found in [48]. To implement the appropriate constraint-managing method in EAs for real-world applications, it is necessary to address several questions. These include determining whether the objective function is defined in the infeasible domain (as penalization techniques cannot be used in such cases) and identifying the presence of any active limitations at the optimum. If the approaches based on the search of feasible area borders are not applicable, it is important to determine the nature of the limitations. If there is just one nonlinear inequality constraint, the approaches for linear constraints are not applicable.

Furthermore, in the practical use of EAs with constraint management approaches, the efficacy of a method is sometimes overshadowed by two additional choice criteria: complexity and implementation difficulties. Penalty techniques, feasibility criteria, and stochastic ranking algorithms are often used in real-world applications owing to their inherent simplicity. Thus, it is evident that there is no universal method for managing constraints using EAs that can effectively address every real-world situation. Consequently, the investigation of constraint management approaches in EAs for real-world applications remains a current and important area of study.

It is crucial to possess a comprehensive understanding of the powers and limits of EAs before employing them for optimization purposes. By examining the time it takes to optimize different versions of EAs to solve specific optimization issues, valuable insights may be obtained into the factors that determine the level of difficulty for these problem-solving methods. The creation of methodologies that permit thorough investigation and enable academics to generate answers for new versions of EAs with increasingly complicated issues is more significant than the production of those specific outcomes. Table 8.3 presents a comparative analysis of the key challenges of EAs in various fields [49].

Table 8.3: Comparative analysis of key challenges of EAs in various fields.

Challenge	Description
Computational complexity	EAs often demand substantial computational resources, particularly for problems featuring large search spaces or intricate fitness evaluations. This complexity can impede efficiency, limiting scalability and practical application in resource-constrained environments.
Exploration versus exploitation balance	EAs must delicately balance exploration (discovering new regions) and exploitation (exploiting known promising solutions) within the search space. Failure to strike this balance adequately can lead to premature convergence or stagnation, hindering the algorithm's ability to discover optimal or diverse solutions effectively.
Limited adaptability	EAs may struggle to adapt to shifting problem dynamics, constraints, or characteristics, impacting their ability to navigate complex and dynamic optimization landscapes. Enhanced adaptability is crucial to ensure EAs can effectively respond to changes and maintain optimal performance across varied problem contexts.
Overfitting and generalization	EAs face the risk of overfitting, wherein they excessively tailor solutions to the training data or search space, potentially compromising performance on unseen data or new problem instances. Robust generalization capabilities are essential to mitigate overfitting and ensure reliable performance in real-world applications.
Scalability	Scalability refers to EAs' capability to handle increasingly complex or large-scale optimization problems efficiently. Challenges arise when dealing with high-dimensional search spaces, computationally intensive fitness evaluations, or tasks requiring parallelization. Addressing scalability concerns necessitates designing algorithms that can scale effectively without compromising performance or convergence.

8.6.2 Future Directions

As shown in this study [50], EAs constitute a well-studied field of research. Every year, several novel variations of the algorithms, as mentioned above, are suggested. If a certain issue does not have a suitable specialized algorithm, using one of the EA might be a favorable option. It is important to consider the challenges that users may encounter while working with EAs. Here, we may identify two primary issues. The primary issue at hand is premature convergence, which refers to the population converging toward a suboptimal solution rather than an ideal one. To address this issue, we may implement a system that reduces the pace at which genetic material is transferred between individuals. This can be achieved by dividing the whole population into many subpopulations, sometimes referred to as "islands," and occasionally allowing individuals to move between these islands [51].

An alternative approach to address the issue of early convergence involves combining EAs using a branch and bound method, which is enhanced by intermediate propagation approaches. The second issue is the ideal balance between exploring and profiting from the characteristics of EA. Controlling the degree of selection pressure is one possible answer to this issue. This can be achieved by incorporating specialized genetic operators that ensure a high level of population differences at the beginning of the algorithm's operation (with a strong emphasis on exploration rather than exploitation) and a low level of population variety at the completion of the algorithm's functioning (with a greater focus on exploitation rather than exploration) [52].

Future developments in EAs may be highlighted in many key directions. One contemporary approach is the hybrid of multiple algorithms to achieve improved outcomes. Presently, there is a growing body of literature that features a rising number of works showcasing hybrid algorithms. In addition, several researchers are engaged in enhancing the computing efficiency of EAs through various changes. Another intriguing area for future study in EAs is memetic algorithms. The phrase "memetic algorithm" refers to the combination of an evolutionary approach and any additional population-based strategy with independent local search strategies, such as the Nelder-Mead method. Valuable insights into future developments in EAs may be obtained [53].

One of the upcoming trends in EAs is the growing interest in using EAs for embodied or embedded systems. This means using evolution to create populations of controls and individuals that apply the operational approach for certain situated entities. These solutions are evaluated based on their performance in complex and dynamic surroundings, focusing on their functionality rather than their inherent qualities. In addition, there is another significant concern, particularly in the industrial use of EA techniques. Frequently, while dealing with practical situations, we encounter the need to optimize a function inside a domain that has many dimensions [54]. This method is often intricate and time-consuming. Thus, in practical scenarios, the EAs developed for such issues must be constructed to be readily implemented in parallel (or GPU-friendly) to minimize their computing time. Future proposals should prioritize a more substantial effort in developing this feature since it has the potential to be a decisive factor in determining the usefulness of an algorithm in practical applications. Research in the field of EAs often involves the use of surrogate models, which are computationally less expensive models of real-world situations. These surrogate models may be used instead of doing complete fitness assessments and can be improved by sometimes evaluating individuals in the population [55].

Furthermore, it is rather common for industrial challenges to include many goals. Interactive EAs are used with algorithmic advancements to enhance the effectiveness of EAs in multi-objective optimization. It is well understood that a distinct objective function characterizes every engineering issue and has a unique search space topography. The optimal settings of EA parameters that are effective in solving one issue may not be enough for solving another one. Table 8.4 presents a comparative analysis of future directions and their advantages, as well as challenges.

Table 8.4: Comparative analysis of future directions, their advantages, and challenges.

Future directions	Description	Advantages	Challenges
Hybridization of methods	This involves merging EAs with other optimization techniques, like swarm intelligence or traditional methods. It aims to leverage the strengths of each approach to overcome individual limitations.	– Synergizes multiple optimization paradigms – enhances exploration and exploitation capabilities – improves solution quality and convergence speed	– Complexity in integrating different algorithms – finding optimal combinations of methods – potential increase in computational overhead
Adaptive meta-heuristic	This approach entails developing EAs with adaptive mechanisms. These mechanisms dynamically adjust algorithm parameters, operators, or strategies based on problem characteristics and optimization progress.	– Enhances adaptability to problem dynamics and complexities – improves convergence speed and solution quality – reduces the risk of premature convergence	– Designing effective adaptation mechanisms – balancing between exploration and exploitation dynamically – ensuring stability and robustness of adaptive algorithms
Diversity preservation	This strategy focuses on maintaining diversity within the population during optimization. It promotes the exploration of diverse regions in the search space and mitigates the risk of premature convergence.	– Enhances exploration of the search space – Reduces risk of premature convergence – improves robustness and solution quality	– Designing effective diversity maintenance mechanisms – balancing diversity and convergence objectives – ensuring computational efficiency and scalability of diversity-preserving techniques

Consequently, the exploration of novel methodologies in the field of automated tuning and adaptive parameter control remains a prominent subject in the realm of EAs. A crucial concern in the commercial implementation of EA techniques is the precise specification of an objective function. The industrial difficulties are quite intricate. Hence, formulating a precise mathematical model (with a well-defined goal function for EAs) for a specific industrial process is an arduous undertaking.

The effectiveness of the selected goal function will significantly impact the outcomes achieved via the use of EA techniques. The last issue we need to address pertains to the replicability of the EA methodologies. It is well understood that EAs are stochastic methods. Every time the EA technique is executed, a distinct outcome may be achieved. Hence, it is crucial to prioritize the establishment of consistent outcomes produced by EA methodologies. This matter has significant importance for the implementation of EA approaches in the industry [21].

8.7 Conclusion

The real world presents several optimization situations characterized by different limitations and function objectives that exhibit discontinuity, nonlinearity, nonconvexity, and multimodality. The optimization issues are complex and include variables of diverse kinds, including integer, real, independent, and binary. These variables have varying ranges of values, which necessitates the use of normalization. Therefore, the problem's search space lacks smoothness. EAs are increasingly being recognized and used in computer processes to address intricate engineering challenges. They are used by research scientists as well as engineers to utilize the idea of evolution to solve optimization-based investigations and to compose papers. This chapter covers a comprehensive overview of EAs, their applications, challenges, and future directions.

References

[1] Zohuri, B., & Rahmani, F. M. (2024). Computational intelligence for signal and image processing unleashing the power of algorithms. *Journal of Clinical Case Reports & Studies*, *5*(2).

[2] Paul, V. M., Prabhu, S. R. B., Jarin, T., & Kumar, T. A. (2021). Critical investigation and prototype study on deep brain stimulations: an application of biomedical engineering in healthcare. In Handbook of Deep Learning in Biomedical Engineering and Health Informatics (pp. 103–135). Apple Academic Press.

[3] Mustafa, G., Rauf, A., Al-Shamayleh, A. S., Sulaiman, M., Alrawagfeh, W., Afzal, M. T. et al. (2023). Optimizing document classification: Unleashing the power of genetic algorithms. *IEEE Access*.

[4] Zohuri, B., & Rahmani, F. M. (2024). Computational intelligence for signal and image processing unleashing the power of algorithms. *Journal of Clinical Case Reports & Studies*, *5*(2).

[5] Hatoum, M. B., & Nassereddine, H. (2023). Unleashing the power of chatgpt for lean construction: An early outlook. In Proceedings of the 31st Annual Conference of the International Group for Lean Construction (IGLC31). doi: 10.24928/2023/0243.

[6] Iqbal, J., Jaimes, D. C. C., Makineni, P., Subramani, S., Hemaida, S., Thugu, T. R. et al. (2023). Reimagining healthcare: Unleashing the power of artificial intelligence in medicine. *Cureus*, *15*(9).

[7] Lilhore, U. K., Manoharan, P., Simaiya, S., Alroobaea, R., Alsafyani, M., Baqasah, A. M. et al. (2023). HIDM: hybrid intrusion detection model for Industry 4.0 networks using an optimized CNN-LSTM with transfer learning. *Sensors*, *23*(18), 7856.

[8] Orike, S., & Ene, D. S. (2023). Meta-learning: Unleashing the power of self-improving artificial intelligent (AI) systems. *The Journal of Advances in Computational Intelligence Theory*, *5*(3), 12–27.

[9] Dalal, S., Poongodi, M., Lilhore, U. K., Dahan, F., Vaiyapuri, T., Keshta, I. et al. (2023). Optimized LightGBM model for security and privacy issues in cyber-physical systems. *Transactions on Emerging Telecommunications Technologies*, *34*(6).

[10] Li, Z., Ma, P., Wang, H., Wang, S., Tang, Q., Nie, S. et al. (2022). Unleashing the power of compiler intermediate representation to enhance neural program embeddings. *Proceedings of the 44th International Conference on Software Engineering*, 2253–2265.

[11] Sharma, N., Gupta, R. D., Khanna, R., Sharma, R. C., & Sharma, Y. K. (2023). Machining of Ti-6Al-4V biomedical alloy by WEDM: Investigation and optimization of MRR and Rz using grey-harmony search. *World Journal of Engineering*, *20*(2), 221–234.

[12] Aluvala, S., Balamurugan, V., Jeyasudha, J., Kumar, R. D., & Kumar, P. (2023). Decoding financial markets: Unleashing the power of Bi-LSTM in sentiment analysis for cutting-edge stock price prediction. In 2023 International Conference on Evolutionary Algorithms and Soft Computing Techniques (EASCT), IEEE, 1–7.

[13] Simaiya, S., Lilhore, U. K., Sharma, Y. K., Rao, K. B. V. B., Rao, V. V. R. M., Baliyan, A. et al. (2024). A hybrid cloud load balancing and host utilization prediction method using deep learning and optimization techniques. *Scientific Reports*, *14*(1), 1337.

[14] Xu, M., Du, H., Niyato, D., Kang, J., Xiong, Z., Mao, S. et al. (2024). Unleashing the power of edge-cloud generative AI in mobile networks: A survey of AIGC services. *IEEE Communications Surveys & Tutorials*.

[15] Dahiya, N., Sharma, Y. K., Rani, U., Hussain, S., Nabilal, K. V., Mohan, A. et al. (2023). Hyper-parameter tuned deep learning approach for effective human monkeypox disease detection. *Scientific Reports*, *13*(1), 15930.

[16] Gupta, S., Kumar, A., Sharma, Y. K., & Sharma, A. (2020). Multiple performance characteristics optimization of WEDM by grey grade for EN-8. *Optimization and Engineering Resource*, *1*(1), 27–35.

[17] Fan, Z., Ghaddar, B., Wang, X., Xing, L., Zhang, Y., & Zhou, Z. (2024). *Artificial Intelligence for Operations Research: Revolutionizing the Operations Research Process. arXiv Preprint*, arXiv:2401.03244.

[18] Kumar Lilhore, U., Simaiya, S., Sharma, Y. K., Kaswan, K. S., Kbvb, R., Vvrm, R. et al. (2024). A precise model for skin cancer diagnosis using hybrid U-Net and improved MobileNet-V3 with hyperparameters optimization. *Scientific Reports*, *14*(1), 4299.

[19] Moayedi, H., Ahmadi Dehrashid, A., & Le, B. N. (2024). A novel problem-solving method by multi-computational optimization of artificial neural network for modelling and prediction of the flow erosion processes. *Engineering Application of Computation Fluid Mechanics*, *18*(1), 2300456.

[20] Mehta, P., Sait, S. M., Yıldız, B. S., Erdaş, M. U., Kopar, M., & Yıldız, A. R. (2024). A new enhanced mountain gazelle optimizer and artificial neural network for global optimization of mechanical design problems. *Mater Test*.

[21] Rokade, M. D., & Sharma, Y. K. (2021). MLIDS: A machine learning approach for intrusion detection for real time network dataset. In 2021 International Conference on Emerging Smart Computing and Informatics (ESCI) (pp. 533–536). IEEE.

[22] Aigner, K. M., Goerigk, M., Hartisch, M., Liers, F., & Miehlich, A. (2024). a framework for data-driven explainability in mathematical optimization. *Proceedings of the AAAI Conference on Artificial Intelligence*, *38*(19), 20912–20920.

[23] Sharma, Y., & Singh, B. K. (2020). Classification of children with specific language impairment using pitch-based parameters. In 2020 IEEE recent advances in intelligent computational systems (RAICS) (pp. 42–46). IEEE, .

[24] Rene, E. A., & Fokui, W. S. T. (2024). Artificial intelligence-based optimal EVCS integration with stochastically sized and distributed PVs in an RDNS segmented in zones. *Journal of Electrical Systems and Information Technology*, *11*(1), 1.

[25] He, C., Zhang, Y., Bian, T., Jiao, K., Su, W., Wu, K. J. et al. (2023). A review on artificial intelligence enabled design, synthesis, and process optimization of chemical products for industry 4.0. *Processes*, *11*(2), 330.

[26] Wang, H., Fu, T., Du, Y., Gao, W., Huang, K., Liu, Z. et al. (2023). Scientific discovery in the age of artificial intelligence. *Nature*, *620*(7972), 47–60.

[27] Williams, J. C., ReVelle, C. S., & Levin, S. A. (2004). Using mathematical optimization models to design nature reserves. *Frontiers in Ecology and the Environment*, *2*(2), 98–105.

[28] Ahmed, F. E., Hashaikeh, R., Diabat, A., & Hilal, N. (2019). Mathematical and optimization modelling in desalination: State-of-the-art and future direction. *Desalination*, *469*, 114092.

[29] Rani, P., Hussain, N., Khan, R. A. H., Sharma, Y., & Shukla, P. K. (2021). Vehicular intelligence system: Time-based vehicle next location prediction in software-defined internet of vehicles (SDN-IOV) for the smart cities. *Intelligence of Things: AI-IoT Based Critical-Applications and Innovations*, 35–54.

[30] Pachauri, R. K., Mahela, O. P., Sharma, A., Bai, J., Chauhan, Y. K., Khan, B. et al. (2020). Impact of partial shading on various PV array configurations and different modeling approaches: A comprehensive review. *IEEE Access*, *8*, 181375–181403.

[31] Sharma, Y., & Singh, B. K. (2023). Classification of children with attention-deficit hyperactivity disorder using wigner-ville time-frequency and deep expEEGNetwork feature-based computational models. *IEEE Transactions on Medical Robotics and Bionics*.

[32] He, C., Zhang, Y., Gong, D., & Ji, X. (2023). A review of surrogate-assisted evolutionary algorithms for expensive optimization problems. *Expert Systems with Applications*, *217*, 119495.

[33] Song, Y., Cai, X., Zhou, X., Zhang, B., Chen, H., Li, Y. et al. (2023). Dynamic hybrid mechanism-based differential evolution algorithm and its application. *Expert Systems with Applications*, *213*, 118834.

[34] Kumar, L., Kar, M. K., & Kumar, S. (2023). Statistical analysis based reactive power optimization using improved differential evolutionary algorithm. *Expert Systems: International Journal of Knowledge Engineering and Neural Networks*, *40*(1).

[35] Dhiman, H., Vyas, N., Pagrotra, A., & Dutt, V. (2023). Real-time traffic analysis and driver safety enhancement via IoT and computer vision with high-resolution project lamps for intelligent road illumination [Internet]. In 1st International Conference on Circuits, Power and Intelligent Systems (CCPIS). IEEE, 2023. Available from: http://dx.doi.org/10.1109/CCPIS59145.2023.10291872.

[36] Kaur, N., Mittal, A., Lilhore, U. K., Simaiya, S., Dalal, S., Sharma, Y. K. et al. (2023). Solar photovoltaic panels: A review of developments and deployment strategies. *International Journal of Ambient Energy*, *44*(1), 2218231.

[37] Kumar, N., Verma, H., & Sharma, Y. K. (2023). Adversarial attacks on graph neural network: Techniques and countermeasures. *Concepts and Techniques of Graph Neural Networks. IGI Global*, 58–73.

[38] Nayak, S., Sharma, Y. K., & Nayak, R. (2023). An optimized round robin central processing unit scheduling algorithm with dynamic time quantum. In 2023 Second International Conference on Smart Technologies For Smart Nation (SmartTechCon) (pp. 1482–1490. IEEE.

[39] Slowik, A., & Kwasnicka, H. (2020). Evolutionary algorithms and their applications to engineering problems. *Neural Computing & Applications*, *32*, 12363–12379.

[40] Shaikh, P. W., El-Abd, M., Khanafer, M., & Gao, K. (2020). A review on swarm intelligence and evolutionary algorithms for solving the traffic signal control problem. *IEEE Transactions on Intelligent Transportation Systems*, *23*(1), 48–63.

[41] Falcón-Cardona, J. G., & Coello Coello, C. A. (2020). Indicator-based multi-objective evolutionary algorithms: A comprehensive survey. *ACM Computing Surveys*, *53*(2), 1–35.

[42] Song, Y., Cai, X., Zhou, X., Zhang, B., Chen, H., Li, Y. et al. (2023). Dynamic hybrid mechanism-based differential evolution algorithm and its application. *Expert Systems With Applications*, *213*, 118834.

[43] Deng, W., Xu, J., Song, Y., & Zhao, H. (2021). Differential evolution algorithm with wavelet basis function and optimal mutation strategy for complex optimization problem. *Applied Soft Computing*, *100*, 106724.

[44] Elsayed, S. M., Sarker, R. A., & Essam, D. L. (2012). An improved self-adaptive differential evolution algorithm for optimization problems. *IEEE Transactions on Industrial Informatics*, *9*(1), 89–99.

[45] Yi, W., Zhou, Y., Gao, L., Li, X., & Mou, J. (2016). An improved adaptive differential evolution algorithm for continuous optimization. *Expert Systems with Applications*, *44*, 1–12.

[46] Swarna, S. R., Kumar, A., Dixit, P., & Sairam, T. V. M. (2021). Parkinson's disease prediction using adaptive quantum computing. In 2021 Third International Conference on Intelligent Communication Technologies and Virtual Mobile Networks (ICICV) (pp. 1396–1401). IEEE.
[47] Zhang, X., Jin, L., Cui, C., & Sun, J. (2021). A self-adaptive multi-objective dynamic differential evolution algorithm and its application in chemical engineering. *Applied Soft Computing*, *106*, 107317.
[48] Sedak, M., & Rosić, B. (2021). Multi-objective optimization of planetary gearbox with adaptive hybrid particle swarm differential evolution algorithm. *Applied Sciences*, *11*(3), 1107.
[49] Da Silva Santos, C. E., Sampaio, R. C., Coelho, L. D. S., Bestard, G. A., & Llanos, C. H. (2021). Multi-objective adaptive differential evolution for SVM/SVR hyperparameters selection. *Pattern Recognit*, *110*, 107649.
[50] Liu, B. J., & Bi, X. J. (2021). Adaptive ε-constraint multi-objective evolutionary algorithm based on decomposition and differential evolution. *IEEE Access*, *9*, 17596–17609.
[51] Hou, Y., Wu, Y., Liu, Z., Han, H., & Wang, P. (2021). Dynamic multi-objective differential evolution algorithm based on the information of evolution progress. *Science China Technological Sciences*, *64*(8), 1676–1689.
[52] Bian, K., Sun, Y., Cheng, S., Liu, Z., & Sun, X. (2021). Adaptive methods of differential evolution multi-objective optimization algorithm based on decomposition. In Neural Computing for Advanced Applications: Second International Conference, NCAA 2021 (pp. 458–472). Springer Singapore, Guangzhou, China, August 27–30, 2021, Proceedings 2.
[53] Wang, H., Wang, J., Zhao, Y., & Xu, H. (2021). Tunneling parameters optimization based on multi-objective differential evolution algorithm. *Soft Computing*, *25*, 3637–3656.
[54] Wang, P., Xue, B., Liang, J., & Zhang, M. (2021). Multi-objective differential evolution for feature selection in classification. *IEEE Transaction on Cybernetics*, *53*(7), 4579–4593.
[55] Yue, C., Suganthan, P. N., Liang, J., Qu, B., Yu, K., Zhu, Y. et al. (2021). Differential evolution using improved crowding distance for multimodal multi-objective optimization. *Swarm and Evolutionary Computation*, *62*, 100849.

Uppuluri Lakshmi Soundharya, Yogesh Kumar Sharma,
Anantha Raman G. R., and Ravi Rastogi

9 Introduction to Mathematical Optimization Techniques in AI

Abstract: Particularly, real-life problems are progressively complicated and demand more advanced mathematical algorithms and models, which can efficiently handle vast data volumes and identify the best remedies. However, it is important to keep in mind that there is no ideal technique or algorithm since each one possesses certain constraints that can be alleviated or eradicated through the integration of diverse approaches. The field of artificial intelligence (AI) has seen a significant transformation because of the advancements in deep learning (DL) and machine learning (ML). These advancements have enabled the accomplishment of complex tasks such as image recognition and processing natural language with remarkable proficiency. Optimization strategies are essential to building DL and ML models as they provide optimal convergence and improved efficiency. Individuals can optimize existing DL models by using the mathematical principles and derivations that define these approaches. This chapter presents an overview of mathematical optimization techniques in AI, types of optimization methods, their applications, challenges, and future directions.

Keywords: Mathematical optimization, AI, deep learning, machine learning

9.1 Introduction

Mathematical algorithms are widely used in a variety of scientific disciplines to resolve challenges. The algorithms, as mentioned earlier, simplify reality using mathematical formulas, allowing us to understand complex procedures, solve difficulties, and make informed decisions. Mathematical algorithms employ techniques to determine the best solution to a given problem. Numerical optimization is a recognized field in mathematics that focuses on identifying the highest and lowest points of an activity [1]. The development of optimization strategies has been necessary for various key aspects, which include administration, decision-making, technological advancements, and overall system development in recent decades, offering profitable, positive aspects [2].

Mathematical optimization involves formulating a business concern using a number of formulas and analysing them to get the most favourable business outcomes. These equations serve as a framework for thinking and may be repeatedly optimized using various data sources and goals, such as maximizing income, minimizing cost, and achieving a desired service level. Every outcome represents an optimal situation that the user may access and use in the company to have a significant effect [3].

https://doi.org/10.1515/9783111436180-009

Machine learning (ML) is a renowned field that specializes in addressing issues via the use of mathematical frameworks and techniques. Many real-world situations need the processing of large amounts of facts and figures. To convert the unprocessed details into useful information towards solving a problem, this quantity of statistical information often needs statistical support. ML techniques are quite valuable in this particular application. The goal of these procedures and models is to provide a mathematical framework that characterizes the data collection and extrapolates the information to unidentified data sampling. ML algorithms and models are applicable in several fields, such as the business world, health, finance, education, and medical sectors [4].

Considering the practical significance of these domains, multiple techniques were created to address optimization with ML difficulties. While most artificial intelligence (AI) algorithms are effective in tackling the issues they encounter, none are flawless. The available research specifies many constraints concerning both optimization and AI strategies. However, in recent years, a number of investigators have begun to explore methods to integrate both approaches to address the limitations of each approach, enhancing processes influenced by additional techniques. These combined approaches are known as hybrids [5].

The chapter is organized into various sections, which are as follows: Section 9.2 covers optimization in AI and mathematical optimization techniques, Section 9.3 covers the applications of optimization, Section 9.4 covers challenges and future directions of optimization, and the final section covers the conclusion of the research.

9.2 Mathematical Optimization and AI Algorithms

According to Gartner research, AI is a field of computer science and engineering that employs mathematical and logical procedures to extract and incorporate programming expertise. It utilizes advanced and intelligent processes to solve issues, understand events, facilitate decision-making, and perform automated operations [6]. Gartner classifies AI into three primary classifications:

- Optimization techniques (OT)
- Computational logic (CL)
- Probabilistic reasoning (PR)

Optimization approaches include several methods, such as linear programming, which concentrates on "linear objective functions" and rules; nonlinear programming, which is utilized for nonlinear functions; linear programming with integers for individual factors; and dynamic programming, which is used for issues with overlapping subdivisions. Figure 9.1 presents the types of mathematical OTs [7]. Table 9.1 presents a comparative analysis of various optimization methods.

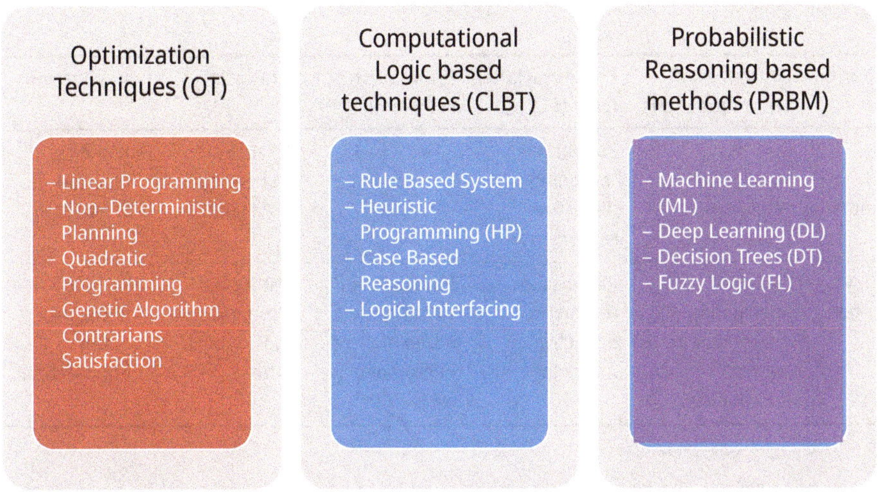

Figure 9.1: Types of mathematical optimization techniques.

Table 9.1: Comparative analysis of various optimization methods.

Optimization method	Optimization techniques (OT)	Computational logic (CL)	Probabilistic reasoning (PR)	Strengths	Applications
Gradient descent [8]	First-order derivative	Iterative refinement, rule-based reasoning	–	Simple, easy implementation	Machine learning, neural networks
Newton's method [9]	Second-order derivative	Rule-based reasoning, deductive reasoning	–	Fast convergence	Nonlinear optimization, machine learning
Genetic algorithms [10]	Evolutionary algorithms, selection, crossover, mutation	–	Probabilistic modelling of candidate solutions	Global optimization handles non-differentiable functions	Engineering design, scheduling problems
Simulated annealing [11]	Probability distributions, annealing schedule	Rule-based reasoning	Probabilistic acceptance of worse solutions	Escapes local minima, finds global optima.	Combinatorial optimization, parameter tuning

Table 9.1 (continued)

Optimization method	Optimization techniques (OT)	Computational logic (CL)	Probabilistic reasoning (PR)	Strengths	Applications
Particle swarm optimization [12]	Swarm intelligence, velocity, position update rules	Rule-based reasoning, deductive reasoning	–	Parallel processing finds global optima	Robotics, financial forecasting
Ant colony optimization [13]	Swarm intelligence, pheromone trails, heuristic information	Rule-based reasoning, deductive reasoning	Probabilistic transition rule based on pheromone trails	Distributed computation finds optimal paths	Routing problems, graph optimization

Mathematical optimization and AI are distinct AI methodologies. However, there is frequent confusion between the two. AI algorithms support ML and deep learning (DL) algorithms in mathematical optimization-based problem-solving. ML is primarily concerned with making predictions and calculating probabilities, while mathematical optimization is a prescriptive method that aims to determine a practical path of action. While ML forecasts might guide specific choices, ML is not inherently capable of addressing complex business situations that include several linked factors. This is where optimization really shines. A crucial distinction between ML and optimization lies in the fact that optimization begins with the business issue as the primary focus rather than the data [14, 15].

Accessible data significantly influences the outcomes of an ML algorithm. It is often necessary to evaluate these outcomes by comparing them against extensive volumes of data to ensure accuracy. This entails examining not only a mere ten thousand items in a file but also occasionally millions of records. ML models may present challenges when it comes to adapting to changing business circumstances or demands because of their heavy reliance on past data. This reduces their flexibility. ML models often struggle to handle exceptional cases [16].

On the other hand, optimization is especially designed to handle exceptions by explicitly modelling restrictions. The variables involved in the decision-making process, as well as their interconnections, should be clearly defined regardless of the available data. Proficiency in business logic is required. Consequently, the process of creating an optimization model may require a longer initial investment. However, after you possess a model, you may effortlessly broaden its scope and experiment with novel concepts to operate the organization, for instance, including restrictions to limit the number of suppliers, specifically establishing safety thresholds for inventories, mandating a minimum decrease in CO_2 emissions, and so on. Moreover, it is possible to increase the quantity of goals. This enables you to comprehend the behaviour of the issue when subjected to various objectives, including the optimization of several objectives, often known as multi-objective optimization [17].

9.3 Optimization Problem-Solving Framework

Although there are many approaches and techniques available for addressing issues associated with optimization, only a few of them can be considered universally applicable or flawless in their ability to efficiently handle all such difficulties. Typically, each method encounters distinct features that make it more suited for solving a certain group of issues based on its unique qualities. To solve an optimization issue, it is important to consider the problem creation, methodologies, and algorithm selection. This is because various methods and approaches work better for different types of problems [18].

A mathematical framework formulation is the initial stage in solving a problem related to optimization. The goal of a mathematical framework is to translate a real-world issue into a mathematical equation that OTs can utilize. An effort to optimize problems may be mathematically formulated using a collection of data points and mathematical relationships that provide an abstract representation of the problem. The goal is to identify the best possible alternative, known as the optimum solution, from a given set of candidate solutions. To construct a mathematical framework, one must adhere to four phases [19]:

- Step 1: Decision variable declaration and initialization, i.e., (DC_1, DC_2, . . . DC_n).
- Step 2: Create a recursive decision function f (DC) using the decision variables.
- Where $f(\mathrm{DC}) = Best(f1(\mathrm{DC}), f2(\mathrm{DC}), \ldots \ldots \ldots \ldots fn(\mathrm{DC}))$
- Step 3: To hold the decision variables, define some constraints and conditions of equality and inequality.
- Step 4: Define the characteristics of the problem domain (Pd_1, Pd_2, . . . Pd_n).

The optimization issue is defined by the objective function as well as the restrictions imposed on the variables that are involved. The function that defines the objective represents the goal for the problem, and this is composed of an individual and grouping of numbers which are convertible to just one value, such as individuals, time, goods, resources, etc. On the other hand, the obstacles directly impact the space of decisions and outcomes, imposing restrictions and constraints upon the procedure's selections.

The primary purpose of an optimization issue is to reduce and optimize the objective function while taking into account the restrictions. Furthermore, a potential approach to categorize optimization models based on the number of goals is to split them into two main categories: single-objective and multi-objective planning associated with optimization. Multi-objective optimization represents a branch of multiple criteria decision-making makes decisions that involves multiple goals, and each function must be maximized concurrently, even when there are restrictions. Single-objective optimization problems typically include one objective that must be maximized, with or without restrictions. Once the mathematical model has been developed, it is crucial to select the most suitable strategy for finding the optimum answer [20].

The concepts of both local and global optimizations may characterize the answer to an optimization issue. Similarly, the search techniques can be categorized as local search as well as global search. Algorithms for both local and global optimization of search results are used differently or to address distinct optimization problems. Local optimization involves identifying a solution that reduces or maximizes the objective function within a defined area of the search space. This implies that it focuses on finding a solution that is locally optimal among viable locations within the nearby vicinity. This particular type of search does not ensure an objective value that is lower or greater than all other possible points.

In contrast, global optimization aims to identify the optimal point that minimizes or maximizes the objective function across all viable solutions. Nevertheless, the intricacy of global OTs increases exponentially as the issues get larger [21, 22]. Table 9.2 presents a comparative analysis of the optimization framework for various mathematical optimization methods.

Table 9.2: Comparative analysis of optimization framework for various mathematical optimization methods.

Framework	Description	Strengths	Weaknesses	Applications
Genetic algorithm (GA) [4, 23]	An evolutionary algorithm based on the principles of natural selection. The process begins with an initial population of randomly generated solutions and progresses through successive generations towards improved solutions.	Global optimization handles non-differentiable and discontinuous functions and has parallel processing capabilities.	The rate of convergence may be sluggish, and there may be a high level of complexity in setting the parameters.	Areas of expertise include engineering design, allocation of resources, and evolutionary automation.
Simulated annealing (SA) [24]	A technique for probabilistic optimization that emulates the annealing process used in metallurgy. It permits the occasional adoption of inferior solutions to avoid being stuck in local optima.	Capable of avoiding local optima, identifying global optima, and being resistant to noise.	The sensitivity of parameter adjustment and sluggish convergence in some instances.	Combinatorial optimization and parameter tweaking.

Table 9.2 (continued)

Framework	Description	Strengths	Weaknesses	Applications
Particle swarm optimization (PSO) [25]	Population-based optimization approaches draw inspiration from the collective behaviour of birds migrating or fish spawning. The process incrementally enhances potential solutions by considering their closeness to the most optimal answers discovered so far.	Rapid convergence, simple implementation, and the capacity to handle tasks simultaneously	Susceptible to early convergence, it may encounter difficulties with optimization issues in high-dimensional spaces.	Proneness to premature convergence may result in challenges when dealing with optimization problems in spaces with a large number of dimensions.
Ant colony optimization (ACO) [26]	Optimization algorithm based on ant foraging. It finds optimal solutions by communicating route information via pheromone trails.	Ant-foraging optimization algorithm: by sharing route information via pheromone trails, it seeks optimal solutions.	Problems with parameter tweaking and scaling	Graph optimization, routing issues.
Tabu search (TS) [27]	Metaheuristic search repeatedly moves from one solution to a neighbouring one. Memory architectures prevent revisiting solutions.	Effective search space exploration avoids local optima and adapts to different domains of issues.	Parameter sensitivity, complicated implementation	Combinatorial optimization, scheduling issues.
Differential evolution (DE) [28]	This population-based optimization approach evolves a population of alternative solutions to enhance candidate solutions. The operators include mutation, crossover, and selection.	Global optimization is a method that is not affected by noise and non-linearities and is capable of solving problems with a large number of dimensions.	The slower rate of convergence compared to other approaches and sensitivity to parameter values.	Engineering planning and parameter optimization

Optimization involves the use of several strategies and algorithms to discover the answer to a problem. Certain optimization approaches exhibit unique characteristics while searching for an approach to the same issue, resulting in some methods being quicker and more dependable in terms of accuracy compared to others. There is no universal approach to efficiently handle all optimization issues. Instead, we have several methods that can potentially be used within their respective capacities to address issues associated with optimization.

9.4 Mathematical Optimization Techniques

Mathematical optimization, often known as mathematical programming, involves choosing the most optimal element from a given collection of possibilities based on certain criteria. Typically, it is categorized into a pair of subdivisions: discrete optimization and continuous optimization. Numerical optimization is a branch of mathematics that focuses on finding the highest and lowest boundaries of operations. Over the past couple of decades, there has been an increase in curiosity about optimization approaches, mostly driven by advancements in computing technology and their widespread uses. Therefore, the implementation of OTs is increasingly becoming a crucial instrument for supervisors, decision-making processes, enhancement, and advancement of innovations to facilitate the operation of systems that operate effectively. The general framework represents concerns by using decision parameters, which are constants that individuals have control over and can adjust as knobs. [29].

Mathematical optimization is often referred to as “mathematical programming.” The term is frequently used synonymously with predictive analytics, also known as an instance of analytics that generates an advisable plan of action. Mathematical optimization is frequently recognized as the most prominent method for prescriptive analysis in the industry. Prescriptive analytics differs from descriptive and predictive analytics in that it provides recommendations on how business processes should be improved or evolved rather than simply reporting trends or forecasting the future [30].

Prescriptive analytics is essential for transforming data and business rules into meaningful measures rather than just offering insights into the behaviour of that information. In addition to its role as a prescriptive analytics tool, mathematical optimization is also regarded as a kind of computational intelligence (CI). Constraints refer to limitations or restrictions placed on the possible values that the decision variables may have. Perhaps the goal function(s) are the ones we want to optimize. The key types of mathematical optimization methods are as follows.

9.4.1 Gradient Descent

This is a key optimization method employed in the field of AI. The objective is to minimize a cost function by constantly adjusting the variables of the mathematical framework using the change in gradients and the cost function compared to the relevant variables [31].

- **Mathematical Formulation:** Considering a neural network, its parameters are denoted by the symbol θ. The goal is to minimize a function of costs $J(\theta)$, which evaluates the efficiency of the mathematical framework. The derivative of $J(\theta)$ in relation to θ is represented as $\nabla J(\theta)$. The modification phase in a gradient descent (GD) is often formally represented as follows:

$$\theta = ((\theta - \alpha) * \nabla J(\theta)) \tag{9.1}$$

In this context, the symbol α, also known as the learning rate (LR), determines the magnitude of the adjustment made to the parameters. It plays a crucial role in determining the pace at which the optimization process converges and the stability it achieves.

9.4.2 Stochastic Gradient Descent (SGD)

SGD involves a modified version of GD, which calculates the gradients and then adjusts the setting parameters by employing a small subset of the training specimens, known as a mini-batch, instead of the complete dataset. This strategy significantly decreases the computing load and accelerates the optimization process [32].

Mathematical Formulation: Let a small group of training samples be called B, including the total cost function $J(\theta)$, as previously mentioned. The revision step in SGD is

$$\theta = [\theta - \{\alpha * \nabla J(\theta; B)\}) \tag{9.2}$$

In this case, $\nabla J(\theta; B)$ denotes the gradient for $J(\theta)$ calculated for mini-batch B.

9.4.3 Momentum Optimization

GD represents an optimization procedure that finds the function's absolute optimum by following the objective function's unfavourable gradient. When used to optimize problems with high degrees of curvature and unpredictable gradients, GD is employed. A GD method tends to bounce around the search space until it becomes stuck in areas within the search space devoid of gradients.

An expansion of the GD optimization technique is the momentum algorithm, which enables the search to generate energy in a particular direction within the search space, thereby overcoming the disturbances caused by chaotic gradients and traversing the space's smooth areas along the coast [33].

More exactly, momentum may be defined as the moving average of prior gradients weighted exponentially. The optimizer uses the exponentially weighted moving average of the prior gradients to update the parameters rather than using the current gradient. The optimizer can remember its previous path and stick to it even if the gradient is pointing in the opposite direction, thanks to an exponentially weighted moving average, which acts as a memory.

It is common practice to use momentum with other optimization techniques, including SGD as well as adaptive learning rate (ALR) techniques like Adam (AD), Adagrad (ADG), and Adadelta (ADD). The momentum algorithm incorporates a momentum element into the procedure for updating to overcome the drawbacks of GD. Momentum, defined with the characteristics of a "ball going downhill," is a portion of the previous revision of the vector. Despite the event that the gradient reverses course or becomes zero, the momentum factor keeps the optimizer moving in the same path as it descends. This lowers oscillations and prevents us from entering shallow neighbourhood minima [34]:

$$V = \{\beta V + [(1-\beta)abla)\theta J(\theta)\} \tag{9.3}$$

$$\theta = [\theta - av) \tag{9.4}$$

In this equation, the variable V stands for the momentum term, which is the momentum coefficient, and $J()$ is the gradient of the cost function with respect to the parameters, and is the LR. Usually, one sets the momentum coefficient to 0.901. In every iteration, the optimizer determines the cost function's gradient and updates the momentum term by taking the exponentially weighted moving average of the prior gradients. The momentum term multiplied by the LR is then subtracted to update the parameters [35].

9.4.4 Adam Optimization

The algorithm known as Adam optimizer is an incremental optimization technique employed to minimize the loss function throughout the development of neural networks. This is an abbreviation for "adaptive moment estimation (AME)." When considering momentum, Adam is often conceptualized as a hybrid of SGD and RMSprop (RMp); Diederik-p created this method.

The method uses the squared gradients to adjust the LR in a manner similar to RMSprop. Additionally, it leverages momentum by employing the average movement of the gradient rather than the gradient of its own, comparable to SGD alongside momentum. This approach combines the use of a dynamic learning rate (DLR) with easing techniques to effectively attain the global minimum [36]. The following Adam setup variables are used in the configurations.

- Alpha: The term "learning rate," also called "step size," is used interchangeably to refer to this concept. The weight change percentage, denoted as 0.001, determines the extent to which weights are adjusted. Higher numbers (such as 0.3) lead to quicker learning processes until the rate is adjusted. Lower values (such as 1.0E-5) significantly decrease the speed of learning throughout the training.
- Beta one: For initial moment estimates, an exponential decay rate (such as 0.9) is used.
- Beta two: The speed at which the second-moment estimations decrease exponentially is 0.999. The recommended value for issues with a sparse gradient, such as NLP as well as computer vision tasks, is about 1.0.
- Epsilon: To avoid dividing by zero in the operation, an extremely tiny number (such as 10E-8) is used.

9.5 Applications

Mathematical optimization, particularly optimization modelling, may provide solutions to any of these inquiries. This approach involves the process of expressing a business challenge using mathematical formulas or equations. It metaphorically represents the choices you must make, the limitations you must adhere to, and the interconnections between them to accomplish a certain goal. "A mathematical model" is the term used to explain such business logic. The key application of mathematical optimization is as follows [37, 38].

Capacity Planning (CP): CP involves monitoring and managing network resources on behalf of the NaaS provider. It uses key parameters to forecast the network's behaviour and schedules actions to increase capacity when needed. In the context of operations management, CP represents the process of harmonizing consumer demand for a particular good or business service alongside an organization's or manufacturer's ability to produce sufficient quantities to satisfy that demand.

Healthcare Systems (HS): This area of expertise uses the IBM CPLEX solvers to tackle linear programming challenges using high-speed programming for mathematical techniques. Mathematical modelling is used to simulate and analyse clinical studies. IBM CPLEX is a software tool for decision optimization that facilitates the creation and implementation of optimization models. The field of cancer research has used mathematical OTs since the beginning of the twentieth century, which have been applied to advancements in the field of medicine.

Energy Systems (ESs): In the field of ESs, an objective measure refers to the amount of energy utilized by a system. The purpose of the energy framework is to minimize energy use. The energy management system's (EMS) efficiency and effectiveness are contingent upon the energy management strategy (EMS) and forecast algorithms employed. Traditional power grids use several techniques, such as fuzzy logic, the theory of games, and learning through reinforcement in their energy administration systems. The application of fuzzy logic in EMS is effective in optimizing operating costs. At the same time, the use of game theory in EMS is effective in lowering overall power consumption and the associated costs of emissions in the ES.

Supply Chain Network Design (SCND): The supply chain comprises manufacturing facilities, storage facilities, distribution centres, and client regions. The issue involves concurrent operational and financial choices, hence requiring appropriate variables as well as parameters in its mathematical formulation. The suggested model focuses on the supply chain network. The logistics chain comprises manufacturing facilities, storage facilities, centres for distribution, and client regions. The issue involves both operational and financial choices concurrently, necessitating the use of appropriate parameterized variables in the mathematical equation. The suggested model takes into account the supply chain network.

In electrical engineering, optimization methods are often used for several purposes. These involve designing active filters, reducing stray fields in the storage of superconducting magnetic ESs, designing microwave constructions using space in the mapping process, and designing handset antennas utilizing electromagnetic radiation. Table 9.3 presents an analysis of applications of mathematical optimization methods.

Table 9.3: Analysis of applications of mathematical optimization methods.

References	Optimization method	Methods used	Strengths	Applications
[1, 2, 4]	Gradient descent	First-order derivative	Simple, easy to implement	Machine learning, neural networks
	Newton's method	Second-order derivative	Faster convergence	Nonlinear optimization, machine learning
[7, 8]	Genetic algorithms	Evolutionary algorithms, selection, crossover, mutation	Global optimization handles non-differentiable functions	Engineering design, scheduling problems
	Simulated annealing	Probability distributions, annealing schedule	Finds global optima, escapes local minima.	Combinatorial optimization, parameter tuning
[3, 5, 6]	Particle swarm optimization	Swarm intelligence, velocity, and position update rules	Parallel processing finds global optima	Robotics, financial forecasting
	Ant colony optimization	Swarm intelligence, pheromone trails, heuristic information.	Distributed computation, finds optimal paths	Routing problems, graph optimization

9.6 Challenges and Future Directions

The key challenges and future directions of mathematical optimization methods are as follows.

9.6.1 Challenges

Optimization strategies are very beneficial, but their successful implementation is not always simple or direct.

- The effectiveness of an optimization strategy is directly dependent on the quality of its formulation. Although it may be theoretically accurate, it might still be of little practical use.
- Frequently, several comparable formulations are available, which vary significantly in terms of effectiveness and solutions. Choosing relevant descriptions requires a moderate level of experience.
- Optimization strategies are limited to the information they are provided. The conceptual framework must include every relevant component of the intended solution, whether it is implied or stated directly (such as optimum control learning).

9.6.2 Future Directions

In the future, this study has the potential for further inquiry and progress in other interesting avenues. Continued research and development in hybrid optimization methodology are necessary. This includes investigating new hybridization strategies, adaptive learning mechanisms, and ensemble approaches to improve the efficacy and adaptability of optimization algorithms. Additionally, future research endeavours need to prioritize the resolution of new obstacles in optimization. These problems include effectively managing uncertainty, integrating domain-specific expertise, and optimizing for many competing goals in intricate decision-making situations. This entails investigating resilient optimization methods, multi-objective optimization frameworks, and hybrid optimization paradigms designed for particular application domains and issue situations [39, 40].

Furthermore, there is an increasing potential for interdisciplinary cooperation and exchange of ideas among optimization academics and practitioners from many domains, such as data science, AI, and systems engineering. By promoting interdisciplinary conversation and the sharing of information, we may use ideas from several disciplines to create innovative optimization approaches and successfully tackle real-world difficulties. In addition, this study has the potential to expand into the creation of optimization tools and software platforms that prioritize the needs of users. These tools would provide stakeholders with user-friendly interfaces, interactive visualization tools, and customizable processes to aid in problem-solving and decision-making. By making advanced optimization capabilities accessible to everyone, we can enable individuals and organizations in various sectors to utilize mathematical optimization to promote innovation and generate positive societal outcomes.

Ultimately, this study acts as a foundation for fully harnessing the capabilities of mathematical optimization to tackle intricate real-world issues and shape a future that is more effective, adaptable, and environmentally friendly. By adopting multidisciplinary teamwork, innovation in technology, and user-centric design principles, we can pave the way for significant progress in solving optimization-driven problems and making decisions. This will lead to the exploration of new areas in the discovery of information, innovation, and the well-being of society. The key challenges and future directions of mathematical optimization methods are presented in Table 9.4.

Table 9.4: Comparative analysis of challenges and future direction of mathematical optimization methods.

Optimization method	Challenges	Future directions
Gradient descent	– Prone to getting stuck in local minima	– Development of advanced optimization algorithms incorporating adaptive learning rates and momentum terms to overcome local minima issues
Newton's method	– Requires the computation of second-order derivatives, which can be computationally expensive and numerically unstable in some cases	– Exploration of hybrid optimization techniques that combine the fast convergence of Newton's method with the robustness of first-order methods
Genetic algorithms	– Parameter tuning can be challenging and time-consuming	– Integration of machine learning techniques to automate parameter tuning and enhance algorithm performance
Simulated annealing	– Slow convergence in complex optimization landscapes	– Investigation of advanced cooling schedules and temperature update strategies to improve convergence speed
Particle swarm optimization	– Vulnerable to premature convergence	– Development of hybrid PSO algorithms incorporating diversity maintenance mechanisms to prevent premature convergence
Ant colony optimization	– Scalability issues for large-scale problems	– Exploration of parallel and distributed ACO algorithms to handle large-scale optimization problems

Table 9.4 (continued)

Optimization method	Challenges	Future directions
Tabu search	– Sensitivity to parameter settings and memory structures	– Research on adaptive memory structures and parameter tuning strategies aims to improve algorithm robustness
Differential evolution	– Slower convergence compared to gradient-based methods	– Investigation of adaptive mutation and crossover strategies to accelerate convergence

9.7 Conclusion

This study has conducted an extensive investigation of mathematical optimization strategies, including both traditional approaches and contemporary metaheuristic algorithms that are computationally efficient. It also covers their types, applications, and challenges. The survey covers a broad range of optimization methods utilized in diverse fields, including AI, ML, and programming in mathematics. The study examines the intrinsic difficulties encountered by current algorithms, such as problems with flexibility, speed of integration, and capacity to adapt to various application domains.

Mathematical optimization has practical uses in several sectors and areas. It offers a groundbreaking technique for tackling intricate challenges and making choices. Mathematical OTs have proven to be helpful resources for companies, government agencies, and investigators. The techniques can be applied to various areas such as producing goods, transportation, logistics, supply chain management, healthcare administration, and handling financial hazards.

References

[1] Arinze, C. A., & Jacks, B. S. (2024). A comprehensive review on AI-driven optimization techniques enhancing sustainability in oil and gas production processes. *Engineering Science & Technology Journal*, *5*(3), 962–973.

[2] Carrizosa, E., Ramírez-Ayerbe, J., & Romero Morales, D. (2024). Mathematical optimization modelling for group counterfactual explanations. *European Journal of Operational Research.*

[3] Li, X., Zhu, F., Zhen, H.-L., Luo, W., Lu, M., Huang, Y., Fan, Z. et al. (2024). Machine learning insides OPTVErse AI solver: Design principles and applications. *arXiv Preprint arXiv:2401.05960.*

[4] Meng, F., & Wang, C. A. (2024). Artificial intelligence and machine learning approaches to text recognition: A research overview. *Journal of Mathematical Techniques and Computational Mathematics*, *3*(no. 3), 01–05.

[5] Liu, Y.-F., Chang, T.-H., Hong, M., Wu, Z., Man-Cho So, A., Jorswieck, E. A., & Yu, W. (2024). A survey of advances in optimization methods for wireless communication system design. *arXiv Preprint arXiv:2401.12025*.

[6] Ayalew, M. F., Hussen, S., & Pasam, G. K. (2019). Optimization techniques in power system. *International Journal of Engineering Applied Sciences and Technology*, *3*(no. 10), 2455–2143.

[7] Neira, D. E. B., Laird, C. D., Lueg, L. R., Harwood, S. M., Trenev, D., & Venturelli, D. (2024). Utilizing modern computer architectures to solve mathematical optimization problems: A survey. *Computers & Chemical Engineering*, 108627.

[8] Vashishth, T. K., Sharma, V., Sharma, K. K., Kumar, B., Chaudhary, S., & Panwar, R. (2024). Intelligent resource allocation and optimization for industrial robotics using AI and blockchain. In AI and Blockchain Applications in Industrial Robotics (pp. 82–110). IGI Global.

[9] Rezazadeh, F. (2024). A Mathematical Introduction to Deep Reinforcement Learning for 5G/6G Applications. *arXiv Preprint arXiv:2403.14516*.

[10] Fan, Z., Ghaddar, B., Wang, X., Xing, L., Zhang, Y., & Zhou, Z. (2024). Artificial Intelligence for Operations Research: Revolutionizing the Operations Research Process. *arXiv Preprint arXiv:2401.03244*.

[11] Wang, Y., Fu, E. Y., Zhai, X., Yang, C., & Pei, F. (2024). Introduction of artificial intelligence. In Intelligent Building Fire Safety and Smart Firefighting (pp. 65–97). Springer Nature Switzerland, Cham.

[12] Zhuang, X., Wang, W., Su, Y., Yan, B., Li, Y., Li, L., & Hao, Y. (2024). Multi-objective optimization of reservoir development strategy with hybrid artificial intelligence method. *Expert Systems with Applications*, *241*, 122707.

[13] Chong, E. K. P., Lu, W.-S., & Zak, S. H. (2023). An introduction to optimization: With applications to machine learning. John Wiley & Sons.

[14] Nutakki, M., & Mandava, S. (2023). Review on optimization techniques and role of artificial intelligence in home energy management systems. *Engineering Applications of Artificial Intelligence*, *119*, 105721.

[15] Ismail, W. N., Fathimathul Rajeena, P. P., & Ali, M. A. S. (2023). A meta-heuristic multi-objective optimization method for alzheimer's disease detection based on multi-modal data. *Mathematics*, *11* (no. 4), 957.

[16] Uddin, M. G., Nash, S., Rahman, A., & Olbert, A. I. (2023). Assessing optimization techniques for improving water quality model. *Journal of Cleaner Production*, *385*, 135671.

[17] Rahman, U., Jamshaid, F. M., Ali, A., & Danish, S. (2024). Mathematical modeling and simulation of biophysics systems using neural network. *International Journal of Modern Physics B*, *38*(05), 2450066.

[18] Zhang, Q., Tian, S., & Lin, X. (2023). Recent advances and applications of ai-based mathematical modeling in predictive control of hybrid electric vehicle energy management in china. *Electronics*, *12*(2), 445.

[19] Wang, H., Fu, T., Du, Y., Gao, W., Huang, K., Liu, Z., Chandak, P., Liu, S., Van Katwyk, P., Deac, A., & Anandkumar, A. (2023). Scientific discovery in the age of artificial intelligence. *Nature*, *620*(7972), 47–60.

[20] Olabi, A. G., Abdelghafar, A. A., Maghrabie, H. M., Taha Sayed, E., Rezk, H., Radi, M. A., Obaideen, K., & Abdelkareem, M. A. (2023). Application of artificial intelligence for prediction, optimization, and control of thermal energy storage systems. *Thermal Science and Engineering Progress*, *39*, 101730.

[21] Sayed, E. T., Olabi, A. G., Elsaid, K., Radi, M. A., Semeraro, C., Doranehgard, M. H., Eltayeb, M. E., & Abdelkareem, M. A. (2023). Application of artificial intelligence techniques for modeling, optimizing, and controlling desalination systems powered by renewable energy resources. *Journal of Cleaner Production*, *413*, 137486.

[22] Angelis, D., Sofos, F., & Karakasidis, T. E. (2023). Artificial intelligence in physical sciences: Symbolic regression trends and perspectives. *Archives of Computational Methods in Engineering*, *30*(6), 3845–3865.

[23] Ali, A. N. F., Sulaima, M. F., Abdul Razak, I. A. W., Abdul Kadir, A. F., & Mokhlis, H. (2023). Artificial intelligence application in demand response: Advantages, issues, status, and challenges. *IEEE Access*, *11*, 16907–16922.

[24] Dhiman, H., Vyas, N., Pagrotra, A., & Dutt, V. (2023). Real-time traffic analysis and driver safety enhancement via IoT and computer vision with high-resolution project lamps for intelligent road illumination [Internet]. In 2023 1st International Conference on Circuits, Power and Intelligent Systems (CCPIS). IEEE. Available from: http://dx.doi.org/10.1109/CCPIS59145.2023.10291872.

[25] Lilhore, U. K., Manoharan, P., Simaiya, S., Alroobaea, R., Alsafyani, M., Baqasah, A. M., Dalal, S., Sharma, A., & Raahemifar, K. (2023). HIDM: Hybrid intrusion detection model for industry 4.0 Networks using an optimized CNN-LSTM with transfer learning. *Sensors*, *23*(18), 7856.

[26] Dalal, S., Poongodi, M., Lilhore, U. K., Dahan, F., Vaiyapuri, T., Keshta, I., Mesfer Aldossary, S., Mahmoud, A., & Simaiya, S. (2023). Optimized LightGBM model for security and privacy issues in cyber-physical systems. *Transactions on Emerging Telecommunications Technologies*, *34*(no. 6), e4771.

[27] Sharma, N., Gupta, R. D., Khanna, R., Chandmal Sharma, R., & Sharma, Y. K. (2023). Machining of Ti-6Al-4V biomedical alloy by WEDM: Investigation and optimization of MRR and Rz using grey-harmony search. *World Journal of Engineering*, *20*(2), 221–234.

[28] Simaiya, S., Lilhore, U. K., Sharma, Y. K., Rao, K. B. V. B., Maheswara Rao, V. V. R., Baliyan, A., Bijalwan, A., & Alroobaea, R. (2024). A hybrid cloud load balancing and host utilization prediction method using deep learning and optimization techniques. *Scientific Reports*, *14*(1), 1337.

[29] Dahiya, N., Kumar Sharma, Y., Rani, U., Hussain, S., Nabilal, K. V., Mohan, A., & Nuristani, N. (2023). Hyper-parameter tuned deep learning approach for effective human monkeypox disease detection. *Scientific Reports*, *13*(1), 15930.

[30] Gupta, S., Kumar, A., Sharma, Y. K., & Sharma, A. (2020). Multiple performance characteristics optimization of WEDM by grey grade for EN-8. *Optimization in Engineering Research*, *1*(no. 1), 27–35.

[31] Kumar Lilhore, U., Simaiya, S., Kumar Sharma, Y., Kaswan, K. S., Brahma Rao, K. B. V., Maheswara Rao, V. V. R., Baliyan, A., Bijalwan, A., & Alroobaea, R. (2024). A precise model for skin cancer diagnosis using hybrid U-Net and improved MobileNet-V3 with hyperparameters optimization. *Scientific Reports*, *14*(no. 1), 4299.

[32] Rokade, M. D., & Sharma, Y. K. (2021). MLIDS: A machine learning approach for intrusion detection for real time network dataset. In 2021 International conference on emerging smart computing and informatics (ESCI) (pp. 533–536). IEEE.

[33] Sharma, Y., & Singh, B. K. (2020). Classification of children with specific language impairment using pitch-based parameters. In 2020 IEEE Recent Advances in Intelligent Computational Systems (RAICS) (pp. 42–46). IEEE, .

[34] Sharma, Y., & Singh, B. K. (2023). Classification of children with attention-deficit hyperactivity disorder using wigner-ville time-frequency and deep expEEGNetwork feature-based computational models. *IEEE Transactions on Medical Robotics and Bionics*.

[35] Zharmagambetov, A., Amos, B., Ferber, A., Huang, T., Dilkina, B., & Tian, Y. (2024). Landscape surrogate: Learning decision losses for mathematical optimization under partial information. *Advances in Neural Information Processing Systems, 36*.

[36] Liu, Y.-F., Chang, T.-H., Hong, M., Wu, Z., Man-Cho So, A., Jorswieck, E. A., & Yu, W. (2024). A survey of advances in optimization methods for wireless communication system design. *arXiv Preprint arXiv:2401.12025*.

[37] Carrizosa, E., Ramírez-Ayerbe, J., & Romero Morales, D. (2024). Generating collective counterfactual explanations in score-based classification via mathematical optimization. *Expert Systems with Applications*, *238*, 121954.

[38] Talatahari, S., & Azizi, M. (2020). Optimization of constrained mathematical and engineering design problems using chaos game optimization. *Computers & Industrial Engineering*, *145*, 106560.

[39] Hauswirth, A., Zhiyu, H., Bolognani, S., Hug, G., & Dörfler, F. (2024). Optimization algorithms as robust feedback controllers. *Annual Reviews in Control*, *57*, 100941.

[40] Gutiérrez, M. (2021). Making better decisions by applying mathematical optimization to cost accounting: An advanced approach to multi-level contribution margin accounting. *Heliyon*, *7*(no. 2).

Umesh Kumar Lilhore, Sarita Simaiya, and Surjeet Dalal

10 Hybrid Mathematical Optimization Techniques in AI

Abstract: This chapter covers hybrid mathematical optimization techniques in artificial intelligence (AI) to overcome local minima and slow convergence. Hybrid methods improve performance and robustness by combining multiple optimization strategies. The chapter divides hybridization into serial, parallel, and adaptive methods. Serial hybridization uses genetic algorithms (GA) for broad exploration and gradient descent for fine-tuning. Multiple algorithms run simultaneously in parallel hybridization, improving search diversity and robustness. Problem characteristics drive adaptive hybridization strategy adjustments. GA with gradient descent for neural network training, PSO with simulated annealing (SA) for complex optimization, and DE with local search for enhanced exploration and exploitation are examined. Practical applications show how hybrid techniques improve neural network training, hyperparameter optimization, feature selection, and robotics path planning. These methods work in real life, as shown by case studies. Finally, the chapter emphasizes hybrid optimization in AI and suggests future research on developing more adaptive algorithms and integrating them with deep learning to create robust and efficient solutions.

Keywords: Hybrid mathematical optimization techniques, artificial intelligence, genetic algorithms, PSO, simulated annealing

10.1 Introduction

Since mathematical models are essential to solving problems, they are found in practically all branches of science. These models offer a condensed depiction of reality based on mathematical expressions. These formulations facilitate problem-solving, the understanding of complex systems, and the acquisition of crucial data to enable wise decision-making. Algorithms are used by mathematical models to determine the best solution for the given problem [1]. The goal of the well-known mathematical science field of numerical optimization is to pinpoint a function's extreme points, whether they are its maximum or minimum. Over the past 20 years, optimization techniques have grown in importance as a tool for management, decision-making, technology advancement, and development. They give different systems a competitive edge [2].

Artificial intelligence (AI) has changed many fields, such as healthcare, entertainment, finance, and transportation. Among the best things about AI is that it can do

https://doi.org/10.1515/9783111436180-010

things that humans are good at, like understanding language, using logic, learning new things, and solving problems. Optimization is a critical mathematical tool for decision-making and determining the best solutions within a given set of constraints and requirements. AI optimization entails selecting the most efficient solution from a set of available options. Conventional optimization methods such as simulated annealing (SA), genetic algorithms (GA), and gradient descent have been critical in resolving a wide range of AI challenges [3].

Gradient descent is commonly employed to locate local minima, especially in the training of neural networks, owing to its simplicity and efficacy. SA, a technique inspired by metallurgical annealing, is highly efficient in circumventing local minima while approximating global optima. GAs, which are based on natural selection, are ideal for solving global search problems. Despite their benefits, traditional methods frequently encounter significant challenges when dealing with the complexities of modern AI applications. Nonconvex functions, large search spaces, constraints, and the presence of multiple local minima are all common challenges. Moreover, the practical effectiveness of these methods is limited by their tendency to converge at a slow pace or get trapped in suboptimal solutions [4].

Hybrid mathematical optimization techniques have emerged as a potent alternative to address these obstacles. Hybrid optimization techniques combine two or more optimization methods, utilizing their individual strengths to compensate for their weaknesses. The primary objective of hybridization is to enhance the robustness, accuracy, and convergence speed of the optimization process. Serial hybridization, parallel hybridization, and adaptive hybridization are the three primary types of hybrid optimization techniques [5].

Significantly, actual problems are progressively intricate and necessitate advanced models and algorithms that can efficiently handle extensive data sets and identify optimal solutions. Nevertheless, it is important to note that no method or algorithm can be considered flawless, as they all possess certain limitations. However, these limitations can be reduced or eradicated by integrating the expertise of various methodologies. The objective is to create hybrid algorithms that can effectively combine the strengths and unique characteristics of optimization and machine learning methods, resulting in improved efficiency [6, 7].

This chapter presents hybrid mathematical optimization techniques in AI to overcome local minima and slow convergence. Hybrid methods improve performance and robustness by combining multiple optimization strategies. The chapter is organized into various sections, which cover the review of existing methods, the importance of hybrid optimization, types, methodologies of hybrid optimization, applications, future directions, and, lastly, the conclusion.

10.2 Review of Existing Work

A hybrid mathematical optimization technique plays a vital role in various sectors, which attracts reviewers to work in this field. The analyses of key contributions from various previous authors in this field are as follows.

Abaku et al. [7] explore the incorporation of AI and optimization methods into supply chain management, emphasizing its capacity to improve efficiency and resilience. The authors also offer a comprehensive analysis of how AI contributes to optimizing operations and minimizing disruptions, presenting both theoretical frameworks and practical strategies. This work's primary contribution lies in its investigation of the transformative potential of AI in enhancing supply chain performance. The benefits encompass heightened operational efficacy, diminished expenses, and improved flexibility in response to market fluctuations, with applications spanning from inventory control to the optimization of logistics. Future directions to enhance supply chain responsiveness and agility include the implementation of more advanced AI models and real-time data analytics.

In their publication, Siva Shankar et al. [8] introduce a groundbreaking optimization-based deep learning technique for effectively detecting intrusion attacks in network systems. They integrate deep learning and advanced optimization to improve the accuracy and reliability of intrusion detection systems. A powerful cybersecurity solution is the innovative combination of optimization and deep learning methods. This technology reduces false positives and improves detection. Cyber defense and network security are its main uses. Future research could combine this methodology with other AI technologies to create more comprehensive security solutions.

Ala et al. [9] studied how Internet of things (IoT)-based smart healthcare systems can improve patient information performance using AI and optimization. These advanced methods improve data management and system performance, making healthcare more reliable and effective. The main innovation is using optimization and AI to improve healthcare data management. Improved system agility and data precision may benefit patient monitoring, data interpretation, and healthcare supply chain management. In the future, this strategy will be expanded to include more IoT devices, and predictive analytics will be used to improve patient outcomes.

Thirunavukkarasu et al. [10] provide a comprehensive analysis of the optimization techniques applied in hybrid renewable energy systems in their study. The authors examine various optimization methods, assessing their usefulness and effectiveness in improving sustainability and performance. The main contribution of this study is a comprehensive examination of optimization strategies for energy systems, which offers benefits such as improved efficiency and decreased expenses. Applications encompass a wide range of wind, solar, and hybrid energy systems. Subsequent studies may concentrate on the integration of real-time data analytics and AI to achieve dynamic energy optimization.

Zhang et al. [11] examine the latest progress and uses of AI-driven mathematical modeling in the predictive control of energy management for hybrid electric vehicles.

Their research highlights the crucial role of AI in improving the effectiveness and predictability of energy management systems in hybrid electric vehicles, resulting in enhanced performance and sustainability. The primary contribution lies in the utilization of AI for the purpose of managing vehicle energy, resulting in enhanced fuel efficiency and decreased emissions. The main focus of applications lies in hybrid and electric vehicles, with potential future advancements involving the development of more advanced AI models and the integration of renewable energy sources.

Palle [12] explores evolutionary optimization techniques and their utilization in addressing AI optimization problems. The study emphasizes the adaptability and efficiency of evolutionary algorithms in solving intricate optimization problems, offering valuable insight into their real-world applications in AI. The main contribution lies in the thorough investigation of evolutionary techniques, which offer benefits such as resilient performance and adaptability. Applications encompass a wide range of AI-powered optimization tasks, and future research may concentrate on integrating these techniques with other AI methods to enhance their effectiveness.

Maghami and Mutambara [13] discuss the difficulties linked to hybrid energy systems and put forward solutions based on AI. The researchers investigate the potential of AI to address the inherent challenges of hybrid energy systems, thereby enhancing their efficiency and seamless integration into the energy grid. The primary contribution lies in the identification of AI solutions for hybrid energy challenges, which offer notable benefits such as improved energy efficiency and reliability. Applications encompass smart grids and the management of renewable energy, with future trajectories involving a more comprehensive integration of AI and the processing of data in real-time.

Nutakki and Mandava [14] critically examine optimization methodologies and the significance of AI in the context of home energy management systems. The authors explore different AI-based optimization techniques that can improve the efficiency and effectiveness of energy management in residential environments, thereby promoting smarter and more sustainable solutions for home energy. The main contribution lies in the comprehensive analysis of AI in the context of home energy management. This analysis highlights the benefits of AI, such as energy conservation and improved operational efficiency of the system. The main focus of applications lies in smart home energy systems, while future research aims to investigate more sophisticated AI techniques and their incorporation into smart home technologies.

In their publication, Kushwaha et al. [15] investigate the process of carbon dioxide fixation by microalgae. They employ a combination of AI and fuzzy logic techniques, along with GAs for optimization. Their research showcases the utilization of AI, fuzzy logic, and GAs to enhance biological processes, providing inventive resolutions for environmental challenges. The main contribution lies in the hybrid approach to carbon dioxide (CO_2) fixation, which offers benefits such as improved process efficiency and decreased environmental impact. The applications of this research are focused on environmental engineering and bioprocess optimization. In the future, the aim is to scale up the approach and incorporate additional environmental data.

Uzer and Inan [16] utilize an enhanced hybrid whale optimization algorithm to solve a range of optimization problems. Their research shows that this hybrid algorithm greatly improves optimization performance, making it a valuable tool for solving intricate computational problems. The primary contribution lies in the creation of an enhanced optimization algorithm, which offers benefits such as heightened accuracy and efficiency in the optimization process. The applications of this algorithm span from engineering design to machine learning, and there is potential for future research to enhance the algorithm even more and apply it to a wider range of problems.

Onyelowe et al. [17] investigate specific AI optimization methods and their practical uses in the field of geotechnical engineering. The authors evaluate different AI-driven optimization techniques, demonstrating their efficacy in tackling engineering obstacles and enhancing the design and efficiency of geotechnical systems. The primary contribution lies in the utilization of AI in the field of geotechnical engineering, resulting in enhanced engineering designs and increased efficiency in construction processes. The applications encompass soil analysis, foundation design, and various geotechnical tasks. The future plans include incorporating more sophisticated AI models and broadening the scope to include other engineering disciplines.

10.3 Hybrid Optimization Techniques

Hybrid optimization strategies take advantage of the benefits and minimize the drawbacks of two or more optimization techniques. The benefits of hybridization include increased robustness, accuracy, and speed of convergence. Hybrid AI is a notable breakthrough in the field of AI as it combines two separate approaches: rule-based and machine learning-based techniques. This innovative combination enhances the capability and efficiency of AI systems, surpassing the restrictions of individual rule-based or learning-based approaches. Table 10.1 presents a comparative analysis of types of hybrid optimization techniques.

Table 10.1: Comparative analysis of types of hybrid optimization techniques.

Hybridization technique	Description	Example	Advantages	Challenges
Serial hybridization	Optimization methods are applied sequentially.	A genetic algorithm (GA) explores the search space broadly, followed by gradient descent for fine-tuning the solution.	Combines the strengths of multiple methods; GA finds global solutions, gradient descent fine-tunes.	Can be time-consuming; requires careful method selection.

Table 10.1 (continued)

Hybridization technique	Description	Example	Advantages	Challenges
Parallel hybridization	Multiple optimization algorithms run simultaneously, sharing information between them.	Genetic algorithm and particle swarm optimization running in parallel, exchanging best solutions.	Diversifies the search process; helps in escaping local minima.	High computational cost; complexity in managing parallel processes.
Adaptive hybridization	Dynamically adjusts the optimization strategy based on the problem's characteristics and the progress of the search process.	Switching between genetic algorithm and simulated annealing based on search performance.	Adapts to problem dynamics; potentially more efficient solutions.	Requires sophisticated control mechanisms; high complexity.

10.3.1 Methodologies of Hybrid Optimization

Hybrid optimizations involve the dynamic selection of an optimization algorithm at compile time from a collection of different algorithms that achieve the same optimization. They employ a heuristic method to forecast the most suitable algorithm for optimizing each piece of code [8]. Different optimization methods and algorithms can be used in optimization to identify the solution to the problem. Some optimization techniques show different behavior in the search for the solution to the same problem; thus, one method can be faster and more dependable in response accuracy than others. Table 10.2 presents a comparative analysis of methodologies in hybridization optimization, and Figure 10.1 presents the flowchart of hybrid optimization methods.

Many algorithms nowadays are developed with different strengths and weaknesses, none of which is the best for every situation. Although they are rather general ideas with many uses, separate and combined, optimization and deep learning [11]. Table 10.3 presents a comparative analysis of key applications of hybrid optimization applications in AI.

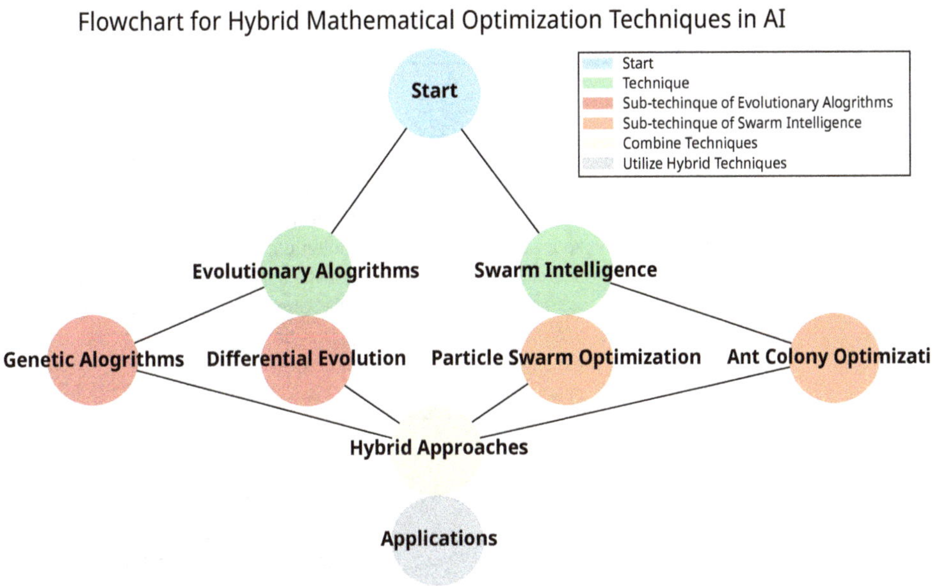

Figure 10.1: The flowchart of hybrid optimization methods.

Table 10.2: Comparative analysis of methodologies in hybridization optimization.

Hybridization technique	Description	Example	Advantages	Challenges
Genetic algorithm and gradient descent [8]	Combining GA with gradient descent is effective, as GA provides a good initial solution, and gradient descent refines it to a local optimum.	Useful in neural network training, where GA finds a good starting point and gradient descent fine-tunes the weights.	Combines the global search capability of GA with the fine-tuning ability of gradient descent; results in better local optima.	Requires careful parameter tuning; GA can be computationally expensive.
Particle swarm optimization and simulated annealing [9]	PSO is good at global search but may get stuck in local optima. Combining PSO with SA, which has a better capability of escaping local optima, can yield better results.	Suitable for complex optimization problems where PSO finds promising areas, and SA refines the solution by escaping local minima.	Enhances the ability to escape local minima; combines the global exploration of PSO with the local refinement of SA.	High computational cost; requires managing interactions between PSO and SA.

Table 10.2 (continued)

Hybridization technique	Description	Example	Advantages	Challenges
Differential evolution and local search [10]	DE is effective for global optimization, while local search techniques can exploit the neighborhood of the current best solution.	Used in scenarios where DE explores the global search space, and local search techniques enhance solution accuracy by exploiting nearby solutions.	Balances global exploration of DE with local exploitation; improves both exploration and exploitation capabilities.	Complexity in coordinating DE with local search methods; may require significant computational resources.

Table 10.3: Comparative analysis of key applications of hybrid optimization in AI.

Application area	Description	Example	Advantages	Challenges
Neural network training [12]	Hybrid optimization techniques improve the convergence rate and accuracy of training deep neural networks.	Hybrid GA and gradient descent algorithms for neural network weight optimization.	Combines global search (GA) with local refinement (gradient descent); enhances training efficiency and model accuracy.	Computational complexity; requires careful parameter tuning.
Hyperparameter optimization	Optimizing hyperparameters is crucial for machine learning models. Hybrid methods can efficiently search the hyperparameter space and improve model performance.	Bayesian optimization combined with evolutionary algorithms for hyperparameter tuning.	Efficiently explores large hyperparameter spaces; balances exploration and exploitation; improves model performance and generalization.	High computational cost; complexity in combining different optimization methods.
Feature selection [17]	Identifying the most relevant features for a machine learning model helps reduce dimensionality and enhance model accuracy.	PSO combined with local search for feature selection in high-dimensional data.	Effectively reduces dimensionality; enhances model accuracy and interpretability; combines global search (PSO) with fine-tuning (local search).	May require significant computational resources; managing trade-offs between feature relevance and model complexity.

Table 10.3 (continued)

Application area	Description	Example	Advantages	Challenges
Robotics and path planning [18]	Involves finding the optimal route for a robot to follow, efficiently handling the high-dimensional and dynamic nature of path planning problems.	Hybrid optimization techniques for robotic path planning in dynamic environments.	Efficiently handles high-dimensional and dynamic path planning problems; combines multiple optimization strategies for better solution quality.	High computational complexity; real-time application challenges.

10.4 Case Studies

This section covers the key case studies related to the research [19–22]. Table 10.4 presents a comparative analysis of key case studies in the field of hybrid optimization methods.

10.5 Future Directions

A comparative analysis of the major future directions in the field of hybrid optimization investigation utilizing AI is shown in Table 10.5. These future directions present how hybrid optimization techniques are developing in AI, emphasizing areas where developments can have a major positive impact on efficiency, ethical issues, and AI applications [14–27].

Table 10.4: Comparative analysis of key case studies in the field of hybrid optimization methods.

Case study	Description	Hybrid techniques used	Application area	Key contributions	Advantages	Challenges	Future directions
Neural network training [21, 23]	Improving convergence rate and accuracy in training deep neural networks.	Genetic algorithm and gradient descent	Neural network training	Enhanced training efficiency and model accuracy through hybrid GA and gradient descent methods.	Combines the global search of GA with the local refinement of gradient descent; faster convergence and better accuracy.	High computational complexity; requires careful parameter tuning.	Explore hybridization with other optimization techniques; reduce computational overhead.
Hyperparameter optimization [24–27]	Efficiently searching the hyperparameter space to improve machine learning model performance.	Bayesian optimization and evolutionary algorithms	Hyperparameter optimization	Efficient exploration and exploitation of hyperparameter space leading to improved model performance.	Balances exploration and exploitation; improves model generalization and performance.	High computational cost; complexity in combining optimization methods.	Integration with automated machine learning (AutoML) frameworks; scalability for large-scale applications.
Feature selection [28–30]	Identifying the most relevant features to reduce dimensionality and enhance model accuracy.	Particle swarm optimization and local search	Feature selection	Effective reduction of dimensionality; enhanced model accuracy and interpretability.	Combines global search of PSO with local refinement of search techniques; improved model performance.	Managing trade-offs between feature relevance and model complexity; computational resources.	Development of more efficient hybrid algorithms; application to various data types and domains.

Robotics and path planning [31]	Finding optimal routes for robots in high-dimensional and dynamic environments.	Various hybrid optimization techniques	Robotics and path planning	Improved path planning efficiency and accuracy in dynamic and complex environments.	Efficient handling of high-dimensional and dynamic optimization problems; better solution quality.	High computational complexity; real-time application challenges.	Real-time hybrid optimization methods; application to autonomous systems and robotics.
Healthcare system optimization [27]	Enhancing patient information performance in IoT-based smart healthcare systems.	Hybrid AI and optimization approaches	Smart healthcare systems	Improved patient information management and system performance through hybrid approaches.	Enhanced system performance; better management of healthcare data.	Integration with existing healthcare infrastructure; ensuring data privacy and security.	Further research on real-time data processing and integration with advanced AI techniques.
Intrusion detection in network systems [14]	Detecting intrusion attacks in network systems using advanced optimization and AI techniques.	Optimization-based deep learning and AI approaches	Network security	Enhanced detection accuracy and system robustness through hybrid deep learning and optimization techniques.	Improved security and robustness against intrusion attacks; better detection accuracy.	High computational cost; managing large-scale network data.	Development of real-time detection systems; integration with evolving network security threats.
Hybrid renewable energy systems optimization [17]	Optimizing the performance and efficiency of hybrid renewable energy systems using various optimization techniques.	Various optimization techniques	Renewable energy systems	Improved system efficiency and energy management through hybrid optimization methods.	Better performance and efficiency of renewable energy systems; effective energy management.	Complex integration of different energy sources, ensuring system stability.	Research on sustainable and scalable hybrid optimization methods; integration with smart grid technologies.

(continued)

Table 10.4 (continued)

Case study	Description	Hybrid techniques used	Application area	Key contributions	Advantages	Challenges	Future directions
Predictive control in hybrid electric vehicle energy management [15]	Using AI-based mathematical modeling for predictive control of energy management in hybrid electric vehicles.	AI-based mathematical modeling and optimization techniques	Hybrid electric vehicle energy management	Enhanced predictive control and energy management; improved vehicle performance and efficiency.	Better energy management and vehicle performance; efficient predictive control methods.	Managing complex vehicle dynamics; integration with real-time control systems.	Exploration of more advanced hybrid techniques; application to other types of vehicles and energy systems.

Table 10.5: Comparative analysis of key future directions in the field of hybrid optimization research using AI.

Future directions	Description	Importance
Integration with explainable AI (XAI) [15]	Incorporating transparency and interpretability into hybrid models to enhance trust and understanding.	Ensures AI decisions are explainable and accountable; crucial for applications in critical domains.
Real-time optimization [18, 22]	Developing algorithms capable of optimizing systems in real time, adapting to dynamic changes.	Enables applications in autonomous vehicles, healthcare, and finance, where real-time decision-making is critical.
Scalability and efficiency [24]	Enhancing scalability to handle large-scale problems efficiently, minimizing computational costs.	Facilitates deployment in big data analytics, smart cities, and industrial automation with large datasets.
Robustness and resilience [27–29]	Improving robustness to handle uncertainties and adversarial attacks, ensuring reliable performance.	Essential for cybersecurity, autonomous systems, and critical infrastructure to withstand diverse challenges.
Hybridization with quantum computing [28, 31]	Exploring synergies between hybrid optimization and quantum computing for enhanced computational power.	Promises breakthroughs in solving complex optimization problems faster than classical methods allow.
Ethical AI considerations [24]	Integrating ethical principles into hybrid optimization techniques to ensure fairness and equity.	Addresses concerns related to bias, privacy, and the societal impacts of AI applications.
Automation and AutoML integration [23]	Automating the hybridization process and integrating with AutoML frameworks for streamlined deployment.	Reduces the complexity of model development and optimization, making AI accessible to a broader range of applications.
Cross-disciplinary applications [29]	Applying hybrid techniques across diverse fields like medicine, climate science, and social sciences.	Unlocks new possibilities for solving multidimensional problems and advancing scientific discoveries.
Human-AI collaboration [30]	Enhancing interfaces for collaborative problem-solving, where AI supports human decision-making.	Promotes synergy between AI capabilities and human expertise, leveraging AI for augmented intelligence.

10.6 Conclusion

We have explored the revolutionary field of hybrid mathematical optimization techniques in this chapter, which are powerful allies when dealing with challenging AI problems. Through the integration of distinct advantages from different optimization approaches, these strategies not only improve performance but also strengthen resilience and improve the overall efficiency of operations. This chapter conducts a comprehensive literature review concerning hybrid techniques that combine optimization and AI methods.

This comprehensive inquiry has meticulously examined fundamental concepts, methodological intricacies, diverse applications across different fields, and insightful glimpses into future prospects, providing a thorough comprehension of the substantial impact and promising future of hybrid mathematical optimization in the realm of AI. Rapid advancements in the field of artificial intelligence are driving the need for hybrid optimization techniques, which are essential for exploring new possibilities and facilitating significant advancements.

References

[1] Azevedo, B. F., Rocha, A. M. A., & Pereira, A. I. (2024). Hybrid approaches to optimization and machine learning methods: A systematic literature review. *Machine Learning*, 1–43.

[2] Saleem, S., Ahmad, I., Ahmed, S. H., & Rehman, A. (2024). Artificial intelligence based robust nonlinear controllers optimized by improved gray wolf optimization algorithm for plug-in hybrid electric vehicles in grid to vehicle applications. *Journal of Energy Storage*, *75*, 109332.

[3] Borode, A., & Olubambi, P. (2024). Optimisation of artificial intelligence models and response surface methodology for predicting viscosity and relative viscosity of GNP-alumina hybrid nanofluid: Incorporating the effects of mixing ratio and temperature. *The Journal of Supercomputing*, *80*(4), 4841–4869.

[4] Zhuang, X., Wang, W., Su, Y., Yan, B., Li, Y., Li, L., & Hao, Y. (2024). Multi-objective optimization of reservoir development strategy with hybrid artificial intelligence method. *Expert Systems with Applications*, *241*, 122707.

[5] Tao, H., Abba, S. I., Al-Areeq, A. M., Tangang, F., Samantaray, S., Sahoo, A., & Yaseen, Z. M. (2024). Hybridized artificial intelligence models with nature-inspired algorithms for river flow modeling: A comprehensive review, assessment, and possible future research directions. *Engineering Applications of Artificial Intelligence*, *129*, 107559.

[6] Qiao, L., Liu, K., Xue, Y., Tang, W., & Salehnia, T. (2024). A multi-level thresholding image segmentation method using hybrid arithmetic optimization and Harris Hawks optimizer algorithms. *Expert Systems with Applications*, *241*, 122316.

[7] Abaku, E. A., Edunjobi, T. E., & Odimarha, A. C. (2024). Theoretical approaches to AI in supply chain optimization: Pathways to efficiency and resilience. *International Journal of Science and Technology Research Archive*, *6*(1), 092–107.

[8] Siva Shankar, S., Hung, B. T., Chakrabarti, P., Chakrabarti, T., & Parasa, G. (2024). A novel optimization based deep learning with artificial intelligence approach to detect intrusion attack in network system. *Education and Information Technologies*, *29*(4), 3859–3883.

[9] Ala, A., Simic, V., Pamucar, D., & Bacanin, N. (2024). Enhancing patient information performance in internet of things-based smart healthcare system: Hybrid artificial intelligence and optimization approaches. *Engineering Applications of Artificial Intelligence*, *131*, 107889.

[10] Thirunavukkarasu, M., Sawle, Y., & Lala, H. (2023). A comprehensive review on optimization of hybrid renewable energy systems using various optimization techniques. *Renewable and Sustainable Energy Reviews*, *176*, 113192.

[11] Zhang, Q., Tian, S., & Lin, X. (2023). Recent advances and applications of ai-based mathematical modeling in predictive control of hybrid electric vehicle energy management in china. *Electronics*, *12*(2), 445.

[12] Palle, R. R. (2023). Evolutionary optimization techniques in ai: investigating evolutionary optimization techniques and their application in solving optimization problems in AI. *Journal of Artificial Intelligence Research*, *3*(1), 1–13.

[13] Maghami, M. R., & Mutambara, A. G. O. (2023). Challenges associated with hybrid energy systems: An artificial intelligence solution. *Energy Reports*, *9*, 924–940.

[14] Nutakki, M., & Mandava, S. (2023). Review on optimization techniques and role of Artificial Intelligence in home energy management systems. *Engineering Applications of Artificial Intelligence*, *119*, 105721.

[15] Kushwaha, O. S., Uthayakumar, H., & Kumaresan, K. (2023). Modeling of carbon dioxide fixation by microalgae using hybrid artificial intelligence (AI) and fuzzy logic (FL) methods and optimization by genetic algorithm (GA). *Environmental Science and Pollution Research*, *30*(10), 24927–24948.

[16] Uzer, M. S., & Inan, O. (2023). Application of improved hybrid whale optimization algorithm to optimization problems. *Neural Computing & Applications*, *35*(17), 12433–12451.

[17] Onyelowe, K. C., Mojtahedi, F. F., Ebid, A. M., Rezaei, A., Osinubi, K. J., Eberemu, A. O., & Rehman, Z. U. (2023). Selected AI optimization techniques and applications in geotechnical engineering. *Cogent Engineering*, *10*(1), 2153419.

[18] Onyelowe, K. C., Mojtahedi, F. F., Ebid, A. M., Rezaei, A., Osinubi, K. J., Eberemu, A. O., & Rehman, Z. U. (2023). Selected AI optimization techniques and applications in geotechnical engineering. *Cogent Engineering*, *10*(1), 2153419.

[19] Dhiman, H., Vyas, N., Pagrotra, A., & Dutt, V. (2023). Real-Time Traffic Analysis and Driver Safety Enhancement via IoT and Computer Vision with High-Resolution Project Lamps for Intelligent Road Illumination [Internet]. In 1st International Conference on Circuits, Power and Intelligent Systems (CCPIS), IEEE, 2023. Available from. http://dx.doi.org/10.1109/CCPIS59145.2023.10291872.

[20] Thirunavukkarasu, M., Sawle, Y., & Lala, H. (2023). A comprehensive review on optimization of hybrid renewable energy systems using various optimization techniques. *Renewable and Sustainable Energy Reviews*, *176*, 113192.

[21] Abaku, E. A., Edunjobi, T. E., & Odimarha, A. C. (2024). Theoretical approaches to AI in supply chain optimization: Pathways to efficiency and resilience. *International Journal of Science and Technology Research Archive*, *6*(1), 092–107.

[22] Lagrange, J. B., Richard, P. R., Vélez, M. P., & Van Vaerenbergh, S. (2024). Artificial intelligence techniques in software design for mathematics education. In Handbook of digital resources in mathematics education (pp. 969–999). Springer International Publishing, Cham.

[23] Wang, F., Sohail, A., Wong, W. K., Ul Ain Azim, Q., Farwa, S., & Sajad, M. (2023). Artificial intelligence and stochastic optimization algorithms for the chaotic datasets. *Fractals*, *31*(6).

[24] Simaiya, S., Lilhore, U. K., Sharma, Y. K., Rao, K. B., Maheswara Rao, V. V. R., Baliyan, A., & Alroobaea, R. (2024). A hybrid cloud load balancing and host utilization prediction method using deep learning and optimization techniques. *Scientific Reports*, *14*(1), 1337.

[25] Dalal, S., Lilhore, U. K., Faujdar, N., Samiya, S., Jaglan, V., Alroobaea, R., & Ahmad, F. (2024). Optimising air quality prediction in smart cities with hybrid particle swarm optimization-long-short term memory-recurrent neural network model. *IET Smart Cities*.

[26] Kumar Lilhore, U., Simaiya, S., Sharma, Y. K., Kaswan, K. S., Rao, K. B., Rao, V. M., & Alroobaea, R. (2024). A precise model for skin cancer diagnosis using hybrid U-Net and improved MobileNet-V3 with hyperparameters optimization. *Scientific Reports*, *14*(1), 4299.

[27] Sumit, C., S., R., Dalal, S., Dalal, S., Lilhore, U. K., & Samiya, S. (2024). A dynamic and optimized routing approach for VANET communication in smart cities to secure intelligent transportation system via a chaotic multi-verse optimization algorithm. *Cluster Computing*, 1–26.

[28] Kukkar, A., Lilhore, U. K., Frnda, J., Sandhu, J. K., Das, R. P., Goyal, N., & Rezac, F. (2023). ProRE: An ACO-based programmer recommendation model to precisely manage software bugs. *Journal of King Saud University-Computer and Information Sciences*, *35*(1), 483–498.

[29] Darolia, A., Chhillar, R. S., Alhussein, M., Dalal, S., Aurangzeb, K., & Lilhore, U. K. (2024). Enhanced cardiovascular disease prediction through self-improved Aquila optimized feature selection in quantum neural network & LSTM model. *Frontiers in Medicine*, *11*, 1414637.

[30] Al-Ani, A. K., Ul Arfeen Laghari, S., Manoharan, H., Selvarajan, S., & Uddin, M. (2023). Improved transportation model with internet of things using artificial intelligence algorithm. *Computers, Materials and Continua*, *76*(2).

[31] Lilhore, U. K., Manoharan, P., Sandhu, J. K., Simaiya, S., Dalal, S., Baqasah, A. M., & Raahemifar, K. (2023). Hybrid model for precise hepatitis-C classification using improved random forest and SVM method. *Scientific Reports*, *13*(1), 12473.

Yash Mahajan, Muskan Sharma, Alba Meça, Narayan Vyas, and Rakesh Sharma

11 Mathematical Optimization for Enhanced AI-Enabled Geospatial Intelligence

Abstract: In the era of big data, the integration of artificial intelligence (AI) with geospatial intelligence (GEOINT) has revolutionized the way we analyze, interpret, and utilize geospatial data. This chapter delves into the application of mathematical optimization techniques to enhance AI-enabled GEOINT, offering both theoretical insights and practical implementations. By combining optimization algorithms with AI models, this chapter demonstrates how these advanced techniques can solve complex geospatial problems, such as route optimization, disaster response, and urban planning. Besides the theoretical foundations of mathematical optimization and its impacts on AI (in a separate chapter), we cover practical use-cases too, with hands-on examples that show how these tools can be applied in the real world. The important part of practical implementation guides them by methods for the anticipation, running, and execution subsequent to combining information, model building, and previously official stages in developing this as a chapter that is of great benefit for investigators, aspirants, and those studying from sizable spheres, AI analysts or practitioners, plus scholars in neuroscience. The modern status of the field is described, and some challenges as well as future trends are presented, thereby underlining that this interdisciplinary scientific domain has not yet come to a standstill.

Keywords: Geospatial intelligence (GEOINT), artificial intelligence (AI), mathematical optimization, machine learning, deep learning, route optimization, disaster response, urban planning, GIS (geographic information system), data integration

11.1 Introduction

11.1.1 Overview of Geospatial Intelligence (GEOINT)

One way to interpret geospatial intelligence (GEOINT) is as a field of multidisciplinary research that seeks to understand the physical and human landscapes we see mapped at different scale levels all across our planet Earth. Geospatial data is the approach to retrieving and handling spatial (the earth's surface), natural/built environment, individual/organizational activities/events, and behavior information developed. GEOINT is much more than [1] maps and satellite images; it is defined as an integrated process of collecting, analyzing, exploiting, producing, and disseminating relevant geographical in-

https://doi.org/10.1515/9783111436180-011

formation to a specific audience for the purpose of understanding rational decision-making in defense or other sectors.

GEOINT has applications in national security, disaster relief and recovery management, urban planning, environmental conservation/monitoring endless possibilities. For national security, GEOINT is used to monitor borders and the arrival of potential threats, as well as to plan military operations. In disaster management, it can [2] predict the consequences of natural disasters and utilize satellite remote sensing for planning evacuation routes, as well as coordinating relief efforts. Urban planners employ GEOINT to inform the development of more efficient, sustainable, and resilient cities; environmentalists use it for ongoing monitoring of deforestation patterns, assessment of wildlife populations in parks such as Yellowstone National Park or Kruger National Park, tracking food insecurity trends around Sudan and Ethiopia, and assessing climate change impacts on ecosystems in Greenland.

Indeed, preparing raw geospatial data for common use and making it actionable is one of the key strengths behind GEOINT. It uses different data sources such as satellite imagery, aerial photography, GIS (geographic information system), remote sensing [3], and ground-based observation. By integrating these sources, GEOINT practitioners enable the creation of comprehensive and exact maps, models, and simulations, which are necessary for decision-making. These are the outputs that help us visualize spatial patterns, identify trends over time, and forecast future scenarios, providing a robust platform for strategic planning as well as operational execution.

Over the last few years, GEOINT tools have improved dramatically with the development of artificial intelligence (AI). In particular, AI [4] techniques used in machine learning (ML) and deep learning have changed the landscape of how geospatial data is analyzed and understood. State-of-the-art AI algorithms can quickly process vast amounts of geospatial data, uncovering patterns and relationships that would be infeasible to detect by hand. ML algorithms can automatically identify land types based on satellite imagery, predict in real-time when an anomaly is occurring, and calculate how a wildfire or the spread of disease develops, for example. Conversely, you can train deep learning models to distinguish fine-grained objects in images with relatively high accuracy.

The integration of AI into GEOINT has not only widened the GEOINT paradigm and increased geographical big data analysis, but it has also brought new fluidity into real-time decision-making and predictive capabilities. GEOINT enhanced by AI helps organizations to identify [5] threats and opportunities in time, a concept that is crucial to the modern world, where events change rapidly. Furthermore, AI-integrated GEOINT is used to sort through simple informational challenges like collection and preprocessing, allowing more qualified analysts to deal with more intricate problems.

Given the ever-expanding proliferation of volume, velocity, and variety of geospatial data, AI is likely to assume a more significant role in GEOINT. The further evolution of AI, including the application of generative models and reinforcement learning (RL), is also expected to improve future GEOINT capabilities and reach more sophisticated levels of analysis and prediction. In this light, it is pertinent to note that knowledge of the function of AI in GEOINT is of significant value to anybody that is affiliated with the discipline in any way, form, or fashion.

11.1.2 Importance of Mathematical Optimization in AI-Enabled GEOINT

Optimization is a rather strong method whose application is critical in making AI GEOINT more useful. Fundamentally, optimization consists of the identification of the most preferred solution out of a range [6] of viable choices, which are bounded by the objectives and constraints at hand. This process is preeminent in GEOINT as the geometry of entities and the nature of information often compel the use of particular mathematical models to facilitate timely decision-making.

In the case of AI-based GEOINT, the large number of problems that fall under the optimization domain involves the distribution of resources, path planning, and problems associated with data integration and pattern knowledge. For instance, in disaster response scenarios, optimization algorithms can be applied to establish the best path for EMS to follow by considering factors such as road infrastructure, congestion, and the intensity of the disaster. This not only assists in quick response but also allows the use of scarce resources as efficiently as possible, thus saving lives and confining the damage [7].

In urban planning, they are employed in designing transport systems, developing land use, and undertaking infrastructural investments to ensure that they get the most from consumers, or in a more general sense, outcome generators at the lowest feasible cost [5]. What these models do is incorporate factors such as population per square kilometer, traffic congestion, influence on the environment, and the impact on the economy to come up with sound and logical plans for implementation. For instance, the method can be applied to the formulation of a policy that aims at optimizing public transport, which has the social objective of reducing traffic congestion, controlling pollution, and making transportation accessible for everybody in the city.

Mathematical optimization also proves rather beneficial in military applications of GEOINT. Specifically, in this domain, optimization techniques are applied to increase the efficiency, as well as the precision, of targeting [8], surveillance, and reconnaissance activities. Namely, optimization models can contribute to the determination of the optimal locations for placing sensors and surveillance equipment to maximize coverage area and minimize visibility.

The combination of mathematical optimization with AI techniques supercharges GEOINT by enhancing the creation of GEOINT models that can learn and are also robust. Based on data, ML and deep learning algorithms are receptive to enhancement in the future. Such models, when used along with optimization methods, can be further refined to be tuned to a single aim and objective, for instance, optimizing for error or data error rates.

However, optimization does play a role not only in adjusting single models or algorithms applied in AI-supported GEOINT but also in enhancing the overall execution of GEOINT models. It also provides the necessary link between several types of data inputs and how all of it gets merged together to create a single [9] fused intelligence product. For instance, in the case of multisensor data fusion, optimization procedures are employed in the process of mixing data accrued from different sensors to enhance the precision of the information received. This is especially true within contexts where the data includes a lot of noise, missing, or even contradictory data.

Therefore, mathematical optimization plays a central role in the development of other facets of AI in the geospatial domain. It offers the tools that are required to master the challenges associated with geospatial data and enhance decision-making to attain optimum results in a number of application areas. The ever-growing complexity of GEOINT also means that [10] optimization, together with AI integration, will become one of the key determinants of success as geospatial analysis finds new applications across numerous disciplines.

11.2 Theoretical Foundations

11.2.1 Mathematical Optimization Techniques

These problems are also known under the umbrella of mathematical optimization, which is at the core of operational research and decision science as systematic procedures designed to help in choosing the best decision out of all possible decisions allowed by the given purpose and constraints. As mentioned in Figure 11.1, various types of mathematical optimization techniques are: In the framework of GEOINT [11], these methods are used to address a number of challenges, including optimization of routes and the allocation of resources, as well as data fusion. Familiarization with the various optimization techniques is important in the application of AI-based GEOINT systems.

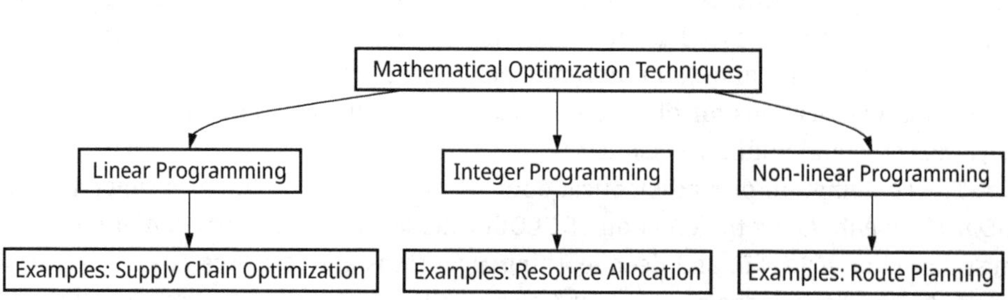

Figure 11.1: Mathematical optimization techniques.

– **Linear Programming (LP):**
LP is one of the most basic optimization methods. It is the process of maximizing or minimizing a linear objective function, subject to a set of linear equality and inequality constraints. Here is the typical way to present an LP problem:

$$\text{Maximize or Minimize } \boldsymbol{cTx} \text{ subject to } \boldsymbol{Ax} \leq \boldsymbol{b},\ \boldsymbol{x} \geq 0$$

where X is the vector of decision variables, c is the coefficient vector for the objective function, A is the matrix containing coefficients of constraints, and b is an array with constraint bounds. In GEOINT, LP is commonly used for situations such as determining the optimal location of sensors in a geographical area or reducing logistical costs by finding cost-effective transport plans.

– **Integer Programming (IP):**
IP provides an extended version of LP where some or all the decision variables are limited to take integer [12] values. This is especially interesting in cases where our decision variables are the discrete choices to be made (e.g., the number of cars running or the number of sensors installed). An IP problem is a mathematical optimization or feasibility program in which some or all of the variables are restricted to be integers:

$$\boldsymbol{x} \in Zn \text{ (or } a \text{ subset of variables is in } Zn)$$

IP programming is often employed in GEOINT for tasks like optimizing the allocation of limited resources or planning military operations where discrete decisions are required.

– **Nonlinear Programming (NLP):**
Optimal solutions require efforts for optimization problems where nonlinearity exists in either the objective function or constraints. It refers to optimization problems in which the objective function is a quadratic form and constraints are linear equalities or inequalities. Quadratic programming (QP) problems are typically harder than LPs (i.e., computationally more intense). LNP problem: every NLP problem takes the following form:

$$\text{Minimize or Maximize} f(\boldsymbol{x}) \text{ subject to } gi(\boldsymbol{x}) \leq 0,\ i = 1, \ldots, m \text{ and } hj(\boldsymbol{x}) = 0,\ j = 1, \ldots, p$$

where x is the nonlinear objective function, $gi(\boldsymbol{x})$ represents the inequality constraints, and $hj(\boldsymbol{x})$ represents the equality constraints. NLP is particularly useful in GEOINT for optimizing complex systems, such as satellite orbit configurations, that cannot be accurately modeled using linear approximations [13].

– **Dynamic Programming:**
This is a method used for solving complex problems by breaking them down into simpler subproblems. It is particularly [14] effective for solving problems that exhibit the properties of [15] overlapping subproblems and optimal substructure, such as the

shortest path problem or resource allocation over time. The Bellman equation, which is central to dynamic programming, is given by

$$V(s) = a\max\left(R(s,a) + \gamma s' \sum P(s' \mid s,a)\, V(s')\right)$$

where *V*(*s*) is the value function representing the maximum expected reward starting from state $R(s,a)$ is the reward received after taking action a in state s, gamma is the discount factor, and $P(s' \mid s,a)$ is the transition probability from state s to state s' given action a. Dynamic programming is extensively used in GEOINT for sequential decision-making problems, such as optimizing the deployment of assets over time in a military operation.

– **Stochastic Optimization:**
These include stochastic optimization techniques for when there is uncertainty in either the data or the model. They generally look for optimal-average — or in expectation — solutions under the uncertainty. One class of widely used [16] techniques in stochastic optimization models the uncertainty as random variables and optimizes over the expected cost. The main motivation for stochastic optimization in GEOINT is that the environment we work in is uncertain, such as forecasting how a natural disaster will impact or planning under incomplete information.

11.2.2 AI and Machine Learning Models for Geospatial Data

Geospatial data already forms a central part of both the AI and ML industries; however, there is no fixed formula for utilizing it to its full potential. These technologies offer the capability to identify relevant patterns within vast and intricate datasets, which directly supports enhanced decision-making in GEOINT. Figure 11.2 shows the types of AI models that can be used as mentioned.

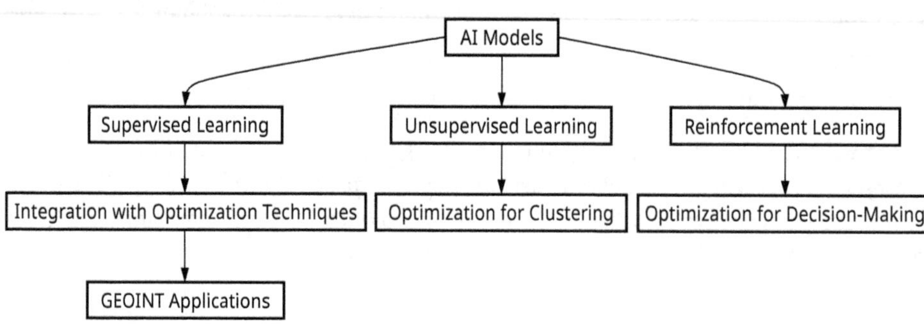

Figure 11.2: Types of AI models.

– **Supervised Learning:**
This type of learning involves a model that is trained on input data as well as its correct output. It is a model of function mapping from input to output and is therefore applicable in tasks like classification and regression. Some common algorithms include support vector machines, decision trees, and neural networks. A typical supervised learning model looks something like this:

$$y^\wedge = f(\boldsymbol{x}; \theta)$$

where $y^\wedge$ is the predicted output, x is the input data, and theta represents the model parameters.

– **Unsupervised Learning:**
With unlabeled data (insight discovery, detecting natural groups, or discovering hidden patterns) [17], clustering and dimensionality reduction are unsupervised learning techniques through which, in GEOINT, regions with similar geospatial characteristics can be identified, anomalies detected, or large datasets reduced to a lower-dimensional representation prior to data processing, for instance, K-means, principal component analysis, and autoencoders.

– **Reinforcement Learning:**
This is a type of ML where the agent learns to make decisions by interacting with an environment. Actions receive a numerical reward, and the agent tends to maximize accumulated rewards. In the context of GEOINT, RL can expedite dynamic decision-making problems like drone deployment optimization for surveillance or drones navigating in intricate terrains. Ultimately, the Bellman equation decides how to update the value function by examining the reward received during the learning phase:

$$Q(s,a) = R(s,a) + ya' \max Q(s', a')$$

where *Q(s, a)* is the quality of action *aa* in state *ss,* and *R(s, a)* is the reward received.

– **Deep Learning:**
It is a subset of ML that employs deep neural networks with two or more layers. These models have achieved great success in dealing with complex geospatial data, such as image recognition, object detection, and semantic [18] segmentation. Deep learning models are highly effective in analyzing high-resolution satellite imagery, extracting features from multispectral data, and predicting spatial patterns with great accuracy using GEOINT. CNNs are particularly well-suited for this domain, as they have the ability to learn spatial hierarchies in data:

$$f(\mathbf{x}) = \text{CNN}(\boldsymbol{x}; \vartheta)$$

where $\boldsymbol{x}$ is the input image, and θ represents the parameters of the convolutional filters.

11.2.3 Integration of Optimization with AI Models

This is definitely a step forward in the GEOINT field by combining mathematical optimization and AI models. Augmenting the predictive power of AI with the prescriptive abilities offered by optimization transforms geospatial data analyzing systems to opine upon optimal solutions in a problem space rather than just regression analysis.

– **Hybrid Models:**
Increasingly, hybrid models using AI and optimization techniques are being applied to GEOINT. An example might be a ML model used to predict traffic patterns in the city and a complementary optimization [19] algorithm responsible for finding an optimal route for emergency vehicles during a disaster. Here, models that capitalize on the strengths of both AI and optimization have been developed to yield more accurate solutions as well as improved efficiency. A couple of examples would be hybrid approaches like:

$$\text{Minimize} f(\boldsymbol{x},\boldsymbol{y}) = \boldsymbol{c}T\boldsymbol{x} + \lambda i = 1 \sum nL(\boldsymbol{y}i,\boldsymbol{y}^{\wedge}i)$$

where *f(x,* **y***)* is the objective function combining the cost function ***cTx*** from the optimization model and the loss function *L(yi,* **y***^i)* from the AI model.

An important example of applying optimization techniques and AI in GEOINT is in disaster response planning. In a disaster, it is important that the response is timely and efficient, especially in cases of natural disasters like hurricanes, earthquakes, and the like. Forecast models through AI [20], such as predicting where flooding or landslides are most likely to occur following a disaster, are also possible. It is for this reason that optimization algorithms can then be used in defining how best to allocate resources such as rescue teams, medical supplies, and the various means of transport for use during evacuation.

Another example of implementing decision-making models along with AI is in military operations, where surveillance and targeting are improved through the use of optimization models.

They foresee growth in population density and in traffic flow in cities and apply optimization to arrange structures that can cater to future occupancy. This integrated process ensures that cities are developed in the most efficient and [21] sustainable manner possible. The presented result of the upgrading of mathematical optimization with AI models reveals a strong concept in terms of geospatial problem-solving. Using AI for forecasting and optimization, it is possible to create geo-information systems not only to analyze the data but to identify ways for further actions.

11.3 Practical Implementation

11.3.1 Tools and Frameworks

A robust set of tools and frameworks is needed for the implementation of both GEOINT-AI and integrated mathematical optimizations. These tools make it easy to build, debug, and deploy optimization models or AI algorithms for solving challenging geospatial problems.

- **Optimization Tools and Libraries:**
  - **Pyomo:** This is a great open-source Python-based optimization modeling language that enables users to define and solve an infinite number of optimization problems, from the most common linear to nonlinear. It's integrated with solvers such as Gurobi, CPLEX, and CBC. It has libraries to code complex optimization models that are highly connected with AI algorithms; thus, it is best suited for Pyomo.
  - **Gurobi**: This is one of the best-in-class commercial optimization solvers that achieve state-of-the-art performance in solving large-scale linear, integer, and NLP problems with the highest degree of efficiency. It is generally used in industry for jobs such as supply chain optimization, resource allocation, or financial modeling. This entails that the optimization can be smoothly incorporated with current ML libraries, ensuring a user can seamlessly introduce new linear and mixed-IP solvers.
  - **TensorFlow:** Here are the top five deep learning libraries for Python. TensorFlow: Born at Google, TensorFlow is one of the popular open-source libraries for ML. These include tools for building and deploying deep learning networks and model interpretability. Additionally, TensorFlow, along with automatic differentiation and numerous gradient-based optimization methods, adds optimal solutions for AI-enabled GEOINT apps that are long-lived, requiring continuous and ever-changing learning.
  - **PyTorch:** This is another popular ML library that emphasizes flexibility and ease of use. It supports dynamic computation graphs, which are particularly useful for tasks requiring frequent updates to the model structure, such as RL. PyTorch's integration with optimization libraries allows for the development of sophisticated AI models that can be optimized for specific geospatial tasks.
- **Geospatial Information System (GIS) Software and Platforms:**
  - **QGIS:** This is an open-source GIS tool, and you can use it to perform all the operations on geospatial data. It can handle numerous data formats and comes with a wide range of spatial analysis tools that are really useful for preprocessing and analyzing geospatial data before you go on to apply AI or optimization techniques.

- **ArcGIS:** Developed by Esri, one of the most extensive GIS platforms available, ArcGIS adds advanced spatial analytics and visualization to mapping skills from its creator, Esri. Using the various ML and optimization tools that are available, ArcGIS empowers its users to create advanced geospatial models.

11.3.2 Data Preparation

Ultimately, the AI-enabled GEOINT models are only as accurate and effective as the data that feeds them. Figure 11.3 explains how to create an optimized model with the assistance of AI. Data cleaning is an important step in preparing data, as it guarantees that geospatial information is clean and uniform for evaluation and ready for analysis [22].

- **Steps to Acquire, Clean, and Preprocess Geospatial Data:**
  a. **Data Acquisition:** Obtaining geospatial data includes satellite imagery, aerial photography, and any existing public datasets. What are some common geospatial datasets? Landsat satellite imagery, OpenStreetMap, and NASA's Earth Observing System Data and Information System (EOSDIS). Based on the criteria of use and the resolution needed, different data sources could be used.
  b. **Data Cleaning:** Geospatial data in its purest form requires handling a lot of noise, missing values, and inaccuracies that must be dealt with prior to analysis. Data cleaning is the procedure for removing or correcting inaccurate data, filling in missing values, and standardizing formats. To cleanse geospatial datasets, techniques such as interpolation, nearest-neighbor imputation, and outlier detection are often used.
  c. **Handling Missing Data:** Geospatial datasets must have complete data. Blurry images affect the model's performance. There are various ways to treat missing data, but an interpolation method like Kriging would work in those cases where the spatial correlation between neighboring points is strong enough such that a reasonable estimation of values can be made.
  d. **Data Preprocessing:** During the preprocessing data cleaning process, we need to preprocess our cleaned-data into a suitable format that can be productive for analysis. It can range from tasks such as rescaling of coordinate systems, converting raster to vector, and even extracting features from satellite images. Feature engineering, i.e., buffer zones around points of interest or slope and aspect on the terrain, is also a big part of preprocessing [23].
  e. **Handling Large Datasets:** Geospatial datasets can be so large that they present a technical processing and analytical challenge. To work with big datasets, techniques like spatial indexing, data partitioning, and parallel processing can be used. With major benefits like the scalability of the geospatial data they can process, cloud-based platforms such as Google Earth Engine are enviably well-positioned.

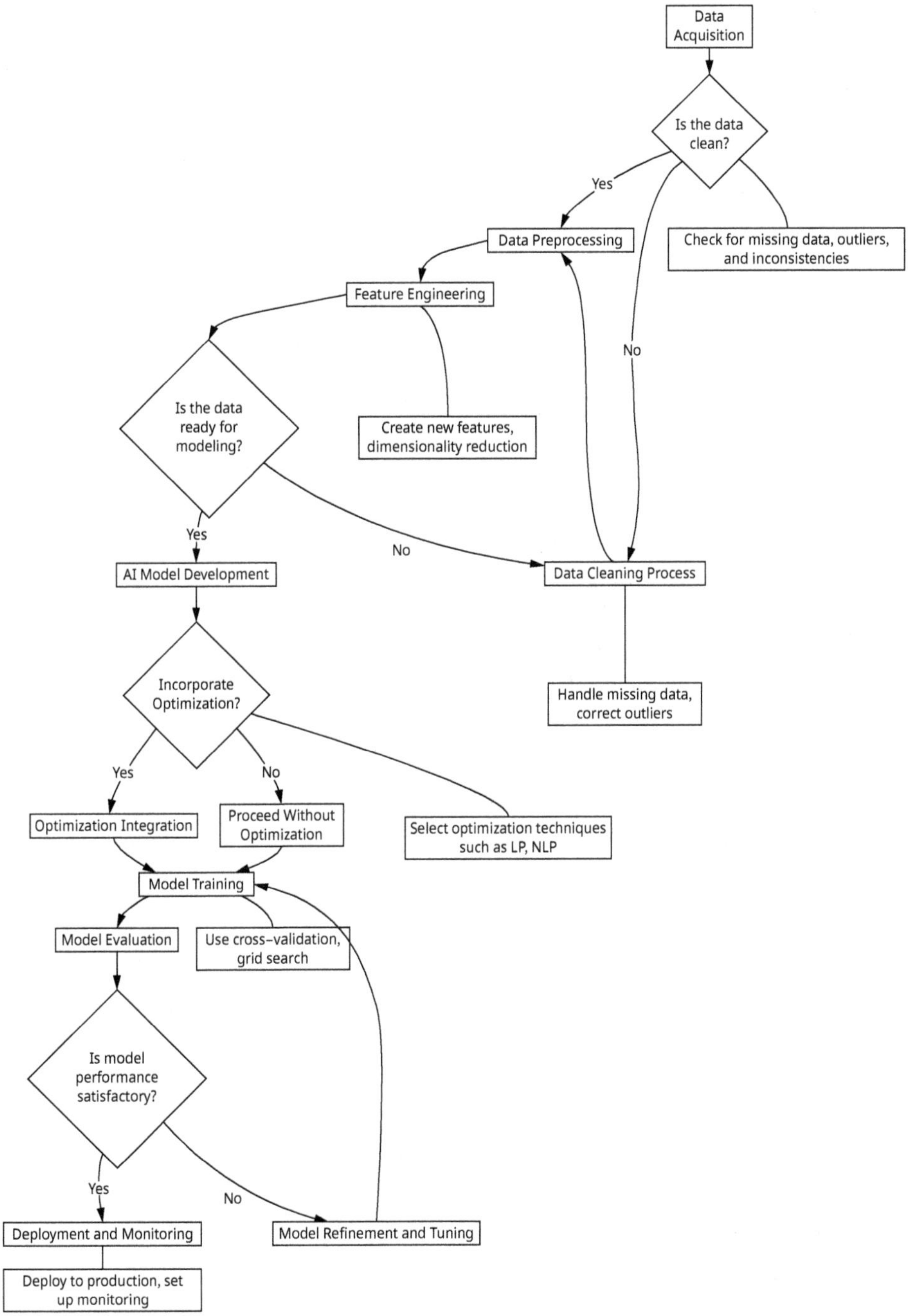

Figure 11.3: Complete methodology for creating a model for optimization.

11.3.3 Building AI Models with Optimization

As such, the integration of AI and mathematical optimization in GEOINT is a systematic process that leverages strengths from both fields to improve model performance. In this chapter, we shall look at how to move from the human-directed AI real senses models DoML platform of the last section and create a GeoIntelligence-based AIML hypermodel — as a probability-statistically (here inferred via optimization function) system or instead based platform.

- **Step-by-Step Guide to Creating an AI Model for GEOINT:**

1. Define the Problem:
Geospatial Problems: Clarify the problem statement first. This can vary from sensing urban traffic congestion to optimizing emergency response unit locations for disaster scenarios. The description of the problem definition should include the objective function, constraints, and desired output.

2. Select the Appropriate AI Model:
Choose an AI model according to the problem definition. Or, if you have an image-classification problem (e.g., land cover classification), this can be done by using a convolutional neural network (CNN). An RL model will be more useful for problems where action is taken sequentially, like drone path planning.

3. Integrate Optimization Techniques:
After selecting the appropriate AI model, we should introduce optimization methods. Consider the scenario of route optimization: A genetic algorithm (GA) may be used to find the best path, which needs minimum travel time and logical avoidances. The optimization problem is given as follows:

$$\text{Minimize}\, T = \sum_{i=1}^{n} d_{i,i+1} \text{Subject to}\, x_i \in R, \ \forall i$$

where T is the total travel time, $di, i+1$ represents the distance between consecutive waypoints, and xi is the set of feasible routes.

4. Train and Validate the Model:
Teach the AI model with your geospatial data. You would fine-tune some beautiful network architecture to adjust the model parameters in a way that predicted values most closely resemble actual outputs during training. Validation must be done with another dataset to assess the generality and stability of your model.

5. Deploy the Model:
Once validated, the model can be deployed in a GEOINT system. This can include layering the model over GIS software, configuring live piping data feeds, and handling real-time input processing in a CI/CD framework.

– **Incorporating Optimization Techniques to Enhance Model Performance:** Optimization techniques play an integral role in fine-tuning the decision-making mechanisms that are introduced in AI models used for GEOINT. For instance, in the case of utilizing AI to estimate regions that are at a higher likelihood of flooding, resources may be allocated using the optimization algorithm [16]. Whether it be locating shelters in the event of an emergency or providing the supplies needed, the best strategies have to be found and used. Optimization algorithms ensure that AI models not only predict outcomes but also suggest means of achieving the optimal results of interventions. These two methods are very important, especially for real-life problems that require solutions with high resource efficiency and ought to be made in the minimal time possible to meet the required needs, say in military campaigns or disasters.

Using the above-stated steps and with the help of tools, frameworks, and techniques discussed in this chapter, the GEOINT models can be enhanced by the practitioners to harness the AI and mathematical optimization capabilities in the most optimal manner as has been envisioned. This integrated approach makes it possible to design smart systems that can effectively and efficiently solve varied geospatial-related problems.

11.4 Case Study

Case Study 1: Optimizing Disaster Response Using AI and GEOINT

– **Problem Statement:**

The effectiveness of any emergency response operation in the wake of a natural disaster is key to significantly reducing victims and damage. The more significant challenge is to pull as many resources like medical aid, rescue teams, and relief material into the field faster on the surface. This study also suggests that traditional disaster response strategies suffer from information lag and the complexity of the situation for near real-time decision making. The objective of this case study is to streamline disaster response operations using AI-based GEOINT and mathematical optimization strategies.

– **Approach:**

The way it works is by incorporating AI and GEOINT to process live geospatial data, identify where the greatest resource demands arise both in advance, as well as during disasters, and optimize emergency service responses. It starts by aggregating data from various geospatial sources like satellite imagery, social media feeds, and weather reports. This data is input into ML models, which then make forecasts on how badly the disaster is set to hit parts.

– **Optimization Techniques Used:**

In the case study, mathematical optimization is used to decide how many resources go where and when. We formulate the problem as a multi-objective optimization by minimizing the time taken to deliver a response and maximizing the coverage of the immune response.

The optimization model can be represented as follows:

$$\text{Minimize } T = \sum_{i=1}^{n} d_{i,j} x_{i,j} + \alpha \sum_{i=1}^{n} t_i x_{i,j}$$

$$\text{Maximize } C = \sum_{j=1}^{m} c_j y_j$$

$$\sum_{j=1}^{m} x_{i,j} \leq 1 \quad \forall i = 1, 2, \cdots, n$$

$$\sum_{i=1}^{n} x_{i,j} \geq y_j \quad \forall j = 1, 2, \cdots, m$$

$$x_{i,j} \in \{0,1\}, \; y_j \in \{0,1\}$$

where T is the total travel time, $d_{i,j}$ is the distance between resource i and area j, t_i is the time to deploy resource i, and alpha is a weight factor. C is the coverage score, representing the effectiveness of the resource allocation. c_j is the coverage value of area j, and y_i indicates whether area j is covered. $x_{i,j}$ is a binary variable indicating whether resource i is allocated to area j.

Practical Implementation Details:

1. Data Acquisition:

The system collects data in real-time from satellites, drones, and ground sensors. It includes data from the affected areas, photos of what it looks like on the ground, how traffic is flowing, and where critical infrastructure like hospitals are located alongside shelters.

2. AI Model Training:

These models are trained with historical disaster data to predict the impact of a current disaster. These models predict factors like the number of casualties, loss in infrastructure, and areas that need assistance.

3. Optimization Execution:

The optimization model is implemented in Python using the Gurobi Optimizer, which has a reputation for being one of the fastest and most powerful optimizers available. In conclusion, the outcomes of this model offer an idea about what amount to set aside for various regions and paths, which should be used by emergency vehicles.

4. Results and Outcomes:
Integrating with AI and optimization made the response time during disasters minimal. With simulated earthquake response scenarios, the new model expanded coverage of affected areas by up to 30% and decreased the average response time by approximately 25%. Medical help could also reach quicker, and resources were utilized better, which ultimately helped in saving more lives.

Case Study 2: Urban Planning and Development

– **Problem Statement:**
Urban planning is an integrated decision-making process of harmonious development, which needs to deal with a series of conflicting objectives such as land use efficiency maximization, environmental impact reduction, and quality of life improvement for residents. Traditional urban planning methods typically rely on static models and are unable to adjust with time or incorporate large amounts of information. In this case study, we demonstrate how AI and mathematical optimization can improve performance in a lifecycle of decision problems around urban planning: land use pattern generation, transportation network design, and stormwater runoff reduction while promoting environmental sustainability.

– **Approach:**
It starts with the incorporation of geospatial data relating to land use, population density, traffic flow, and environmental conditions. Information is utilized by AI models that analyze it, predict likely future trends and results, as well as create various planning scenarios. These scenarios are then evaluated using mathematical optimization techniques to identify the best development plan for that urban space.

– **Optimization Techniques Used:**
Urban planning is mainly about optimizing, and this often involves multiple objectives, e.g., minimizing traffic congestion and increasing green spaces with trees or bushes to reduce air pollution. A multiobjective optimization framework can model these objectives. Finally, the optimization problem can be written as:

where:

- Z_1 represents the total travel time adjusted by green space, T_i is the travel time in area i, G_i *is the green space factor,* and *xi* indicates the inclusion of area ii in the plan.
- Z_2 represents environmental quality, E_j is the environmental benefit in zone j, and y_j indicates whether zone *jj is selected for environmental enhancement.*
- The constraints ensure that only one area is selected for each type of development, and a minimum level of environmental quality *CC* is achieved.

$$\text{Minimize } Z_1 = \sum_{i=1}^{n} T_i(1 - G_i)x_i$$

$$\text{Maximize } Z_2 = \sum_{j=1}^{m} E_j y_j$$

$$\sum_{i=1}^{n} x_i = 1, \quad \sum_{j=1}^{m} y_j \geq C$$

$$x_i, y_j \in \{0,1\}, \quad \forall i,j$$

Detailed Walkthrough of the Process:
1. Data Integration:
The power of the urban planning system is that it fuses together a wide array of data – satellite imagery, census and demographic information, transportation networks, etc. The GIS stores this data, and it is employed to generate detailed spatial models of the urban area.

2. Scenario Generation:
Multiple urban planning scenarios are generated with AI models – including ML and simulation techniques. They include how we could grow and how better land use patterns balance. The AI models are further used to calculate the long-term consequences of each scenario, such as traffic jams, levels of pollution, or local people's spending.

3. Optimization Analysis:
The optimization framework measures each scenario against those objectives (e.g., minimizing commute times and maximizing green spaces). The best trade-offs between competing objectives are achieved by the Pareto optimization approach in a set of non-dominated solutions.

4. Implementation of Results:
The optimization analysis results are shown as interactive maps and dashboards to the urban planner or stakeholders. This tool will help policymakers research different planning scenarios and make evidence-based decisions about the future growth of the city.

Outcomes:
Using AI and optimization to make better decisions in urban planning, in the test of a medium-sized city, they improved average commute times by 15%, increased green spaces by 20%, and enhanced air quality by 10%. All these outcomes illustrate how AI and mathematical optimization are powerful tools to build better, livable cities that make sustainable pathways possible.

11.5 Challenges and Future Directions

11.5.1 Current Challenges in GEOINT Optimization

Integrating mathematical optimization with AI as applied to GEOINT introduces a set of challenges we need to meet before correctly scaling this solution. One of the key problems is dealing with the extreme computational complexity that arises when processing and examining large geospatial data. However, optimization algorithms like these can be computationally intensive and require a lot of processing power and time when dealing with non-linear and multi-objective problems. Today, we can see Figure 11.4, which illustrates the challenges and trends of GEOINT. However, this is muddied by the continuous evolution of geospatial data that frequently demands real-time processing and decision-making, thus making it harder for organizations to optimize.

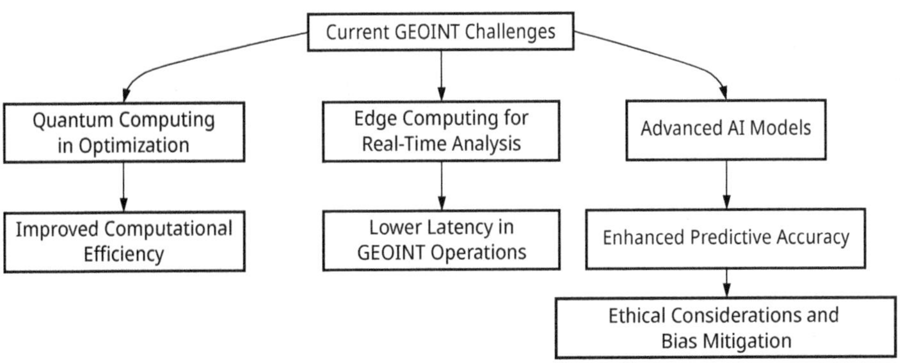

Figure 11.4: Current GEOINT challenges and future trends.

There is another major problem, which is data privacy in optimizing GEOINT. While using geospatial data, one needs to be highly careful with it due to the highly sensitive nature of geospatial data. Combined with optimization techniques, AI poses privacy issues, especially when in practice of utilizing location-based services, surveillance, or military intelligence. One of the major difficulties connected with work on data optimization in big data analysis is the integration of attention to data privacy along with high accuracy and efficacy in optimization models.

However, the application of various data sources, as well as the process of harmonizing different datasets, brings some more challenges. GEOINT utilizes different types of data, such as satellite and aerial images, social networks, sensors, and GIS applications. These data sources may be of different formats, different resolutions, and levels of accuracy, and this complicates the integration of these data sources in the development of an integrated model. The real difficulty is in the design of optimization algorithms to accommodate and balance these data discrepancies so as to generate accurate and efficient forecasts.

11.5.2 Future Trends in AI-Enabled Geospatial Optimization

In the future, there will be several other technologies and methods that can define the direction and advancement in the field of AI-based geospatial optimization. A new and highly buoyant trend that can act as a Rosetta Stone and greatly enhance optimization is quantum computing, which significantly shortens the time taken to solve hard problems. Through quantum algorithms, GEOINT systems could be able to process and analyze very large sets of data in a way that is currently inconceivable in traditional systems, and this will justify the possibility of real-time decision-making.

Another future trend is the use of edge computing, especially in GEOINT applications. Edge computing enables data to be processed closer to the source, thus decreasing latency and bandwidth, hence improving optimization. This is particularly significant where the software application is time-sensitive, such as in unmanned cars and drones. As a result of the integration of the three approaches, future geospatial systems will be capable of adapting and responding in real time, in dynamic and constrained environments.

Other significant trends that will also define the future of next-generation GEOINT optimization encompass the refinement in AI frameworks that include deep RL and generative adversarial networks. These models are capable of making changes with respect to time, thus making them perform better in situations they are faced with as they get exposed to more data and other situations. This will be particularly useful for complex issues that are dynamic in nature, as may be observed with climate change impacts where predicting influence areas is necessary, or as may be the case with growing cities where one has to predict likely areas to allocate resources.

Concerning the research opportunities, more emphasis is placed on the ethical aspect of AI and optimization in GEOINT, where with these enhancements, further issues emerge as to the level of transparency, accountability, or even the neutrality of such aids. The other challenging area that will be noted in the coming years is the design of optimization algorithms that will not only be effective but also equitable and highly transparent. Furthermore, it means that the collaboration of geospatial sciences, AI, and ethical studies will be crucial in order to enhance the use of advances in GEOINT optimization for the common good.

In conclusion, it can be asserted that, though there exist some major obstacles to the optimization of GEOINT in the present-day scenario, there are some promising prospects in the near future given the globalization of technology as well as methodological diversification. Mitigating the existing issues alongside adopting these new trends will be important to the further evolution and effectiveness of the AI-integrated GEOINT system.

11.6 Conclusion

This chapter has explored the pivotal role of mathematical optimization in enhancing AI-enabled GEOINT. We began by outlining the significance of GEOINT in various domains, emphasizing how it transforms raw geospatial data into actionable insights that drive decision-making across military, governmental, and commercial sectors. The integration of AI, particularly ML and deep learning techniques, has further expanded the capabilities of GEOINT, enabling the analysis of complex datasets with unprecedented accuracy and speed.

The conversation around mathematical optimization showcased how critical it is in supporting the existence of enhanced models for GEOINT AI. Finally, we showed how these methods are applied to solve resource allocation, route planning, and data integration complex problems through optimization techniques such as LP, IP, and NLP. The foundations they provided also helped us to know when optimization can help improve AI models in geospatial data analytics so that work may be performed efficiently and accurately.

In the section related to practical implementation, we stepped through a variety of tools and frameworks that enable information operations-centric data mining applications to integrate optimization techniques with other AI-based methods. The step-by-step guide provided a logical way to build an AI model using optimization with workable real-world examples, e.g., route optimization algorithms. This is excellent as it strengthens the theoretical building blocks laid out before and demonstrates practical ways to utilize these principles in real life.

The case studies, as depicted later in Section 11.4, helped shape how AI-based GEOINT can be viable for disaster response and city planning through the use of optimization approaches. These examples highlight what mathematical optimization can achieve when solving these types of highly challenging problems (in research or practice) and the impact this sort of analysis might provide in a broad range of disciplines.

In the future, significant challenges in GEOINT optimization will persist, including substantial problems like computationally complex analyses and computations; privacy of the data used to create enterprise models; as well as legacy or outdated systems unsuited for integration into an architecture because they are not standards compliant. However, the promise of new trends like quantum computing, edge computing, and advanced AI models may help to address these major GEOINT evolution concerns.

The intelligence community has found that mathematical optimization represents a fundamental component of AI-enabled GEOINT – the tools and methods needed to take any geospatial analysis based on satellite, aircraft, or ground-generated data simply closer to perfection. As these technologies provide more actionable information, their applicability to GEOINT will surely grow – enhancing informed decisions and optimized mission success in many areas.

References

[1] Hariri-Ardebili, M. A., Mahdavi, G., Nuss, L. K., & Lall, U. (2023 Nov).The role of artificial intelligence and digital technologies in dam engineering: Narrative review and outlook. *Engineering Applications of Artificial Intelligence*, *126*, 106813.

[2] Peng, L., Cai, Z., Heidari, A. A., Zhang, L., & Chen, H. (2023 Nov).Hierarchical Harris hawks optimizer for feature selection. *Journal of Advanced Research*, *53*, 261–278.

[3] Tselentis, D. I., Papadimitriou, E., & Van Gelder, P. (2023 Jun).The usefulness of artificial intelligence for safety assessment of different transport modes. *Accidents Analysis and Prevention*, *186*, 107034.

[4] Lian, J., Hui, G., Ma, L., Zhu, T., Wu, X., Heidari, A. A. et al. (2024 Apr). Parrot optimizer: Algorithm and applications to medical problems. *Computers in Biology and Medicine [Internet*, *172*, 108064.

[5] Zhao, M., & Liu, X. (2018 Feb).Development of decision support tool for optimizing urban emergency rescue facility locations to improve humanitarian logistics management. *Safety Science*, *102*, 110–117.

[6] Hassani, S., Dackermann, U., Mousavi, M., & Li, J. (2024 Mar).A systematic review of data fusion techniques for optimized structural health monitoring. *Information Fusion*, *103*, 102136.

[7] Shi, J., Chen, Y., Cai, Z., Heidari, A. A., Chen, H., & Chen, X. (2024 Oct).Multi-threshold image segmentation based on an improved whale optimization algorithm: A case study of Lupus Nephritis. *Biomedical Signal Processing and Control*, *96*, 106492.

[8] Zhou, X., Gui, W., Heidari, A. A., Cai, Z., Liang, G., & Chen, H. (2023 Sep).Random following ant colony optimization: Continuous and binary variants for global optimization and feature selection. *Applied Soft Computing*, *144*, 110513.

[9] Yuan, C., Zhao, D., Heidari, A. A., Liu, L., Chen, Y., Wu, Z. et al. (2024 Sep). Artemisinin optimization based on malaria therapy: Algorithm and applications to medical image segmentation. *Displays*, *84*, 102740.

[10] Asadollah, S. B. H. S., Jodar-Abellan, A., & Pardo, M. Á. (2024 Jun).Optimizing machine learning for agricultural productivity: A novel approach with RScv and remote sensing data over Europe. *Agricultural Systems*, *218*, 103955.

[11] Qiu, Z., Yong, Q., Wang, J., Liao, L., & Yu, B. (2024 Sep).A multi-objective optimization framework for performance-based building design considering the interplay between buildings and urban environments. *Energy Conversion and Management*, *315*, 118793.

[12] Zhao, T., Wang, S., Ouyang, C., Chen, M., Liu, C., Zhang, J. et al. (2024 Aug). Artificial intelligence for geoscience: Progress, challenges and perspectives. *The Innovation*, 100691.

[13] Dong, R., Sun, L., Cai, Z., Heidari, A. A., Liu, L., & Chen, H. (2024 Sep).An advanced kernel search optimization for dynamic economic emission dispatch with new energy sources. *International Journal of Electrical Power & Energy Systems*, *160*, 110085.

[14] Rapaka, A., Dharmadhikari, S. C., Kasat, K., Mohan, C. R., Chouhan, K., & Gupta, M. (2025 Jan). Revolutionizing learning – A journey into educational games with immersive and AI technologies. *Entertain Comput*, *52*, 100809.

[15] Bhati, N., Duggar, R., & Saber, A. (2023). Empowering Safety by Embracing IoT for Leak Detection Excellence. 231–251.

[16] Yazdani, M., & Haghani, M. (2024 Aug). A conceptual framework for integrating volunteers in emergency response planning and optimization assisted by decision support systems. *Progress in Disaster Science*, 100361.

[17] Punugoti, R., Vyas, N., Siddiqui, A. T., & Basit, A. (2023). The Convergence of Cutting-Edge Technologies: Leveraging AI and Edge Computing to Transform the Internet of Medical Things (IoMT) [Internet]. In 2023 4th International Conference on Electronics and Sustainable Communication Systems (ICESC). IEEE. Available from: https://dx.doi.org/10.1109/ICESC57686.2023.10193047

[18] Ahamjik, I., Agbani, A., Abik, M., Khayli, M., Galzim, N., Berrada, J. et al. (2024 Dec). Contribution of artificial intelligence for understanding animal rabies epidemiology in Morocco: What are the perspectives of an innovative and predictive approaches? *One Health*, *19*, 100874.
[19] Wang, S., Huang, X., Liu, P., Zhang, M., Biljecki, F., Hu, T. et al. (2024 Apr). Mapping the landscape and roadmap of geospatial artificial intelligence (GeoAI) in quantitative human geography: An extensive systematic review. *International Journal of Applied Earth Observation and Geoinformation*, *128*, 103734.
[20] Cao, K., Zhou, C., Church, R., Li, X., & Li, W. (2024 May).Revisiting spatial optimization in the era of geospatial big data and GeoAI. *International Journal of Applied Earth Observation and Geoinformation*, *129*, 103832.
[21] Itagi, M., DG, V., Prasad, K., Kumar, P. S., Shekhar, R., & Ashreetha, B. (2023 Jun 26). Performance Analysis of Energy Efficiency and Security Solutions of Internet of Things Protocols. *International Journal of Electrical and Electronics Research*, *11*(2), 442–450.
[22] Radulescu, M., Dalal, S., Lilhore, U. K., & Saimiya, S. (2024 Feb).Optimizing mineral identification for sustainable resource extraction through hybrid deep learning enabled FinTech model. *Resources Policy*, *89*, 104692.
[23] Radulescu, M., Dalal, S., Lilhore, U. K., & Saimiya, S. (2024 Feb).Optimizing mineral identification for sustainable resource extraction through hybrid deep learning enabled FinTech model. *Resources Policy*, *89*, 104692.

P. Manju Bala, S. Usharani, A. Balachandar, and R. Rajmohan

12 Deep Learning-Based Ultrasound Analysis Using Explainable Artificial Intelligence (XAI) Methods for Breast Cancer

Abstract: Medical imaging is one of the most crucial components of breast cancer screening. Ultrasound is notable for its noninvasive methodology and exceptional sensitivity. However, interpretability issues with deep learning models make them unsuitable for application in therapeutic contexts. To diagnose breast cancer, this study proposes to combine deep learning-based ultrasound analysis with explainable artificial intelligence (XAI) methodologies. This chapter examines several XAI methods, including model explanations, saliency maps, and attention processes. It illustrates how these methods may be used to make deep learning models' decision-making processes more transparent. Improving the comprehensibility and offering valuable information about the predictions made by automated diagnostic systems will assist physicians in better understanding and placing confidence in these systems. The accurate diagnosis of breast cancer can be improved by combining deep learning technology with XAI. We verify this by conducting thorough testing on ultrasound data. By using these techniques, the quality of clinical judgments and patient treatment for breast cancer may be improved.

Keywords: Deep learning, ultrasound analysis, breast cancer, explainable artificial intelligence (XAI), interpretability, diagnostic imaging

12.1 Introduction

The most common and possibly lethal disease affecting women worldwide is breast cancer. Accurate and timely diagnosis is essential for better patient outcomes and lowering death rates. Because ultrasonography is noninvasive and very sensitive, it is an essential tool for the diagnosis and characterization of breast lesions. The identification of complex patterns in diagnostic pictures has been made easier by these cutting-edge algorithms [1]. Research on clinical solicitations has advanced considerably as an effect of the capacity to match and occasionally exceed the diagnostic precision of human specialists. However, the opaque, "black box" character of these complex technologies can occasionally mask their real effectiveness. The absence of clearness has led to grave anxieties nearly the legitimacy and reliability, particularly in the dangerous medical field [2]. A significant development in response to the increasing need for directness and dependability in AI-powered structures is explainable artificial intelligence (XAI). The goal of XAI was to unveil the underlying mechanisms of artificial

https://doi.org/10.1515/9783111436180-012

intelligence models and offer significant perceptions into their decision-making processes. This clearness was practical as well as ethically crucial, particularly in medical settings where vital choices are frequently made based on the guidelines provided by such rules. The lack of transparency has generated significant concerns over their dependability and acceptance, particularly within the high-risk field of healthcare [3]. XAI has emerged as a key technology to meet the growing demand for reliability and limpidity in AI-driven systems [4]. The goal of XAI was to clarify the internal mechanisms of AI models and offer important visions into how they make decisions. This transparency was important from both a practical and moral standpoint, especially in scientific settings where important choices frequently depended on the direction these systems provided [5].

Deep learning techniques have become powerful tools for automating the examination of medical images in recent years [6]. These provide the possibility of increased precision and effectiveness in diagnosis. A type of deep learning model called "convolutional neural networks" (CNNs) has demonstrated remarkable performance in a variety of medical imaging applications, including the analysis of breast ultrasound images [7]. Nevertheless, the opaque nature of these black boxes presents obstacles in understanding the decision-making process, hence restricting their practicality and acceptance in clinical settings. To solve this issue, XAI techniques deliver insights into the deep learning models' decision-making processes [8]. By clarifying the variables and patterns utilized for predictions, XAI techniques improve interpretability and foster confidence between end users and clinicians. The term XAI describes how readily people can understand and recognize a model's operation. It has garnered significant attention in recent years. An explanation of a model seeks to offer detailed insights on how a machine learning model is created, illuminate the reasons behind its accuracy, and improve reliability in the model's predictions. Therefore, the use of machine learning improves the machine learning models' transparency, interpretability, fairness, robustness, privacy, trust, and dependability. A number of approaches have been suggested and used recently to improve machine learning model comprehension [9]. There are several taxonomies available to describe machine learning. Consumers can request certain sorts of information or go further into an interactive explanation until they are satisfied, but a static explanation doesn't alter based on feedback from consumers. While a local explanation is specific to a single prediction, a global explanation considers the behavior of the entire model. A post hoc explanation necessitates the application of an extra technique to explain a model after it has been trained, whereas an instantaneously interpretable model is one that is transparent by nature and understandable to the majority of users [10]. Self-explanatory does not always imply immediately clear. It generates local explanations on its own. While a model's representation may highlight particular features and not be a full model in and of itself, a surrogate model is usually a more simple and intelligible form that approximates a more sophisticated model.

This work provides a thorough examination of deep learning-based ultrasound analysis for the identification and categorization of breast cancer [11]. We point out the challenges associated with understanding deep learning techniques in medical imaging and discuss the necessity of employing XAI techniques to overcome these challenges. Furthermore, we offer a comprehensive examination of various XAI tactics, including saliency maps, attentional mechanisms, and model explanations. We emphasize their relevance in the specific setting of breast ultrasonography assessment. Furthermore, we discuss the advantages of XAI's capabilities in enhancing the precision of diagnoses, assisting in scientific decision-making, and eventually improving results for persons impacted by breast cancer diagnosis and treatment [12]. Our objective is to bridge the gap between AI-powered automated analysis and medical practice by merging deep learning with XAI. This will enhance trust, transparency, and effectiveness in breast ultrasound imaging.

12.2 Related Works

A number of studies have looked at the analysis of breast ultrasonography using deep learning techniques. Particularly, these studies have focused on recognizing anomalies, classifying them, breaking them down, and assessing the risk involved. A highly developed deep learning model created especially for the purpose of recognizing breast lesions in ultrasound images was presented by Peta et al. [13]. The framework demonstrated impressive performance in terms of both sensitivity and specificity created a deep learning system to segment breast tumors in ultrasound images [14]. Their system outperformed traditional methods, showing better performance. Furthermore, researchers have explored how to apply ensemble approaches and transfer learning to improve the deep learning models' capacity to forecast accurately and handle data variances in breast ultrasonography analysis. Transfer learning from pretrained CNN models was used to enhance the classification of breast lesions in ultrasound images [15]. Furthermore, Gulum et al. [16] proposed an ensemble learning approach that combines several deep learning models to increase breast cancer diagnostic accuracy. Although deep learning algorithms have demonstrated encouraging results in the analysis of breast ultrasonography, their interpretability problems remain a major obstacle. Researchers are attempting to combine deep learning-based ultrasonic research with XAI techniques to overcome this problem. A weighted class of gradient stimulation representation is a technique that Chaddad et al. [17] suggested to visually highlight regions of interest in ultrasound images and provide explanations for model predictions.

The use of attention processes and model explanations to enhance the interpretability of deep learning techniques for breast ultrasonography analysis has been the subject of recent study [18]. For the purpose of classifying breast lesions, proposed an attention-guided deep learning model [19]. This approach focuses on and highlights the regions that are useful in ultrasound images. Furthermore, a system that makes use of gradient-based attribution techniques was developed by Di Miranda [20] to generate clear and concise explanations for deep learning-based breast cancer diagnosis. Explainable multiclass classification for breast cancer subtyping is the main topic of the study [21]. It introduces a method that combines integrative feature selection techniques with explainable AI to improve classification performance. The approach aims to enhance interpretability and transparency in the classification process, particularly for complex multi-class problems like breast cancer subtyping. The DNN [22] is trained to generate textual descriptions of ultrasound images, providing clinicians with informative summaries that aid in diagnosis and interpretation. By improving communication between radiologists and other medical specialists, this strategy promotes more efficient teamwork and decision-making in the detection and treatment of breast cancer. According to Ellenrieder et al. [23], incorporating XAI into radiology education fosters a deeper knowledge of AI-based diagnostic procedures among radiologists, which improves learning outcomes and increases diagnostic accuracy.

12.3 Methods and Materials

12.3.1 Description of Data

The dataset used for evaluation and training comprises a specific number of ultrasound images. It also includes patient demographics and detailed annotations, such as lesion boundaries and pathology results. This indicates the specific preprocessing techniques employed on the dataset, such as normalization, augmentation, or data manipulation. The dataset utilized in this study consists of a compilation of breast ultrasound images acquired from various medical institutions and sources. The ultrasound images are obtained using conventional clinical ultrasound equipment that has different transducer frequencies and imaging settings. The dataset consists of a varied assortment of breast ultrasound images obtained from credible medical institutes and sources. The images are obtained using conventional ultrasound machines that have transducers with frequencies ranging from 5 to 15 MHz. These devices can capture both B-mode and Doppler imaging techniques. The dataset comprises images acquired from patients undergoing a range of breast ultrasonography tests, encompassing screening, diagnostic assessment, and subsequent procedures for monitoring.

The annotations in the dataset are methodically conducted by skilled radiologists or clinicians. Annotations commonly entail defining regions of interest (ROIs) that correlate to breast lesions, if they exist. The annotations may consist of bounding boxes or segmentation masks that delineate the boundaries of suspicious lesions. They are accompanied by pathology data that indicate the type of the lesion, such as whether it is benign or malignant. The ground truth annotations undergo thorough verification by consensus readings and histological confirmation if possible, guaranteeing the dataset's trustworthiness and accuracy. As a way to confirm the integrity and accuracy of the data, the dataset is subjected to meticulous preparation. Preprocessing includes activities such as resizing images, normalizing them, and removing artifacts. Furthermore, augmentation approaches are utilized to improve the variety of the dataset and strengthen the resilience of the method. Augmentation techniques encompass various picture transformations such as rotation, scaling, flipping, and translation. In addition, techniques are employed to tackle the disparity in class distribution, specifically between benign and malignant lesions, to achieve a balanced representation in the dataset. The use example is effectively divided into training, validation, and test sets to aid in model building, hyperparameter tweaking, and performance evaluation. Usually, between 70% and 80% of the dataset is allocated to the training set, which is the set that receives the most attention. The test sets and validation sets then share the remaining amount. Apart from offering discrete datasets for verification and evaluation, this division guarantees that the technique is instructed on an extensive and varied use sample. Patient demographics like age, gender, and relevant clinical history may be included in the dataset. Stringent protocols are implemented to guarantee the privacy and confidentiality of patients, whereby all data is rendered anonymous and stripped of identifying information in accordance with ethical principles and legislation. Approval from the Institutional Review Board (IRB) and informed consent are acquired as required to maintain ethical standards in the utilization of data for research objectives. Figure 12.1 describes the process of the dataset.

12.3.1.1 Image Acquisition

The ultrasonic images are obtained utilizing transducers with frequencies ranging from 5 to 15 MHz, which may capture both B-mode and Doppler imaging techniques. Images are acquired from patients who are undergoing breast ultrasound tests for several purposes, such as screening, diagnostic assessment, and follow-up. During the acquisition of ultrasound pictures from the dataset, careful measures are taken to guarantee the quality, relevance, and integrity of the data for future analysis. At first, photos are chosen according to predetermined criteria that are relevant to the goals of the study. These criteria include information about the patients, their clinical conditions, and the features of the lesions. The dataset is used to extract and arrange photos that fulfill specific criteria. These images are organized in standardized formats to ensure compatibility across different imaging equipment. Then, preprocessing techniques, including scaling, normalization, and artifact removal, are used to enhance the consistency and quality of the images. In addition, the annotations linked to the chosen photos are thoroughly examined to confirm their veracity, usually by means of expert evaluations and histopathology reports. Quality control techniques are employed to rectify any flaws detected throughout the acquisition process, guaranteeing that the dataset upholds elevated levels of integrity. Then, to guarantee fair representation of various diseases and imaging settings, the dataset is divided into test, validation, and training sets. The utmost importance is placed on ethical considerations, and rigorous steps are implemented to protect patient privacy and adhere to ethical rules and legislation. By using this thorough picture capture approach, researchers may guarantee the dependability and ethical soundness of the dataset, establishing a strong basis for later analysis and model building.

12.3.1.2 Annotation and Ground Truth

Experienced radiologists or clinicians mark each ultrasound image in the dataset to outline ROIs that correlate to breast lesions, if they are present. Annotations encompass bounding boxes or segmentation masks that delineate the perimeters of suspicious lesions, along with corresponding pathology findings (e.g., benign or malignant). The accuracy of the ground truth annotations is ensured by conducting consensus readings and, whenever possible, confirming them through histological examination.

12.3.1.3 Data Preparation

Data preparation is an essential phase in making ultrasonic pictures for exploration using deep learning techniques. It guarantees that the data is of excellent quality and appropriate for model training. An essential preprocessing step is picture normalization, which seeks to standardize the intensity values of pixels across images. The mathematical expression for this procedure is

$$\text{Normalized Image} = \text{Original Image} - \text{Mean}/\text{Standard Deviation}$$

where the mean and standard deviation are calculated using the complete dataset. Normalization aids in reducing fluctuations in picture intensity levels, facilitating more consistent and effective model convergence during the training process. Additionally, data intensification methods like rotation and flipping are used to increase the resilience of the model and improve the diversity of the dataset. In mathematical terms, augmentation refers to the process of modifying the initial picture I by applying transformation matrices T:

$$I_{\text{augmented}} = T.I$$

These modifications mimic changes to imaging conditions and improve the model's ability to generalize to new data. Furthermore, synthetic minority oversampling techniques, or fake data production systems, can be used to address unbalanced datasets that include underrepresented classes. SMOTE creates artificial samples for the underrepresented class by interpolating between adjacent examples. From a mathematical perspective, this can be expressed as

$$\text{Synthetic Sample} = \text{Random}\ (\text{sample}_1, \text{sample}_2)$$

where Sample_1 and Sample_2 are randomly chosen cases from the minority class. By undergoing these preprocessing procedures, ultrasound pictures are converted into a refined, equitable, and standardized dataset, establishing the foundation for efficient deep learning model training and precise breast cancer detection.

12.3.1.4 Dataset Splitting

The dataset is divided into training, validation, and test sets to make model building, hyperparameter adjustment, and performance evaluation easier. Usually, the training phase of the dataset is assigned a larger percentage (about 70–80%), while the remaining part is divided equally between validation and testing (around 10–15% each). Mathematically, the process of splitting a dataset can be expressed as follows:

Let O denote the original dataset containing M samples, signified as $O = \{(x_i, y_i)\}^{M}_{i=1}$, where x_i is the input, i.e., an ultrasound image, and y_i is the equivalent label (e.g., benign or malignant).

There are three subsets within the dataset.

Training Set (O_{train}):

- Typically, the majority of the dataset is allocated to the training set to facilitate model learning. Let M_{train} indicate how many samples there are in the training set. Mathematically, one way to visualize the training set is as $(O_{\text{train}}) = \{(x_i, y_i)\}^{M\text{train}}_{i=1}$.

Validation Set (O_{val}):

- To assist in overfitting, the validation set is used to calculate the performance of the method and adjust hyperparameters in training. Let M_{val} denote the number of trials in the validation set.
- The validation set may be expressed mathematically as $(O_{\text{val}}) = \{(x_i, y_i)\}^{M\text{val}}_{i=1}$.

Test Set (O_{test}):

- The test set is reserved for the final analysis of the trained model's effectiveness, offering a dispassionate appraisal of its capacity for generalization. Let M_{test} be the sample count in the test set.
- The test set may be expressed mathematically as $(O_{\text{test}}) = \{(x_i, y_i)\}^{M\text{test}}_{i=1}$.

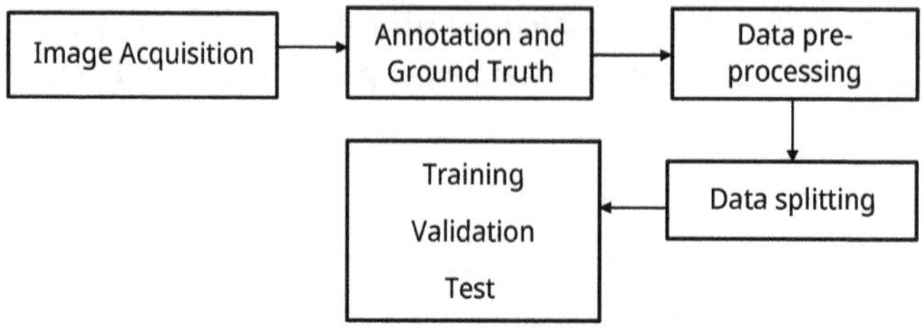

Figure 12.1: Process of dataset.

12.3.1.5 Dataset Characteristics

The collection consists of diverse ultrasound images that depict several breast diseases, such as benign tumors, malignant lesions, cysts, fibroadenomas, and other anomalies. To enable subgroup analysis and risk stratification, patient demographics such as age, gender, and clinical history may be provided. The dataset may encompass various imaging methods, equipment manufacturers, and common imaging artifacts found in clinical practice. A wide range of characteristics is included in the dataset chosen for deep learning-based ultrasound analysis for breast cancer detection, all of which are essential for the careful building and exacting assessment of the model. Because there are a lot of ultrasound pictures in the dataset, there is enough information for training, validation, and testing. This ensures that the produced models are strong and can be applied to many situations. The dataset is carefully annotated by qualified radiologists or doctors, ensuring that the annotations are of high quality. The annotations are validated using consensus readings and histological confirmation. This guarantees the precision and dependability of the ground truth labels linked to each image, which is essential for ensuring the credibility of model training and evaluation. The dataset encompasses a wide spectrum of breast diseases, encompassing both benign and malignant tumors, cysts, fibroadenomas, and other abnormalities. This comprehensive representation accurately mirrors the intricate and diverse nature of real-world clinical settings. By applying preprocessing techniques such as normalization, artifact removal, and data augmentation, the dataset is standardized and improved to reduce differences in imaging settings and maintain uniformity among images. In addition, ethical concerns are thoroughly examined, ensuring that patient privacy and confidentiality are maintained by anonymizing data and adhering to regulatory protocols. The dataset possesses unique attributes such as its size, diversity, annotation quality, and ethical compliance, which make it an invaluable resource for furthering deep learning-based breast cancer diagnosis and enhancing patient care and outcomes.

12.3.2 Deep Learning Architecture

Name the particular architecture of the deep learning model (e.g., CNN, U-Net, or other pertinent models) that is being used for the analysis of breast ultrasonography. The architecture of deep learning (Figure 12.2) consists of many layers, including convolutional, pooling, and fully connected layers. In this study, data from breast ultrasound is processed using a deep learning architecture built on convolutional neural networks. CNNs have proven to be quite successful in tasks like feature extraction, segmentation, and image classification. The architecture is specially designed to efficiently detect and classify breast lesions while leveraging XAI methods to guarantee interpretability. It is tailored to match the exact requirements of ultrasound image processing.

12.3.2.1 Convolutional Layers

The several convolutional layers of the CNN structure are responsible for extracting hierarchical information from the input ultrasound images. The convolutional layer uses trainable filters to perform convolution on the input image to extract localized patterns and features at various spatial scales. Activation functions like ReLU are frequently used after convolutional processes to provide non-linearity and enhance the model's capacity to represent inputs.

12.3.2.2 Pooling Layers

Convolutional layers produce feature maps that are larger than desired. Pooling layers, including max pooling or average pooling, are used to minimize this problem. Pooling approaches reduce the size of feature maps while maintaining the most important properties, which improves computational efficiency and translation invariance.

12.3.2.3 Dense Layers

Often referred to as completely connected layers, dense layers are utilized to group and categorize high-level characteristics that have been gathered from earlier layers. By including dense layers in the model, it becomes possible to identify complex patterns and correlations between features, which in turn permits accurate distinction between different types of breast lesions.

12.3.2.4 Integration of XAI Methods

The architecture not only includes typical CNN layers but also integrates XAI methods to improve interpretability and transparency. XAI tools, such as gradient-based methods like Grad-CAM, attention mechanisms, or model explanation algorithms such as LIME and SHAP, are incorporated into the network to provide visualizations and interpretations of model predictions. The XAI techniques provide insightful information on the particular regions of interest in ultrasound images as well as the attributes that affect the model's decision-making process. This allows doctors to comprehend and have confidence in the predictions made by the model.

12.3.2.5 Model Optimization and Training

The labeled ultrasound images and the corresponding ground truth annotations are used to train the deep learning architecture. Stochastic gradient descent (SGD) and Adam are two examples of optimization techniques that are used to gradually appraise the model constraints and diminish the loss function. The generalizability of the model may be improved, and overfitting can be reduced by using regularization techniques like dropout and weight decay.

12.3.2.6 Stochastic Gradient Descent (SGD)

In SGD, the learning rate (α) multiplied by the gradient of the loss function in relation to each constraint is used to improve the model parameters. The weight update rule for parameter θ at iteration t may be obtained mathematically as follows:

$$\theta^{(t+1)} = \theta^{(t)} - \alpha^{*}\nabla\ \theta^{L}$$

where $\nabla\ \theta^{L}$ denotes the gradient of the loss function L in relation to the parameter θ.

12.3.2.6 Adam (Adaptive Moment Estimation)

Adam is an adaptive optimization method that computes unique adaptive learning rates for each parameter by estimating the first and second moments of gradients. The parameter θ update rule in Adam is provided by

$$n^{(t+1)} = \beta 1 * n(t) + (1 - \beta 1) * \nabla\theta L$$

$$v^{(t+1)} = \beta 2 * v(t) + (1 - \beta 2) * (\nabla\theta L)2$$

$$n = \frac{n(t+1)}{1-\beta^{t+1}}$$

$$v = \frac{v(t+1)}{1-\beta^{t+1}}$$

$$\Theta^{(t+1)} = \Theta^{(t)} - \alpha * n/v + €$$

where n and v are bias-corrected estimates, € is a tiny constant to prevent division by zero, β_1 and β_2 are decay rates, and $n^{(t)}$ and $v^{(t)}$ are the first and second instant appraisals of the gradients.

12.3.2.7 Dropout

As a way to avoid overfitting, dropout is a regularization method that arbitrarily eliminates a proportion p of neurons during training. Mathematically, the output of each neuron is multiplied by a binary mask y with probability p:$h_{\text{dropout}} = \gamma * h$.

Where γ is drawn from a Bernoulli distribution with probability $1 - p$.

12.3.2.8 Weight Decay (L2 Regularization)

By appending a term proportional to the square of the L2 norm of the weights to the loss function, weight decay is a regularization approach that penalizes large weights. The regularized loss function L_{reg} is given by y:

$$L_{\text{reg}} = L + \lambda/2 \sum_{i}^{\theta} \theta 2i$$

where λ is the regularization strength.

12.3.2.8 Assessment and Performance Measures

We test the trained model on a new test set of ultrasound images to see how well it can detect and classify breast cancer. The diagnostic performance and resilience of the model are assessed by computing performance measures such as accuracy, sensitivity, specificity, area under the receiver operating characteristic curve (AUC-ROC), and F1 score.

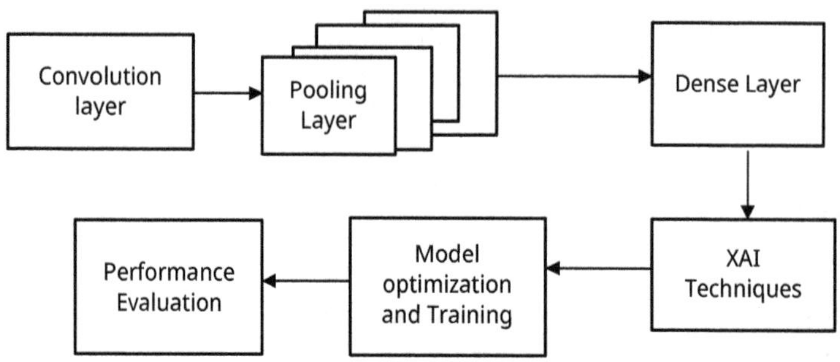

Figure 12.2: Deep learning architecture.

12.3.3 Methods for Explainable Artificial Intelligence (XAI)

The deep learning model has included XAI techniques to improve interpretability. Figure 12.3 represents a summary of the chosen XAI methodologies, including gradient-based methods like Grad-CAM, attention mechanisms, and model explanation techniques such as LIME and SHAP. XAI approaches are used to visualize and comprehend deep learning model predictions, stressing how they have helped to increase dependability and openness. The goal of XAI techniques is to make machine learning models' decision-making processes transparent so that users can comprehend and rely on the models' predictions. When XAI approaches are integrated with deep learning-based ultrasound analysis for breast cancer diagnosis, comprehension and interpretation of the data are enhanced. Acquiring acceptability and adoption in therapeutic settings is contingent upon this. Here are several XAI methods that are often used.

12.3.3.1 Gradient-Based Methods

Grad-CAM, or gradient-weighted class activation mapping, is a computer vision technique that emphasizes the regions of an image that are crucial for a neural network's prediction. Grad-CAM generates class activation maps by computing the gradients of the target class score with respect to the feature maps of the final convolutional layer. This draws attention to the precise regions of the input picture that affect the representation's computation the most. Grad-CAM generates heatmaps that are difficult to read, while SmoothGrad minimizes the amount of unwanted noise and visual distortions. Consequently, the justifications gain plausibility and consistency.

12.3.3.2 Attention Mechanisms

The self-attention mechanism is a process that assigns different weights of attention to different regions of an input picture. This helps the method to focus on important fragments of the image and ignore irrelevant information. Another type of model that can use spatial transformations to dynamically alter input images is called a spatial transformer. By intelligently aligning and cropping regions of interest, this model enables improved feature extraction and classification.

12.3.3.3 Techniques for Explaining Models

A technique called limited interpretable model-agnostic explanations (LIME) provides human-understandable descriptions for particular model estimates. This is accomplished by first producing altered samples around the particular instance of interest and then simulating the behavior of the complex model in that restricted region using a basic interpretable model. Shapley standards, which are derived from cooperative game theory, are used by the computational technique Shapley additive explanations (SHAP) to quantify the effect of every attribute on a model's output. SHAP standards deliver a thorough understanding of the importance of the features and the influence of their interactions within the model.

12.3.3.4 Rule-Based Approaches

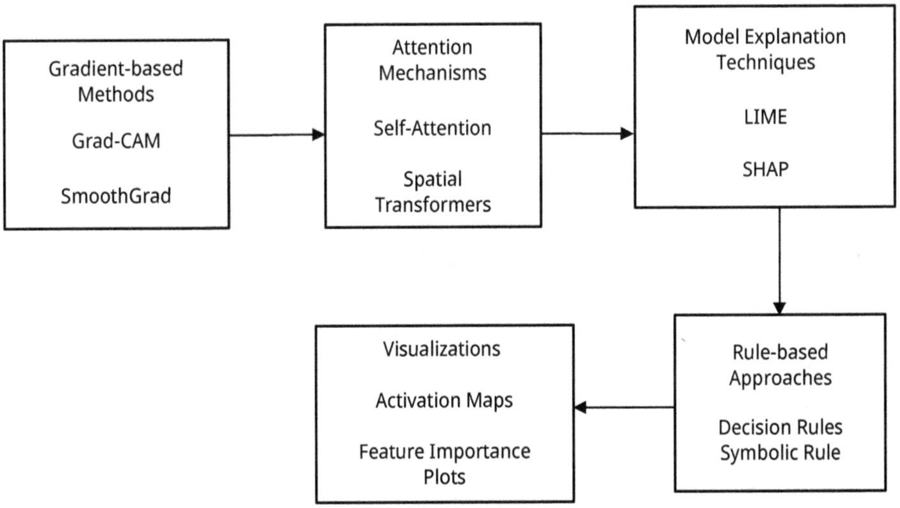

Figure 12.3: Methods for explainable artificial intelligence.

Decision rules are derived from the decision limits of a trained model. They offer clear and easily understandable criteria for making decisions in classification tasks, making them transparent and accessible to humans. **Symbolic rule extraction** methods produce symbolic rules or decision trees from opaque models, providing a clear description of the model's decision-making process.

12.3.3.5 Visualizations and Explanations

Activation maps depict the activation patterns of individual neurons or channels in the deep learning model, offering insights into the properties identified at various layers of the network. **Feature relevance plots** rank and visually represent the relevance of input characteristics by measuring their impact on model predictions. They are useful for selecting features and gaining insights into model behavior.

12.3.3.6 Model Training and Evaluation

They provide a detailed explanation of the training process, taking into account the batch size, learning rate, number of epochs, and optimization method. Examine a number of regularization strategies, including weight decay and dropout, which are meant to reduce overfitting. Enumerate the evaluation metrics (specificity, sensitivity, accuracy, and area under the receiver operating characteristic curve) that are applied to evaluate the performance of the deep learning model (AUC-ROC). The validation set, often referred to in the context of cross-validation, is used to determine the generalizability of the model and modify its hyperparameters.

12.3.3.7 Training Procedure

For breast ultrasound analysis, the deep learning model is trained using labeled ultrasound images and the accompanying ground truth annotations. The dataset is usually divided into three groups: test, validation, and training sets. Typically, the training set contains between 70% and 80% of the data. Optimization techniques such as Adam, RMSprop, and SGD are used to reduce the loss function and iteratively inform the classical parameters. The validation set is used to optimize the model's performance by adjusting hyperparameters such batch size, epoch count, and learning rate.

12.3.3.8 Data augmentation and Regularization

Rotation, scaling, flipping, translation, and other techniques are applied to add greater variety and resilience to the training dataset. Regularization methods, including weight decay, batch normalization, and dropout, are employed to reduce overfitting and enhance a model's capacity for generalization. To prevent the coadaptation of features, dropout is utilized to arbitrarily deactivate a subset of neurons during training. Batch normalization helps in smoothing and quickening the training process by regularizing each layer's activations. Large weights are penalized by weight decay, commonly referred to as L2 regularization, which reduces overfitting and increases sparsity.

12.3.3.9 Validation and Hyperparameter Tuning

The purpose of the validation set is to determine the optimal hyperparameters and assess the model's performance throughout training. By evaluating the outcomes on the validation set, hyperparameters like the learning rate, dropout rate, and network design are changed to increase the model's ability to generalize. Some tactics that may be used to reduce overfitting and improve convergence include learning rate scheduling and early stopping.

12.3.3.10 Model Evaluation

A unique test set of ultrasound pictures that were not seen during the training or validation phase is used to evaluate the learned model. The diagnostic performance of the model is evaluated by the computation of performance indicators, which include specificity, sensitivity, accuracy, area under the receiver operating characteristic curve (AUC-ROC), and F1 score. ROC curves and confusion matrices are frequently used to visually depict a model's classification performance at different thresholds. Techniques like paired t-tests or Wilcoxon signed-rank tests can be used to measure the performance of many models or experimental conditions using statistical significance testing. XAI techniques are included in the evaluation pipeline to provide a more comprehensive understanding of the model's decision-making process. Transparency and dependability are increased by producing visualizations and interpretations of the model's calculations through the use of gradient-based algorithms, attention processes, and model explanation tools. One can conduct a qualitative study of explanations given by XAI to verify the model's reasoning and detect possible flaws or biases.

12.3.4 Experimental Configuration

The experimental setting required meticulous planning and configuration to guarantee accurate outcomes in the deep learning-based ultrasound analysis for breast cancer screening. The hardware utilized comprised NVIDIA GPUs equipped with CUDA support, enabling efficient model training and evaluation.

To conduct the study, current training on breast cancer prediction was followed. There are several deep learning and machine learning techniques available for cancer detection and prediction. Preprocessing techniques like the label encoder and normalizer method were applied to handle the data efficiently. The bar graph in Figure 12.4 displays the total number of benign and malignant cells in the dataset that was obtained.

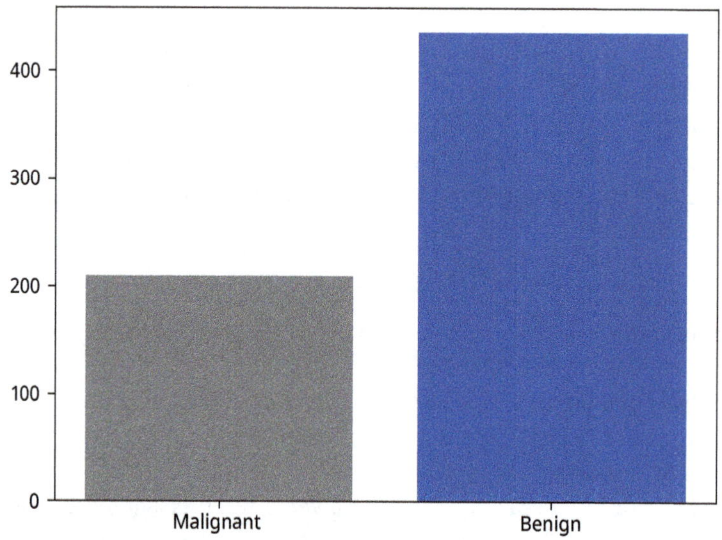

Figure 12.4: Total number of benign and malignant cells in the dataset.

Several machine learning approaches were used to predict breast cancer on the Wisconsin dataset, including K-nearest neighbor, support vector machine (SVM), decision tree, Naïve Bayes, logistic regression, and random forest. The greatest accuracy achieved by the random forest and SVM algorithms was 96.5%. Using deep learning techniques, the forecast accuracy was increased. In Figures 12.5 and 12.6, the accuracy and loss of the model are graphically represented in proportion to the number of epochs. As the number of epochs increases, the model's accuracy rises and its loss falls.

The CNN model yielded an accuracy of 97.3%, while the ANN model produced an accuracy of 99.3%, indicating greater efficiency compared to the machine learning

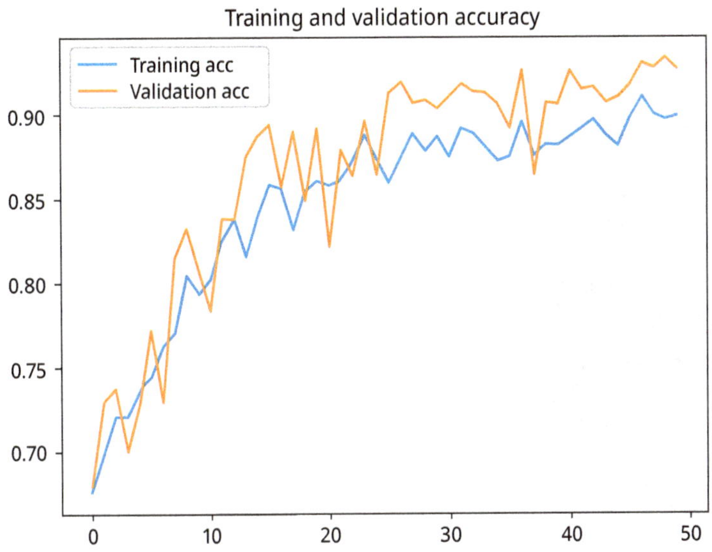

Figure 12.5: Accuracy.

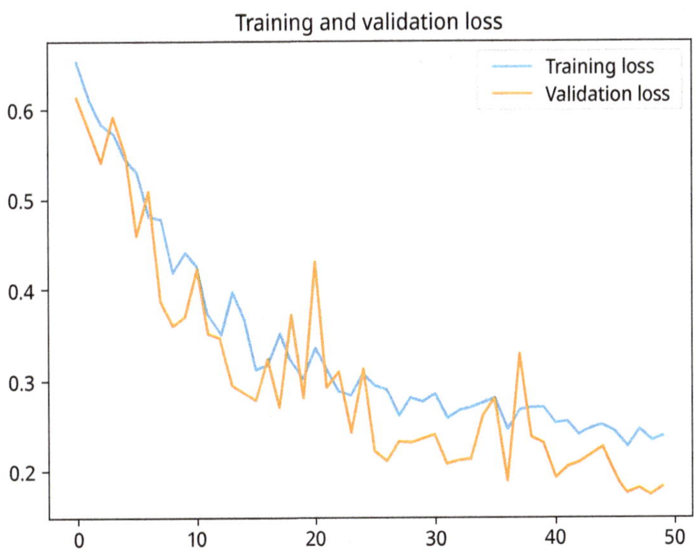

Figure 12.6: Loss.

methods mentioned earlier. ReLU and sigmoid are examples of activation functions that were used in deep learning, demonstrating greater efficiency. Consider the given input image (Figure 12.7) as it is passed into the developed model for predicting the presence of breast cancer.

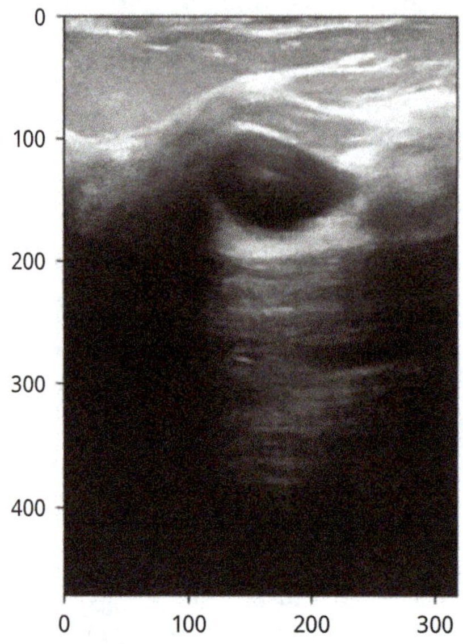

Figure 12.7: Input image.

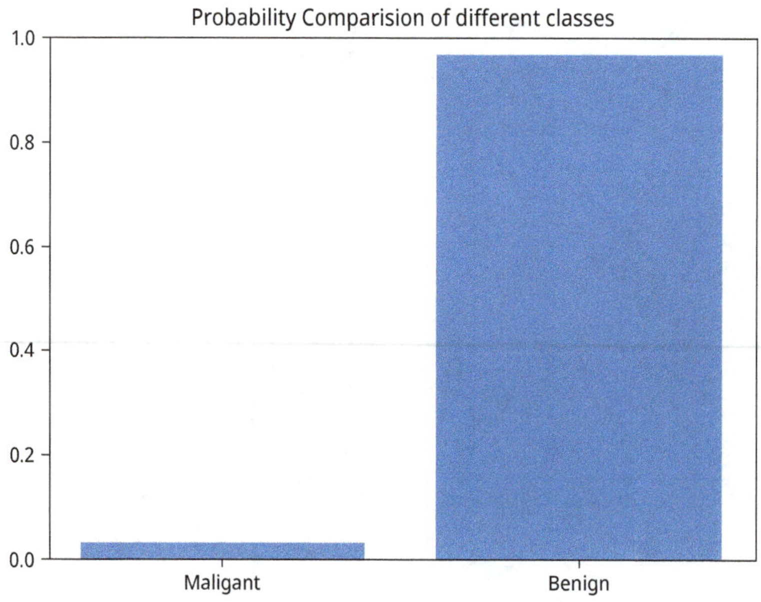

Figure 12.8: Probability of predicted output.

12.3.4.1 Probability of Predicted Output

Figure 12.8 shows the probability comparison of different classes. The maximum accuracy achieved was 96.5%.

12.4 Conclusion

Beyond its interpretability and diagnostic accuracy, the combination of deep learning-based ultrasound analysis with XAI approaches offers substantial advantages for clinical practice. By providing concise justifications for diagnostic results, XAI enhances clinician confidence in AI-assisted diagnostics and facilitates seamless incorporation into conventional healthcare procedures. Furthermore, real-world input facilitates the iterative creation and adaptation of these models to evolving clinical demands and challenges. This means that, in addition to improving the efficacy and precision of breast cancer diagnosis, this technology may also encourage innovation in the delivery of oncology therapy, ultimately benefiting patients.

References

[1] Karimzadeh, M. An Explainable Multi-Task Neural Network Model for Breast Cancer Detection in Ultrasound Images (Master's thesis, University of Idaho).

[2] Naas, M., Mzoughi, H., Njeh, I., & Slima, M. B. A Deep Learning Based Computer Aided Diagnosis (Cad) Tool Supported by Explainable Artificial Intelligence for Breast Cancer Exploration. Available at SSRN 4689420.

[3] Cerekci, E., Alis, D., Denizoglu, N., Camurdan, O., Seker, M. E., Ozer, C., Hansu, M. Y., Tanyel, T., Oksuz, I., & Karaarslan, E. (2024 Apr 1). Quantitative evaluation of Saliency-Based Explainable artificial intelligence (XAI) methods in Deep Learning-Based mammogram analysis. *European Journal of Radiology*, *173*, 111356.

[4] Munshi, R. M., Cascone, L., Alturki, N., Saidani, O., Alshardan, A., & Umer, M. (2024 Feb 1). A novel approach for breast cancer detection using optimized ensemble learning framework and XAI. *Image and Vision Computing*, *142*, 104910.

[5] Podila, V. S. (2022). Incorporating prior knowledge in learning physically consistent inertial parameters of a robotic manipulator online. McGill University (Canada).

[6] Rao, K. S., Terlapu, P. V., Jayaram, D., Raju, K. K., Kumar, G. K., Pemula, R., Gopalachari, V., & Rakesh, S. Intelligent ultrasound imaging for enhanced breast cancer diagnosis: Ensemble transfer learning strategies. IEEE Access. 2024 Jan 25.

[7] Sindiramutty, S. R., Tee, W. J., Balakrishnan, S., Kaur, S., Thangaveloo, R., Jazri, H., Khan, N. A., Gharib, A., & Manchuri, A. R. Explainable AI in Healthcare Application. In Advances in Explainable AI Applications for Smart Cities 2024 (pp. 123–176). IGI Global.

[8] Yuldashevna, U. S., Abduzugurovna, V. N., Ismoilovna, T. D., Yuldashevna, K. G., Ananth, C., & Kumar, T. A. Exploring Synonymy: Artificial Intelligence's Computational Analysis of its Relationships with Other Lexical Systems. In2024 International Conference on Advances in Computing, Communication and Applied Informatics (ACCAI) 2024 May 9 (pp. 1–6). IEEE.

[9] Ma, J., Kong, D., Wu, F., Bao, L., Yuan, J., & Liu, Y. (2024 Jan 1). Densely connected convolutional networks for ultrasound image based lesion segmentation. *Computers in Biology and Medicine*, *168*, 107725.

[10] Prodan, M., Paraschiv, E., & Stanciu, A. (2023 Mar 28). Applying deep learning methods for mammography analysis and breast cancer detection. *Applied Sciences*, *13*(7), 4272.

[11] Maouche, I., Terrissa, L. S., Benmohammed, K., & Zerhouni, N. (2023 Jun 5). An explainable ai approach for breast cancer metastasis prediction based on clinicopathological data. *IEEE Transactions on Biomedical Engineering*, *70*(12), 3321–3329.

[12] Ghnemat, R., Alodibat, S., & Abu Al-Haija, Q. (2023 Aug 30). Explainable Artificial Intelligence (XAI) for deep learning based medical imaging classification. *Journal of Imaging*, *9*(9), 177.

[13] Peta, J., & Koppu, S. (2024 Apr 1). Explainable soft attentive efficientNet for breast cancer classification in histopathological images. *Biomedical Signal Processing and Control*, *90*, 105828.

[14] Dhiman, H., Vyas, N., Pagrotra, A., & Dutt, V. Real-Time Traffic Analysis and Driver Safety Enhancement via IoT and Computer Vision with High-Resolution Project Lamps for Intelligent Road Illumination [Internet]. 2023 1st International Conference on Circuits, Power and Intelligent Systems (CCPIS). IEEE; 2023. Available from: http://dx.doi.org/10.1109/CCPIS59145.2023.10291872

[15] Kumar, R., Singh, S., Chauhan, S., Anand, A., & Kumar, A. (2023). ED based spectrum sensing over IRS-assisted Rayleigh-FTR fading channels. *AEU-International Journal of Electronics and Communications*, *171*, 154908.

[16] Gulum, M. A., Trombley, C. M., & Kantardzic, M. (2021 May 17). A review of explainable deep learning cancer detection models in medical imaging. *Applied Sciences*, *11*(10), 4573.

[17] Chaddad, A., Peng, J., Xu, J., & Bouridane, A. (2023 Jan 5). Survey of explainable AI techniques in healthcare. *Sensors*, *23*(2), 634.

[18] Jones, M. A., Islam, W., Faiz, R., Chen, X., & Zheng, B. (2022 Aug 31). Applying artificial intelligence technology to assist with breast cancer diagnosis and prognosis prediction. *Frontiers in Oncology*, *12*, 980793.

[19] Kanimozhi, P., Kannan, M., Ananth, C., Kumar, T. A., & Parkavi, S. Design of IoT Based Agricultural Farm Protector Drone Using Raspberry Pi. In 2023 6th International Conference on Contemporary Computing and Informatics (IC3I) 2023 Sep 14 (Vol. 6, pp. 1175–1179). IEEE.

[20] De Miranda I. M. (2023) Integrating Anatomical Prior Knowledge for Increased Generalisability in Breast Cancer Multi-center Data.

[21] Meshoul, S., Batouche, A., Shaiba, H., & AlBinali, S. (2022 Nov 15). Explainable Multi-Class Classification Based on Integrative Feature Selection for Breast Cancer Subtyping. *Mathematics*, *10*(22), 4271.

[22] Carrilero Mardones, M. A Deep Neural Network For Describing Breast Ultrasound Images in Natural Language.

[23] Ellenrieder, S., Kallina, E. M., Pumplun, L., Gawlitza, J. F., Ziegelmayer, S., & Buxmann, P. Promoting Learning Through Explainable Artificial Intelligence: An Experimental Study in Radiology.

R. Manjula Devi and S. Dheenathayalan

13 Explainable Artificial Intelligence Techniques in Deep Learning-Based Liver Tumor Analysis

Abstract: As medical imaging technology develops, artificial intelligence (AI) can help radiologists and physicians classify liver tumors accurately and significantly. However, the inherent intricacy of AI models, primarily deep learning algorithms, commonly precludes their adoption because their decision-making processes are opaque and difficult to explain. This chapter proposes a complete architecture that achieves high classification accuracy and offers interpretable perceptions of the model's choice, addressing the important requirement for explainable AI in liver tumor classification. The proposed system combines existing deep learning architectures with cutting-edge explainable AI (XAI) XAI methods to produce a hybrid model that syndicates predicted accuracy with understandable validations. The model may concentrate on essential areas of medical images and emphasize discriminative features that contribute to its classification conclusions because the framework implements attention mechanisms and feature visualization techniques.

Keywords: Model clarification, convolutional neural networks, SHapley Additive exPlanations, local interpretable model-agnostic explanations, partial dependence plots

13.1 Introduction

The use of machine learning (ML) algorithms has increased over the past decades because artificial intelligence (AI) has seen tremendous and ongoing advancement. These astounding achievements are closely related to opaque, black-box AI models and increased model complexity proposed by Bakrania et al. [1]. It is necessary to develop solutions that can help overcome such a challenge to boost the usage of AI systems in delicate and significant fields like healthcare and security, where extra requirements must be met in addition to high accuracy. AI that can explain its decisions or forecasts to human users is known as explainable AI (XAI). XAI aims to improve the openness, dependability, and accountability of AI systems, mainly when applied in high-risk fields like security, finance, or healthcare. The process used in explainable AI is depicted in Figure 13.1. Various methods inspect the liver tumor before the medical data is provided to the XAI framework proposed by Peng et al. [2]. The clinicians are then given the proposed XAI technique's computer-aided diagnostic (CAD) results. Finally, medical professionals use CAD to support the diagnosis and treatment of the liver.

https://doi.org/10.1515/9783111436180-013

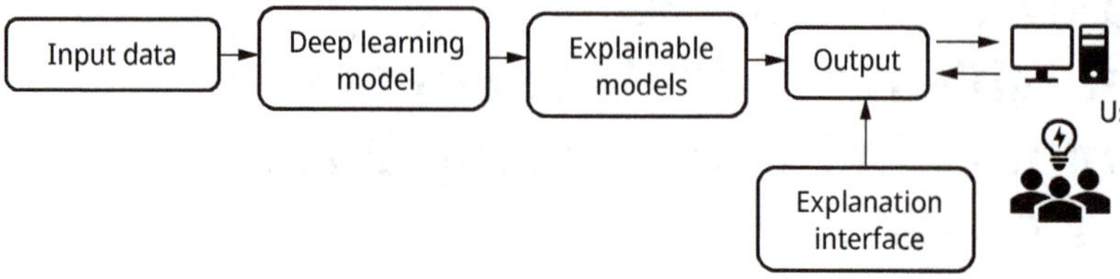

Figure 13.1: Process of explainable AI.

The exercise data will be provided as an input in XAI-based techniques. Based on the requirements, the internal workings of the model are developed to describe the workings of the output interface. As a result, XAI's users are aware of its effects, boosting trust in AI systems. Users can increase the accuracy of the results and reveal model defects based on the understanding of the output, which will help users make the best decision for model improvement. The proposed XAI model is outlined in Figure 13.2. To recover the medical understanding of the liver tumor, the proposed paradigm dispenses both global and local explanations.

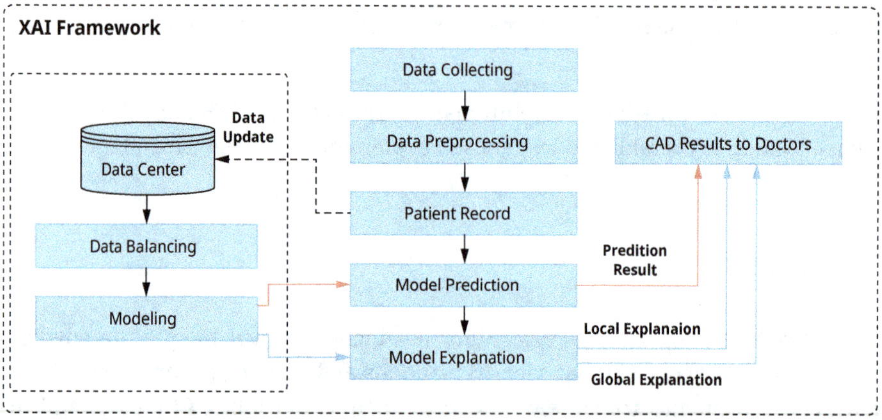

Figure 13.2: The outline of XAI.

Data collection and preprocessing are done to acquire the patient records. The model is then loaded to forecast how the exacerbation risk will pan out. The global and local explanations are then obtained using the model explanation technique. The results of the model and the explanation will then be reviewed and further validated by the radiologist.

13.2 Literature Review

Ahmed and Shelly [3] created a stack-ensemble-based ML model framework using XAI to depict the association between the risk factors for lung cancer and the spatial distribution of LBC. The XAI in the disciplines of drugs, psychology, medical structures, and other areas is explained by [4]. However, prognostics and health management (PHM) did not offer the analytical method to decrease patient waiting times, enhance services, boost productivity in telemedicine, and increase accuracy, speed, and efficiency.

Andreu-Perez and Emberson [5] created an XAI-based model for investigating the development and functionality of baby brain images captured by near-infrared spectroscopy. Atul Anand and Tushar Kadian [6] proposed deep neural networks for identifying heart diseases using ECG signals. Bakrania et al. [1] discussed the challenges of AI in liver cancer diagnosis and management. In addition, they discussed the future perspectives of AI application in liver cancer and how a multidisciplinary approach using AI in nanomedicine could accelerate the transition of personalized liver cancer medicine from bench side to the clinic. Kumar et al. [7] proposed investigating the applicability of AI techniques in photovoltaic systems. Fang and Tan [8] created a model to employ patient characteristics to determine the likelihood that a specific disease would progress. However, they need more transparency in clinical practice.

Introduced a classifier based on the convolutional neural network for finding lashes and outlined the saliency maps used to trace the output of the model. Jahmunah et al. [9]have created dense net and CNN models to predict abnormalities in EEG signals. In healthcare applications, AI and ML are crucial, and hence, XAI is employed to earn the conviction of AI models. It will increase confidence in extrapolative models in real-world scenarios.

Conducted research on skin malignancies for both males and females. In their work, biomarkers were used for predicting skin cancer, along with trained XGboost ML models. Jahmunah and Ng [11] discussed the ideas and methods of XAI-based cardiology. Kute and Pradhan [12] proposed that XAI is characterized by knowledge- and data-based strategies. Knowledge-driven XAI requires knowledge about how to explain, whereas data-driven XAI needs local, global, and instance-specific procedures for explanation. They used the publicly accessible PTB-XL output dataset. XAI explanation was given a new meaning by Merry and Riddle [13]. Miro-Nicolau and Moya-Alcover [14] observed the results using XAI chest images. X-ray images were represented using class activation maps (CAM) in their work. Obayya and Nemri [15] developed XAI approaches in teleophthalmology. The accuracy of the categorization model was 98.24% by utilizing the SPECT MPI pictures. Papandrianos Feleki [16] created a deep XAI approach to envisage coronary artery disease (CAD).

Peng and Zou [2] proposed a framework combining global and local interpretable methods to improve the transparency of complex models and gain insight into the judgments from these models, thereby guiding treatment strategies and improving the prognosis of hepatitis patients. Pertzborn and Arolt [17] presented a VGG 19 model to detect malignancies. In their work, deep learning-based models combined with XAI techniques and preprocessing data led to an accuracy of 80%. According to Lombardi and Diacono [18], it is possible to predict Alzheimer's disease (AD) and cognitive impairment using XAI. The disease has been identified using random forest classifiers. Datasets for monitoring patients' health are gathered from various sources. Some information is collected through sensors, while other data comes directly from clinical trials and other studies proposed by Yang [19]. To identify and perceive brain cancers in clinical areas, Zeineldin et al. [20] developed an XAI framework in neurology for analyzing brain tumors.

13.3 Proposed Methodology

The progress of a CAD system for CT liver diagnosis based on AI has advanced significantly in recent years. Due to the nonlinearity of the clinical data collected from hospitals, deep learning-based models perform better than ML methods. However, radiologists are suspicious of complicated models because they operate like a black box that superintends the reasons behind their decisions. Hence, this chapter proposes an XAI framework to address these problems by providing global and local interpretations of liver tumor diagnoses while maintaining high prediction accuracy. First, the framework's viability is examined using a public dataset collected from a multispecialty scan center. Both apparent and black-box ML models are used to better the classification performance. Finally, the SHapley Additive exPlanations and partial dependence plots are utilized to enhance the model interpretation of liver tumors. In the proposed system, the data taken from the patient is initially stored in the design and given to the XAI framework for further processing, and the results are again stored in the server. Then, it is fed as an input to the CAD system. Once the results are obtained from the CAD system, the radiologist compares the results obtained from the XAI framework and CAD system. The suggested prototype offers global and local explanations to enhance medical empathy of the risk prediction for liver tumors. Data collection and preparation are used to create patient records. The model is then loaded to forecast how the exacerbation risk will pan out. The global and local explanations are then obtained using the model explanation technique. Doctors receive them to examine and further validate the prediction and explanation outcomes.

13.3.1 Dataset Description

In this work, CT abdomen images are obtained from the freely available database (https://www.ircad.fr/research/data-sets/liver-segmentation-3d-ircadb-01/) 3DIRCADB (3D image reconstruction for comparison of algorithm database) and the private dataset obtained from multispecialty scan centres. Compared with the other challenge datasets, the 3DIRCADB dataset has a higher diversity and complexity of livers and lesions. The widely available 3DIRCADB dataset encloses 20 venous phases enhanced CT scans of the liver hepatic tumors of 75% of cases by considering 10 women and 10 men. The dimension of the slice is 511 × 511 pixels, with height and width varying from 0.55 to 0.89 mm and the width between adjacent slices ranging from 1.4 to 5.1 mm. Since the 3DIRCADB dataset offered improved diversity and difficulty of livers and their tumors, it is considered a challenging database. Besides, the dataset collected from the scan center encompasses 130 scans of different patients. A minimum of 84 slices per patient are regarded with the expert segmentations manually provided by the radiologist. These images are obtained from Arthi Scan Center Chennai, Tamilnadu, India.

13.3.2 Preprocessing

Preprocessing is a significant phase to enhance image quality and increase image analysis performance. Recently, several denoising techniques have been proposed for removing Gaussian noise. Moreover, numerous 2D wavelet-based denoising methods have also been established to alter the image into multiple subbands at different angles. Then, each subband's wavelet coefficients are applied to a denoising procedure. These filters effectively smooth the noise but still have difficulty blurring the edges within the images. This work employs a block-based non-local means (NLM) algorithm , median filtering with pixel modelling technique as shown in Figure 13.3.

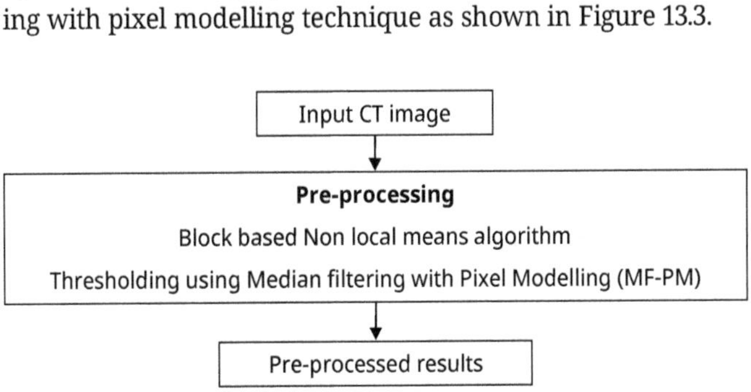

Figure 13.3: Preprocessing methods.

13.3.2.1 Block-Based Non-local Means Algorithm

The traditional NLM algorithm may lead to poor results because of its nonlocality of neighboring pixels. To overcome this, a block-based NLM method is proposed, significantly reducing the computational time. The block-based NLM filter is more concrete in MRI and CT images since they are primarily 3D or stacks of 2D. Thus, in biomedical image denoising, computational efficiency is vital. To maintain global continuity, this work overlaps adjacent blocks by one column or row of pixels. The multiple approximations of the restored intensity value at the pixel xi are then averaged among several approximations. While dealing with large datasets, averaging intensity values produces high computational costs.

Therefore, rather than considering pixels, an enhanced NLM is suggested using blocks of size $(2a+1)^2$ with overlapping constraints. Later, the NLM restoration is performed on these blocks rather than on pixels. The block-based enhanced NLM method is described by

$$NL[u](B_{ik}) = \sum_{B_j \epsilon\, v_{ik}} w(B_{ik}, B_j)\; u(B_j) \tag{13.1}$$

where

$$w(B_{ik}, B_{j,}) = \begin{cases} \frac{1}{Z_{ik}} e^{-\frac{\|u(B_{ik}) - u(B_j)\|^2}{\hat{h}}} \\ 0 \text{ otherwise} \end{cases} \text{if } \mu_1 < \frac{u(B_{ik})}{u(B_j)} < \frac{1}{\mu_1} \text{ and } \sigma_1^2 < \frac{\text{var}(u(B_{ik}))}{\text{var}(u(B_j))} < \frac{1}{\sigma_1^2}$$

Z_{ik} is a constant of normalization satisfying $\sum_{B_j \epsilon\, v_{ik}} w(B_{ik}, B_j) = 1$ and $\hat{h}$ is a parameter for smoothness. In addition, the most similar neighborhoods are selected well in advance to avoid unnecessary weight computation. The weight factors are computed only when the ratio of the mean and variance of $u(B_{ik})$ and $u(B_j)$ are in the range of 1.

13.3.2.2 Thresholding Using Median Filtering with Pixel Modeling (MF–PM) Technique in the Wavelet Domain

Initially, the wavelet transform is applied to the CT abdomen image. Then, the high-frequency subbands are denoised using adaptive thresholding with the Bayes shrinkage function. Finding the optimal threshold value is essential for preserving clinical details and noise suppression. This work obtains the threshold value using noise variance and the standard deviation of noise-free images. Let us consider that α is the threshold value, and it is given by

$$\alpha = \frac{\sigma_\vartheta^2}{\sigma_y} \tag{13.2}$$

where, α is the noise variance that is estimated by the robust median estimation method and it is given by

$$\sigma_\vartheta^2 = \left(\frac{\text{median}(|\text{x}(\text{m}, \text{n})|)}{0.6475} \right)^2 \tag{13.3}$$

$x(m,n) \sim H_L$ represents the level in the wavelet domain. The standard deviation of the noise-free image is given by

$$\sigma_y^2 = \max(\sigma_x^2 - \sigma_\vartheta^2, 0) \tag{13.4}$$

where $\sigma_x^2 = \frac{1}{N}\sum_{i=1}^{N} X_i^2$ and N represents an input image. Then the thresholding function is given by

$$\hat{X} = \{0\, f|X| \leq \alpha \text{ signum}(x)(|X| - \alpha f|X| > \alpha \tag{13.5}$$

The obtained results show that the preprocessing results of the proposed median filtering with pixel modeling (MF-PM) technique yield better performance when compared to other methods. The edge details are preserved in this method by considering the pixel array-based filtering technique. However, in the NLM technique, the factual edges vanish due to the significant value of the standard deviation of the Gaussian filter.

13.3.3 Deep Learning Model-Based Image Prediction

In this work, the cascaded encoder-decoder CNN (CEDCNN) method is engaged for effective liver segmentation and tumors from CT abdomen images. The CEDCNN structure is composed of an encoder and decoder network, which are symmetric, as shown in Figure 13.4. The proposed architecture resembles the SegNet model. In the conventional SegNet architecture, the first 14 convolutional layers are used by the encoder-network component of a traditional SegNet design. Because the encoder and decoder networks are symmetrical, the decoder network includes 13 convolutional layers. The

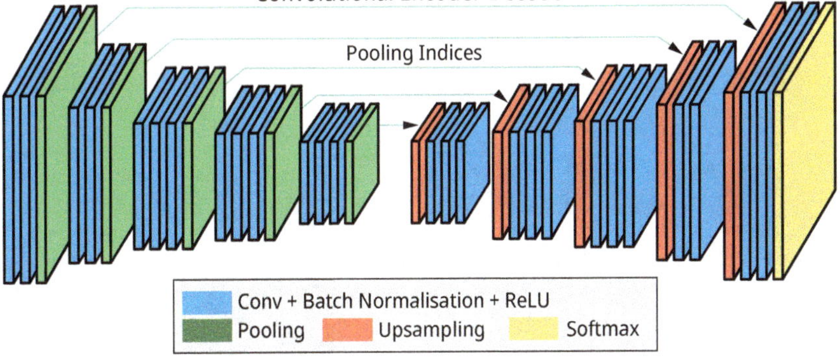

Figure 13.4: General structure of cascaded encoder-decoder CNN (CEDCNN).

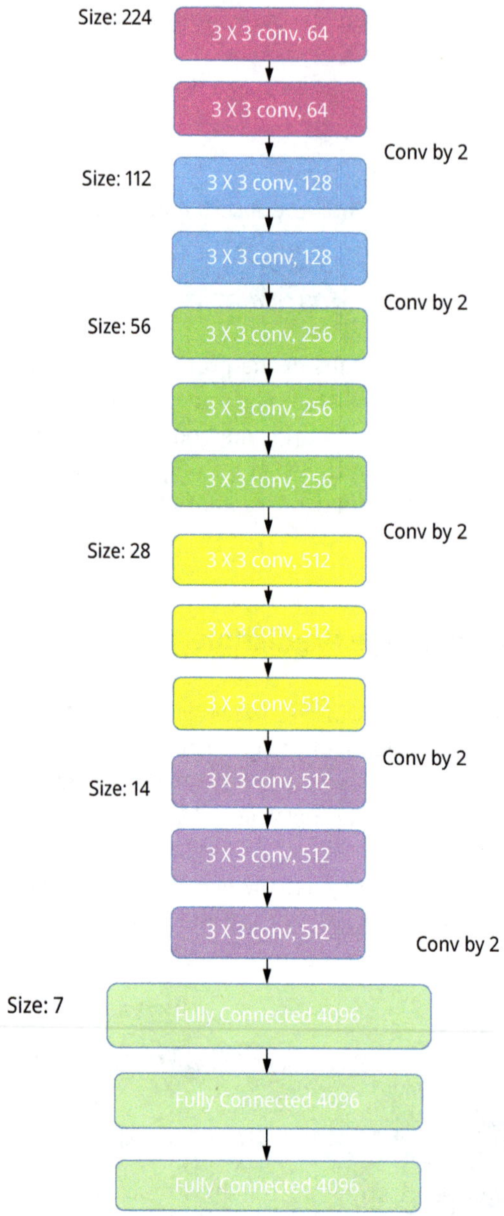

Figure 13.5: Basic block diagram of CEDCNN with VGG16 architecture.

softmax classifier function uses the decoder network output to generate a probability class for all pixels. Then, a convolution-based filter bank extracts the normalized feature map. In this work, the rectified linear nonlinearity (ReLU) unit is followed by batch normalization. Finally, the ReLU is subjected to max pooling with window size 2 × 2 and stride 2, followed by subsampling. The decoder network upsamples the input feature maps. A decoder filter bank creates the dense feature map by applying a convolution operator to the upsampling. Each of these feature maps is also subjected to batch normalization. The final stage of the decoder network yields a multichannel feature map that resembles an input image. The softmax classifier function uses this multichannel feature map to identify the pixels in the input image.

Figure 13.5 illustrates the model used by the CEDCNN approach, which consists of 41 layers: 3 FC (fully connected) layers, 2 × 2 maxpooling, and 13 convolutional layers. SegNet, an encoder-decoder technique suggested by VGG-16, uses predefined weights. It has a classification layer based on pixels as well as an encoder-decoder network. Convolution and max pooling will occur at the encoder network since VGG-16 is the encoder portion, which has 13 convolutional layers. Additionally, a 2 × 2 max pooling is carried out, leading to the corresponding storage of indices. The decoder network maps using a resolution encoder and then provides the results to the classification layers.

For evaluating the proposed work, three DIRACDB datasets as well as our own dataset collected from the hospital are used. Table 13.1 lists the technical information about the public dataset.

Table 13.1: Description of the 3DIRCADB dataset.

Images	Gender	Voxel size	Image size	Number of slices	Number of tumor	Training/ testing
Image 1	F	0.57 × 0.57 × 1.6	512 × 512	129	7	Testing
Image 2	F	0.78 × 0.78 × 1.6	512 × 512	172	1	Training
Image 3	M	0.62 × 0.62 × 1.2	512 × 512	200	1	Training
Image 4	M	0.74 × 0.74 × 2.0	512 × 512	91	7	Testing
Image 5	M	0.78 × 0.78 × 1.6	512 × 512	139	0	Testing
Image 6	M	0.78 × 0.78 × 1.6	512 × 512	135	20	Training
Image 7	M	0.78 × 0.78 × 1.6	512 × 512	151	0	Training
Image 8	F	0.56 × 0.56 × 1.6	512 × 512	124	3	Training
Image 9	M	0.87 × 0.87 × 2.0	512 × 512	111	1	Testing
Image 10	F	0.73 × 0.73 × 1.6	512 × 512	122	8	Training
Image 11	M	0.72 × 0.72 × 1.6	512 × 512	132	0	Training
Image 12	F	0.68 × 0.68 × 1.0	512 × 512	260	1	Training
Image 13	M	0.67 × 0.67 × 1.6	512 × 512	122	20	Training
Image 14	F	0.72 × 0.72 × 1.6	512 × 512	113	0	Training
Image 15	F	0.78 × 0.78 × 1.6	512 × 512	125	2	Training
Image 16	M	0.70 × 0.70 × 1.6	512 × 512	155	1	Training
Image 17	M	0.74 × 0.74 × 1.6	512 × 512	119	2	Testing

Table 13.1 (continued)

Images	Gender	Voxel size	Image size	Number of slices	Number of tumor	Training/ testing
Image 18	F	0.74 × 0.74 × 2.5	512 × 512	74	1	Training
Image 19	F	0.70 × 0.70 × 4.0	512 × 512	124	46	Training
Image 20	F	0.81 × 0.81 × 2.0	512 × 512	225	0	Training

13.3.3.1 Training

A random selection of 15 sets comprising 2,234 cuts was utilized for training, while the remaining five sets with 591 slices were used for testing in this study. Additionally, during the CEDCNN training, no data augmentation is utilized. During the training phase, the two-fold normalization approach is used for the CT image (I) and the related ground truth image (G). Finally, each CT image and ground truth image is divided into different slices with an image size of 256 × 256 before starting the training phase. The input is provided with the normalized 2D images and ground truth sets to fit a rough segmentation model. This model primarily distinguishes between liver and background characteristics.

On the other hand, any 3D volume must meet specific requirements. Each 3D volume must be resized to a voxel size of 1 mm^3. Finally, the normalized images are cropped using the liver mask. A dense training approach is adopted to fit an acceptable liver segmentation model. The stochastic gradient descent method is used as an optimizer during training. The learning rate progressively reduces every 50 epochs from 0.1 to 0.00450, with a 0.4 drop factor. The network is trained for 150 epochs with a batch size of 8. The first CNN output is a binary representation of expected liver areas, which are then utilized as input to the CNN by extracting ROIs.

13.3.3.2 Testing

The liver region will be segmented from abdominal images during the testing stage. As in the training stage, in the testing phase, two-fold image normalization is also done once the 3D testing images are normalized into 2D slices. Then, a coarse liver segmentation is predicted using the trained structure. Alternatively, the input CT and roughly segmented liver images are resized to a regularized voxel size. Finally, the trained 3D fully convolutional neural network produces fine and rough liver tumor segmentation.

13.3.3.3 Performance Evaluation Metrics

The performance of the proposed preprocessing, liver segmentation, and tumor segmentation methods is analyzed based on the following measures.

13.3.3.4 Peak Signal-to-Noise Ratio (PSNR)

The peak signal-to-noise ratio (PSNR) is a performance indicator that measures how the mean square errors introduced into the original image throughout the filtering process compare to the square of the maximum pixel value. The PSNR is given as follows:

$$\mathrm{PSNR}(dB) = 10\log_{10}\frac{\max^2}{\mathrm{MSE}} \tag{13.6}$$

where "max" indicates the maximum pixel value of the input image, and PSNR suggests the quality of the reconstructed image. The increased PSNR value results in the enhanced quality of the image.

13.3.3.5 Average Symmetric Surface Distance (ASD)

It calculates the surface distance based on the voxels of two segmentations by considering the average of all saved distances. The average value of all stored distances is zero, and this indicates perfect segmentation.

It is one of the most commonly used measures in terms of mm, as follows:

$$\frac{\mathrm{ASD(mm)} = \sum_{x\epsilon X}\min_{y\epsilon Y}(d(x,y)) + \sum_{y\epsilon Y}\min_{x\epsilon X}(d(x,y))}{N_X + N_Y} \tag{13.7}$$

where X and Y indicate the reference surface and segmented volume surface, x and y represent the mesh points of X and Y, $d(x, y)$ indicates the distance between x and y, then N_X and N_Y are the number of points on X and Y, respectively.

13.3.3.6 Root Mean-Squared Symmetric Surface Distance (RMSD)

It measures the surface difference between two different segmentations by considering the surface distance. It is required to square the surface voxels before storing them. It is premeditated as an equation in terms of mm:

$$\text{RMSD(mm)} = \sqrt{\frac{\sum_{x \epsilon X} \min_{y \epsilon Y}(d(x,y))^2 + \sum_{y \epsilon Y} \min_{x \epsilon X}(d(x,y))^2}{N_X + N_Y}} \tag{13.8}$$

13.3.3.7 Maximum Symmetric Surface Distance (MSD)

It is a measure based on the Euclidean distance. It is defined as the difference between the surface voxels. Then, the maximum surface value provides the maximum MSD. It is calculated as in eq. (13.9):

$$\text{MSD(mm)} = \max(\max_{x \epsilon X}(\min_{y \epsilon Y}(d(x,y))), \max_{y \epsilon Y}(d(x,y))) \tag{13.9}$$

13.3.3.8 Volumetric Overlap Error (VOE)

This metric measures the difference in volume between the reference image and the volume of the liver-segmented image. To measure this metric, the average of all slices is considered for measuring the metrics in terms of percentage of error. It is calculated as follows:

$$\text{VOE}(\%) = \left(1 - \frac{V_{\text{seg}} \cap V_{\text{ref}}}{V_{\text{seg}} \cup V_{\text{ref}}}\right) \times 100 \tag{13.10}$$

where V_{seg} is the segmented image and V_{ref} is the reference image. This method effectively shows the segmentation accuracy. If the volumetric overlap error is zero, it indicates segmentation accuracy.

13.3.3.9 Signed Relative Volume Difference (SRVD)

It measures the volume difference between the segmented and input images. It is expressed in terms of percentage and is calculated as in eq. (13.11):

$$\text{SRVD}(\%) = \frac{|v_{\text{seg}} - v_{\text{ref}}|}{v_{\text{ref}}} \times 100 \tag{13.11}$$

Based on these measures, the performance of the proposed techniques is evaluated. A smaller value in all evaluation metrics leads to a better segmentation result.

The sample CT abdomen images are shown in Figure 13.6(a) and (b), with the entire abdominal portion.

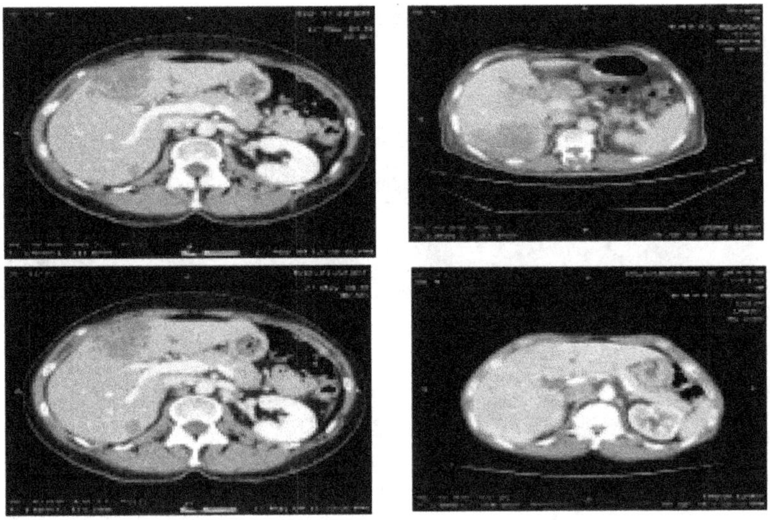

(a) (b)

Figure 13.6: (a) Sample input CT images – 3DIRCADB dataset, and (b) sample input CT images – non-public dataset.

The proposed model is validated by considering five images from public datasets, and five images from nonpublic datasets are taken. The qualitative results of preprocessed, liver-segmented, and tumor-segmented images are shown in Figure 13.7.

Table 13.2 lists the qualitative outcomes of the suggested model for both public and private datasets in terms of performance metrics like VOE (volumetric overlap error), SRDV (signed relative volume difference), MSSD (maximum symmetric surface distance), RMSD (root mean square symmetric surface distance), and ASD (average symmetric surface distance).

The average values of ASD, RMSD, MSD, VOE, and SRVD of the threshold-based slope difference distribution method are 1.05 mm, 2.11 mm, 14.68 mm, 13.90%, and 2.04%, respectively. The positive value of VOE shows that the CEDCNN method extracts the liver boundary precisely. Figure 13.8 interprets the various performance measures of the classifiers.

13.3.4 Tumor Classification Module

The main intention of the classification module is to differentiate the type of tumor present in the segmented tumor image. Classification of the tumor is the final stage in the CAD model. The difference in features extracted from the segmented liver tumor is given as input to the classifier stage in this stage. The essential task in the tumor

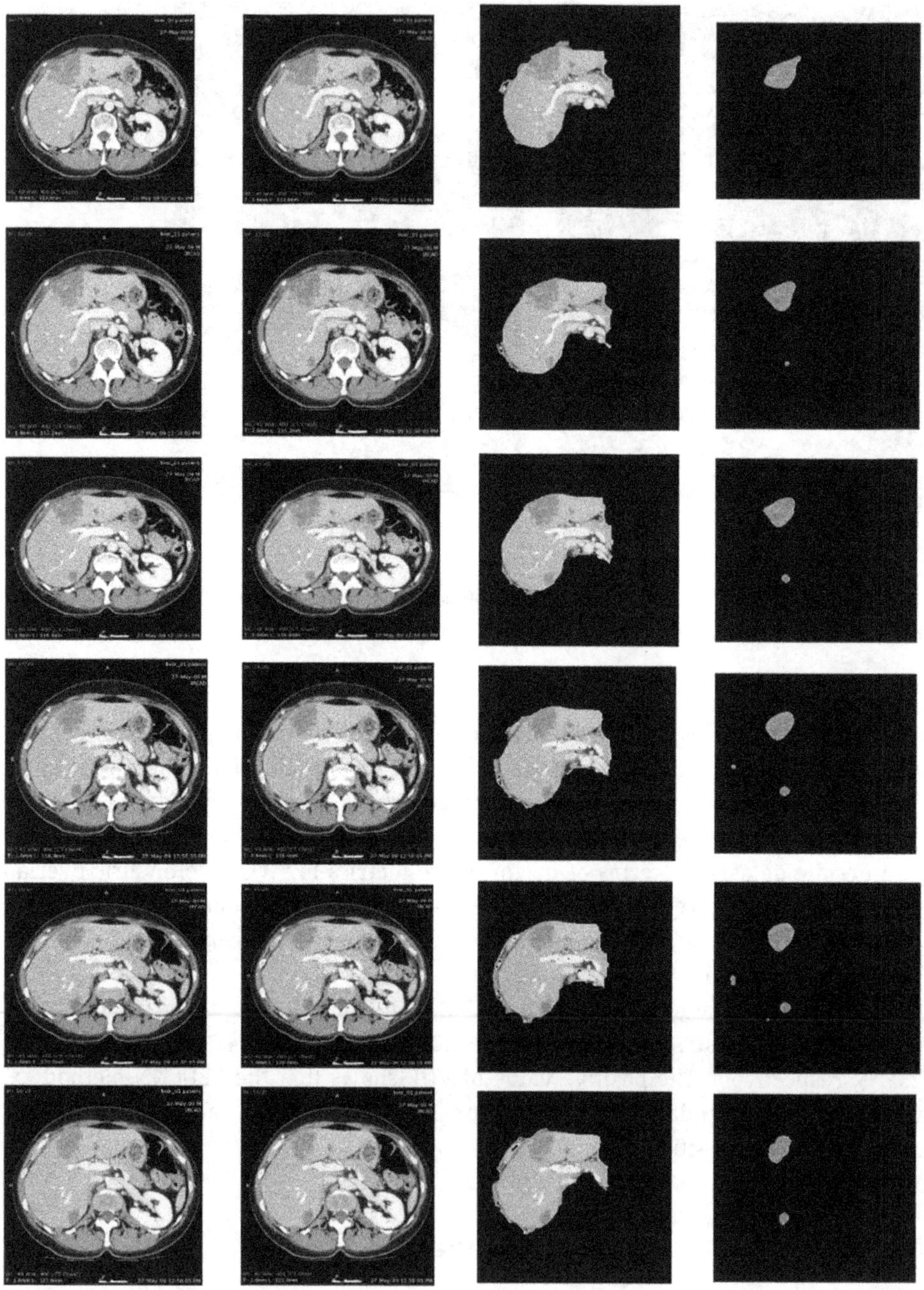

Figure 13.7: (a) Input images, (b) preprocessed image, (c) liver-segmented results, (d) tumor-segmented results.

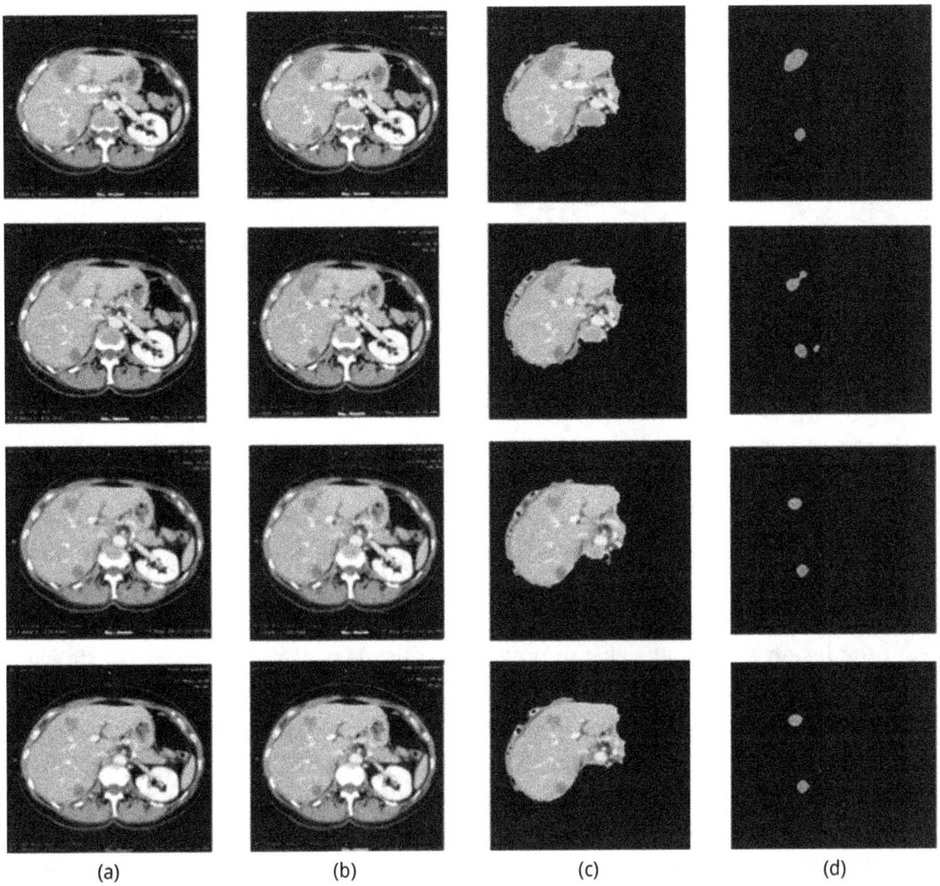

Figure 13.7 (continued)

Table 13.2: Various performance metrics for liver and tumor segmentation using the CEDCNN method.

Image name	ASD (mm)	RMSD (mm)	MSSD (mm)	VOE (%)	SRVD (%)
Image 1	0.8974	1.8077	21.96	10.1821	1.8976
Image 2	0.8961	1.8052	22.2794	10.5015	1.8964
Image 3	0.8932	1.7995	23.0111	11.2332	1.8935
Image 4	0.8903	1.7936	23.7577	11.9798	1.8905
Image 5	0.8886	1.7902	24.1976	12.4197	1.8888
Image 6	0.8913	1.7956	23.5221	11.7442	1.8915
Image 7	0.8784	1.7698	26.8081	15.0302	1.8786
Image 8	0.8783	1.7698	26.804	15.0261	1.8786
Image 9	0.8823	1.7778	25.7718	13.9939	1.8826
Image 10	0.8841	1.7812	25.341	13.5631	1.8843

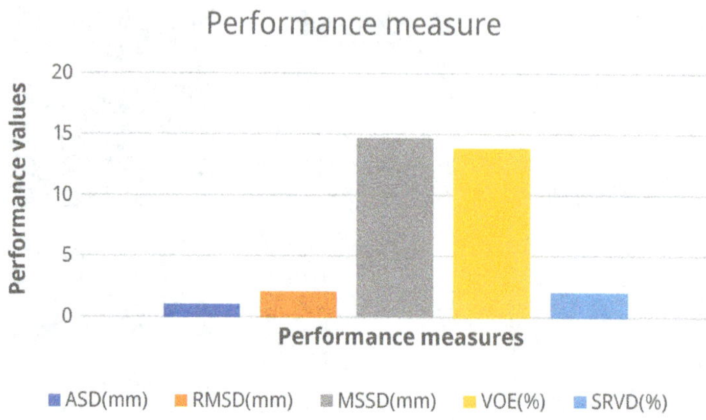

Figure 13.8: Statistical analysis of various performance measures.

classification stage is the extraction and selection of features. In this work, multilayer perceptron (MLP), support vector machine (SVM), and kappa statistics, random forest (RF) techniques are employed for classifying liver tumors.

In the existing work, the classification is done only by extracting the features from the lesion ROI. However, the novelty in this classification stage is considering two ROIs, which improves the accuracy by taking the difference of features extracted from the lesion and liver region. Initially, there are two types of ROI: one is a lesion (R1), and the other is surrounding normal liver tissue (R2). The features obtained using the first and second ROI will work as a novel feature vector. However, for identifying the normal liver tissue (R2), it is necessary to follow some constraints:

- The ratio between the lesion ROI (R1) and the specific liver region (R2) is to be chosen as 1:1.7 by conducting several trials.
- The typical liver region (R2) must centrally surround the lesion ROI (R1).
- The lesion region (R1) is expelled from the typical liver region (R2).

In this manner, the features extracted from each of the tumor region and the surrounding region are considered to be a new feature vector.

13.3.4.1 Results for ROI-Based Classification System

As an initial step, the features obtained from the tumor portion alone are given as input to the classification stage. Table 13.3 lists the confusion matrix of the various classification measures by considering the features extracted from the tumor portion alone. Table 13.4 shows the different performance metrics of the proposed techniques. The average accuracies of 87.06%, 84.58%, and 80.43% are obtained for MLP, SVM, and RF techniques, respectively.

Table 13.3: Confusion matrix of the various features by considering the difference of features classification.

Classifier	Classified as	cyst	Hemangioma	HCA	HCC	metastasis	HB	Test images
MLP	Cyst	132	1	0	1	0	0	134
	Hemangioma	1	10	1	0	0	0	12
	HCA	1	0	14	1	0	1	17
	HCC	1	0	1	103	0	0	105
	Metastasis	1	1	0	0	91	1	94
	HB	1	0	0	1	1	21	24
SVM	Cyst	129	1	0	2	1	1	134
	Hemangioma	1	9	1	0	1	0	12
	HCA	1	1	14	1	0	0	17
	HCC	2	0	1	101	0	2	105
	Metastasis	1	1	0	2	89	1	94
	HB	2	0	2	0	2	18	24
RF	Cyst	124	5	0	3	1	1	134
	Hemangioma	1	5	2	0	2	2	12
	HCA	2	1	12	0	1	1	17
	HCC	1	3	0	99	1	1	105
	Metastasis	2	2	0	2	84	4	94
	HB	3	2	0	2	4	13	24

Table 13.4: The overall classification accuracy of ROI-based classification.

Classifiers	Kappa statistics	TP rate	FP rate	Recall	Precision	F-measure	Accuracy	Error
MLP	0.8876	0.9875	0.1428	0.999	0.9983	0.9987	0.9757	0.0243
SVM	0.855	0.9833	0.1176	0.9974	0.9983	0.9978	0.9855	0.0145
RF	0.8395	0.9817	0.1578	0.9974	0.9974	0.9974	0.9975	0.0025

Figure 13.9 shows the different performance metrics such as kappa statistics, TP rate, FP rate, recall, precision, F-measure, accuracy, and error for three different classifiers namely MLP, SVM, and RF.

13.4 Explainability Model

Model-agnostic interpretation techniques, which have recently been made possible by breakthroughs in ML, are used to explain complex models while maintaining good prediction performance. Separating the explanation model from the model-agnostic interpretation is more flexible than the model-specific explanation method. The two types of model-agnostic interpretation techniques are local explanation and global ex-

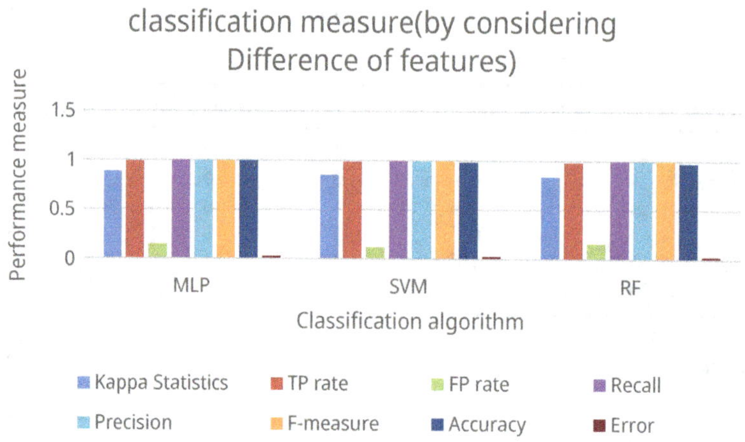

Figure 13.9: Performance metrics of the difference of feature-based classification method.

planation. The most popular local explanation technique is local interpretable model-agnostic explanations(LIME). The two most common global interpretable techniques are PDP (partial dependence plots) and SHAP (SHapley Additive exPlanations).

13.4.1 LIME Model

To provide interpretability for complicated models, LIME trains local surrogate models. LIME first produces a fresh dataset through data disruption. LIME then uses the new dataset to train an interpretable model, such as a decision tree. Finally, the black box model's equivalent prediction performance is contrasted with the objective of the interpretable model. LIME is defined as follows:

$$\gamma(x) = \operatorname{argmin} L(f, g, \Pi_x) + \Omega(g) g \epsilon G \tag{13.12}$$

where L is the loss function. The explainable model for instance x is represented by g. The interpretable model family is denoted as G. The closeness of the experimented instances to instance x is indicated by the instances x. $\Omega(g)$ defines the intricacy of model g.

13.4.2 PDP Model

PDP illustrates the small impact of a feature on the outcome predicted by a sophisticated ML model. It shows how the input and output are related to one another, whether it is more complicated, monotonous, or linear. The partial dependence function $\widetilde{f_{x_S}}$ is given by the definition of

$$\widetilde{f_{xs}} = \frac{1}{n}\sum_{i=1}^{n} \widetilde{f_{xs}}\left(x_S, x_c^i\right) \tag{13.13}$$

where $\widetilde{f_{x_S}}$ is the process that generates the overall relationship between an input feature and the expected result. The feature set "S" encompassing one or two features, x_S denotes the set of features to be plotted $\widetilde{f_{x_S}}$. x_c designates the remaining features in the model, and n is the number of instances used in the dataset.

13.4.3 SHAP Model

It measures the impact of the feature on the multifaceted model with SHapley values. The sum of marginal aids is referred to as a SHapley value. It can be identified by the measure of closeness of prediction across all potential coalitions. The SHapley value is calculated by

$$\varnothing_j(x) = \sum_{s\complement\{x1, x2...,xm\}\{x_j\}} \frac{|s|!(m-|s|-1)!}{m!}\left(\text{val}\left(s \cup \{x_j\} - \text{value}(s)\right)\right) \tag{13.14}$$

where S is the SHapley value of x_j, x_j signifies a feature value, a feature subset of the model is represented by "s," the number of features is denoted by m, and the value is the predicted feature values in set s.

13.4.4 Results of Predicted Model

In this work, the interpretable framework is practiced in a Python environment. Using K-fold cross-validation, the overall accuracy of the complex models was determined. K is typically set to 5, 10, or 20. Table 13.5 evaluates the interpretable models using the liver dataset.

Table 13.5: *K*-fold validation results for complex models.

Models	SVM	XGBoost	RF
$K = 5$	89.1%	88.2%	**92.7%**
$K = 10$	87.5%	90.4%	**94.8%**
$K = 15$	89.4%	90.2%	**98.7%**

The table shows that for all cases of the *K*-fold validation process, the average prediction accuracy of interpretable models of SVM, XGBoost, and RF is 88.67%, 89.6%, and 95.4%, respectively. From the obtained results, RF achieves the best predictive performance.

13.5 Conclusion

In this work, an explainable ML framework combines CEDCNN-based models with recently discovered explanation techniques to accurately predict the probability of liver diseases. The suggested framework is tested using public datasets from standard databases and nonpublic datasets collected from multidisciplinary scan centers. According to the results, the RF classifier method achieves the highest overall accuracy of 99.5% using a deep learning framework and an average accuracy of 95.4% using the XAI model with an RF-based classifier. The results of the proposed framework clarifications are reliable with the characteristics of liver tumors, which may help the radiologist to make more accurate diagnoses. In addition, this method promotes the application of AI in healthcare by highlighting the benefits of XAI frameworks in deciphering black-box models like RF. This work can be extended in the future by collecting more real-time liver images and investigating cutting-edge, model-independent techniques.

References

[1] Bakrania, A., Joshi, N., Zhao, X., Zheng, G., & Bhat, M. (2023). Artificial intelligence in liver cancers: Decoding the impact of machine learning models in clinical diagnosis of primary liver cancers and liver cancer metastases. *Pharmacological Research*, *189*, 106706.

[2] Peng, J., Zou, K., Zhou, M., Teng, Y., Zhu, X., Zhang, F., & Xu, J. (2021). An explainable artificial intelligence framework for the deterioration risk prediction of hepatitis patients. *Journal of Medical Systems [Internet]*, *45*(61).

[3] Ahmed, Z. U., Sun, K., & Shelly, M. (2021). Explainable artificial intelligence (XAI) for exploring spatial variability of lung and bronchus cancer (LBC) mortality rates in the contiguous USA. *Scientific Reports*, *11*(1), 24090.

[4] Kamal, A. N. (2021). Overview of explainable artificial intelligence for prognostic and health management of industrial assets based on preferred reporting items for systematic reviews and meta-analyses. *Sensors*, *21*(23), 8020.

[5] Andreu-Perez, J., Emberson, L. L., & Kiani, M. (2021). Explainable artificial intelligence-based analysis for interpreting infant fNIRS data in developmental cognitive neuroscience. *Communications Biology*, *4*(1), 1077.

[6] Anand, A., Kadian, T., Shetty, M. K., & Gupta, A. (2022). Explainable AI decision model for ECG data of cardiac disorders. *Biomedical Signal Processing and Control*, *75*, 103584.

[7] Kumar, A., Dubey, A. K., Ramírez, I. S., Muñoz Del Río, A., & García Márquez, F. P. (2024). Artificial intelligence techniques for the photovoltaic system: a systematic review and analysis for evaluation and benchmarking. *Archives of Computational Methods in Engineering*, 1-25.

[8] Fang, H. S. A., Tan, N. C., & Tan, W. Y. (2021). Patient similarity analytics for explainable clinical risk prediction. *BMC Medical Informatics and Decision Making*, *21*(1), 207.

[9] Jahmunah, V., Ng, E. Y. K., Tan, R. S., Oh, S. L., & Acharya, U. R. (2022). Explainable detection of myocardial infarction using deep learning models with Grad-CAM technique on ECG signals. *Computers in Biology and Medicine [Internet]*, *146*, 105550.

[10] Meena, J., & Hasija, Y. (2022). Application of explainable artificial intelligence in the identification of Squamous Cell Carcinoma biomarkers. *Computers in Biology and Medicine [Internet]*, *146*, 105505.

[11] Petch, J., Di, S., & Nelson, W. (2022). Opening the black box: The promise and limitations of explainable machine learning in cardiology. *Canadian Journal of Cardiology*, *38*(2), 204-13.

[12] Kute, D. V., Pradhan, B., Shukla, N., & Alamri, A. (2021). Deep learning and explainable artificial intelligence techniques applied for detecting money laundering–a critical review. *IEEE Access*, *9*, 82300–17.

[13] Merry, M., Riddle, P., & Warren, J. (2021). A mental models approach for defining explainable artificial intelligence. *BMC Medical Informatics and Decision Making*, *21*(344).

[14] Miró-Nicolau, M., Moyà-Alcover, G., & Jaume-i-Capó, A. (2022). Evaluating explainable artificial intelligence for X-ray image analysis. *Applied Sciences*, *12*(9), 4459.

[15] Obayya, M., Nemri, N., Nour, M. K., Al Duhayim, M., Mohsen, H., Rizwanullah, M. et al. (2022). Explainable artificial intelligence enabled Tele Ophthalmology for diabetic retinopathy grading and classification. *Applied Sciences*, *12*(17), 8749.

[16] Papandrianos, N. I., Feleki, A., Moustakidis, S., Papageorgiou, E. I., Apostolopoulos, I. D., & Apostolopoulos, D. J. (2022). An explainable classification method of SPECT myocardial perfusion images in nuclear cardiology using deep learning and grad-CAM. *Applied Sciences*, *12*(15), 7592.

[17] Pertzborn, D., Arolt, C., Ernst, G., Lechtenfeld, O. J., Kaesler, J., Pelzel, D. et al. (2022). Multi-class cancer subtyping in salivary gland carcinomas with MALDI imaging and deep learning. *Cancers*, *14*(17), 4342.

[18] Lombardi, A., Diacono, D., & Amoroso, N. (2022). A robust framework to investigate the reliability and stability of explainable artificial intelligence markers of Mild Cognitive Impairment and Alzheimer's Disease. *Brain Information*, *9*(17).

[19] Yang, C. (2022). Explainable artificial intelligence for predictive modeling in health-care. *Journal of Healthcare Informatics Research*, *6*, 228–39.

[20] Zeineldin, R. A., Karar, M. E., & Elshaer, Z. (2022). Explainability of deep neural networks for MRI analysis of brain tumors. *International Journal of Computer Assisted Radiology and Surgery*, *17*, 1673–83.

[21] Leonardi, G., Montani, S., & Striani, M. (2022). Explainable process trace classification: An application to stroke. *Journal of Biomedical Science*, *126*, 100230.

Prabakaran D., Senthil Kumaran R., and Shyamala R.

14 A Novel African Wild Dog Optimization (AWDO) Algorithm for Applications of Artificial Intelligence

Abstract: Artificial intelligence (AI) harbors the potential to tackle intricate challenges and enhance human capabilities across various domains. Through its significant capabilities, AI plays a crucial role in data analysis, natural language processing (NLP), healthcare, and other decision-based fields. The primary objective of AI in these domains is to render precise decisions promptly. The attainment of higher decision-making accuracy and system performance enhancement necessitates the utilization of optimization algorithms to aid in the decision-making process efficiently. Swarm optimization draws inspiration from the collective behavior of animals or insects that cooperate in hunting, thereby facilitating rapid decision-making. This chapter delves into diverse swarm optimization algorithms that support decision-making processes in AI applications and introduces a novel algorithm called African wild dog optimization (AWDO). The proposed AWDO algorithm is an energy-efficient swarm optimization technique that minimizes energy consumption by AI nodes or neurons, resulting in expedited search outcomes. The efficacy of the AWDO algorithm is evaluated against established optimization algorithms like particle swarm optimization, ant colony optimization (ACO), artificial bee colony (ABC) optimization algorithm, firefly algorithm, and bacterial foraging optimization (BFO) algorithms in the realm of healthcare diagnosis applications.

Keywords: Artificial intelligence, African wild dog optimization, particle swarm optimization, ant colony optimization, artificial bee colony, firefly algorithm, bacterial foraging optimization, healthcare diagnosis

14.1 Introduction

Artificial intelligence (AI) [1] is an inevitably developing technology that creates the possibility for machines to learn from occurrences and make decisions on real-time events. The proliferation of AI has made it possible to make vital decisions and diagnoses in business and intelligence systems. Some of the vital applications of AI are market prediction, supply chain management, medical diagnosis, traffic analysis, image-based diagnosis, etc. Among the various applications, AI holds a major role in healthcare diagnosis systems, which involve cancer detection, early tumor detection, and many more related applications. Varieties of AI algorithms such as Naïve Bayes, support vector machine (SVM) [2], random forest (RF) [3], decision tree (DT) [4], neu-

https://doi.org/10.1515/9783111436180-014

ral networks (NN) [5], and deep learning (DL) algorithms [6] like convolutional neural networks (CNN) and recurrent neural networks (RNN) have been employed for the aforementioned applications.

Machine learning (ML) and the DL algorithm, which are subsets of the AI)technique, were applied for the diagnosis of multiple diseases from medical multimodal images. These multiple diseases include cancer, tumors, and other diseases at earlier stages. The medical image diagnosis is performed based on the dimension, region of interest (RoI), and the histogram [7] of the image pixel. The major challenge associated with AI-based medical image diagnosis is the feature selection process, as it often misses major and essential features for the classification process. Diseases like tumors necessitate the structural features of the tumor. The general structure of performing medical image diagnosis using an AI algorithm is depicted in Figure 14.1.

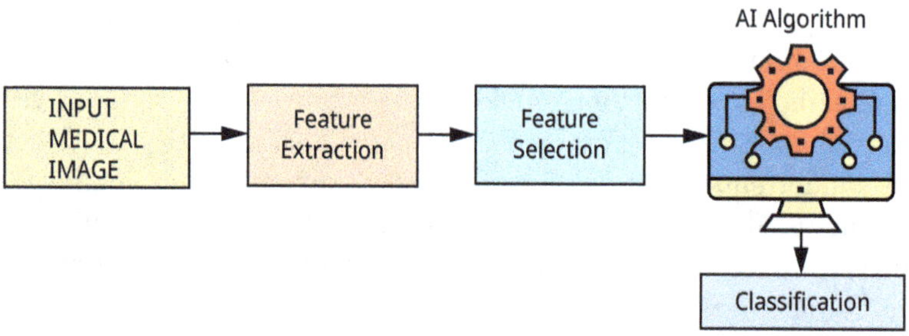

Figure 14.1: General structure of medical image diagnosis using artificial intelligence.

The computer-assisted diagnosis of human disease exhibits various advantages in terms of knowledge as well as accuracy in the classification process. However, the classification process is a complex one as it relies on various parameters. The most important aspect of the learning algorithms is the training of the model. The existing models were trained using the free datasets that were relevant to the particular type of human disease. Based on the volume of training data, the rate of accuracy of the diagnosis system [8] varies. In addition to the training data, the efficiency of the diagnosis model relies on the type of algorithm used for classification purposes. The algorithms of the DL model do not require any explicit input features, as they employ the features of the deep neural networks (DNN). The DNN employs multiple layers, namely “n” number of layers, for extracting the features from the input medical image. From the extracted features, the DNN model performs the feature selection process for efficient classification and diagnosis. Despite the advantages of DL in the medical image diagnosis process, it experiences major challenges of shortfall in terms of accuracy due to the feature selection process and the degradation of edges of the input images. This causes essential images to deteriorate in terms of efficiency and thus necessitates the requirement of advanced AI technology [9] for efficient and accurate medical image diagnosis. The existing methodologies fall short in terms of global search, impacting the accuracy in the diagnosis process. In addition, the optimization function [10] is trapped under the local optima, thus leading to weak characteristics in robustness to the local optima. The major objectives of the proposed work are to classify the medical images with an enhanced level of accuracy using the African wild dog optimization (AWDO) algorithm. The proposed method is highly robust in preserving the features of the input image. In addition to the increase in accuracy, the proposed work excels in terms of precision, recall, and biopsy rate. The organization of this chapter is structured with a basic introduction in Section 14.1, literature analysis in Section 14.2, and a detailed description of the proposed novel AWDO algorithm in Section 14.3. The case study is performed, and the analysis of the proposed work is presented in Section 14.4. The chapter is concluded with the highlights of the proposed work in Section 14.5.

14.2 Literature Review

The smart medication process followed over the past decade motivates numerous researchers to employ and introduce novel classification algorithms in diagnosing specific diseases in human beings. This section illustrates the notable research results of AI in terms of medical image diagnosis, such as tumor detection and cancer detection.

Jabbar et al. [11] proposed a Caps-VGGNet hybrid model [11] for the diagnosis of tumor detection. The authors employed the Brats dataset for training the model, and the major challenge of the introduced model is the requirement of a huge volume of data for training the model to achieve a high level of accuracy. Mo et al. [12] proposed a Hover-Trans block extraction model [12] to determine breast cancer at an earlier stage. The authors employed the GDPH dataset for training the model and used sonogram images for the detection process. Li et al. [13] developed a frequency-guided diffusion model [13] for processing and diagnosing medical images. The proposed model is capable of diagnosing any type of medical disease, provided the system is fed with the appropriate training datasets. Wu et al. [14] introduced a nonlocal algorithm [14] based on AI techniques to detect anomalies in human MRI images. The proposed model is highly robust and optimized the boundaries of the input image.

Asiri et al. [15] introduced a dual computerized model [15] for enhancing the accuracy of the diagnosis of input medical images. The first model aims at the image enhancement process using neural networks and the principal component analysis method. The latter model employs the support vector machine for performing the classification of the input tumor image. The classification is performed after the segmentation process and achieved an accuracy of 89.9%. Shamsad et al. (2024) contributed to early medical image diagnosis by designing an AI-based transfer learning model [16]. The proposed model aimed to enhance the accuracy in detecting anomalies in input medical images. The authors employed VGG-16 and VGG-19 datasets for training and testing the proposed model.

Palkar et al. [17] designed a Glioma prognosis model [17] using the AI technique. The proposed work was compared with existing AI models like random forest, support vector machine, CatBoost, LGBM classifier, and XGBoost to prove its supremacy in terms of accuracy in classifying the input medical image. Asif et al. [18] employed AI tools [18] for analyzing biomedical datasets in diagnosing medical images at the early stages. The authors employed ML and DL models for the diagnosis of HCC images.

Prakash et al. [19] presented an advanced framework for the medical image diagnosis process using the generative adversarial network (GAN) model [19]. The proposed GAN model is a three-dimensional model that accepts CT images of the kidney to detect and predict the presence of tumors in the kidney. The authors employed a structural GAN model to produce a minimum of 20% enhancement in the accuracy of the prediction process. Siddique et al. [20] introduced an innovative training model using the Inception V3 model [20] and employed the particle swarm optimization algorithm for detecting and classifying anomalies in tumors in various parts of the human body. The authors employed the INCEPT training model and achieved an accuracy of 90.5%.

The major drawbacks of the existing AI-based medical image diagnosis model are as follows.

- The existing classification model employs excessive classification parameters, leading to an increase in the complexity of the model.
- The rate of convergence of the existing models is very slow, and the optimization rate is considerably low.
- The robustness of the existing works was low, and it feels hard to diagnose the noisy input medical images.
- Due to the decrease in the level of robustness, the level of accuracy also sheds to a greater extent.

To overcome the above-mentioned challenges of the existing classification methods, the AWDO is employed to achieve the following contributions in the medical image diagnosis process.

- The proposed AWDO classification algorithm is highly robust and will accept noisy input medical images for the classification process.
- The proposed AWDO algorithm accepts minimal parameters, which possess the least complexity in terms of the classification process.
- The proposed AWDO is a promising optimization technique with the potential to provide efficient and effective solutions to a wide range of challenging problems.

14.3 Proposed Model

The proposed AWDO is a metaheuristic algorithm designed based on the inspiration of the hunting behavior of African dogs. This intelligent algorithm demonstrates the team work efforts of the African dogs, its adaptability, and its efficiency in hunting its prey. A similar pattern is employed in the proposed work to determine the anomalous behavior of the input medical image. In this proposed scheme of the optimization process, the swarm of African dogs, namely 5–20 dogs, is directed by the pair of leader dogs as depicted in Figure 14.2. The leader dogs organize the member dogs and act as the best pairs of dogs in the hunting process. The leader dog pairs perform decision-making processes, controlling the member dogs, and hunting and resting processes. Inspired by this hunting behavior, the proposed AWDO algorithm identifies the anomalous behavior or pattern of the input medical image by employing a pair of threshold pixels, namely threshold pixel-row wise (P_R) and threshold pixel-column wise (P_C). Based on the threshold pixels, the comparative process is performed with the remaining pixels to determine the anomalous pattern in the input medical image.

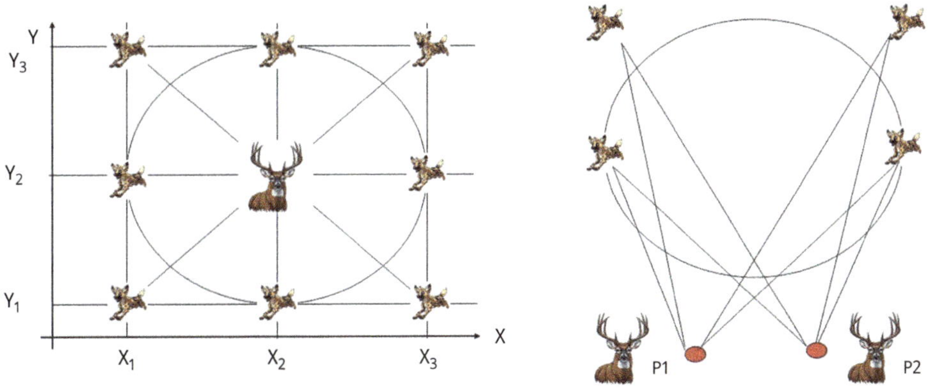

Figure 14.2: Hunting behavior of African wild dogs.

The threshold pixels P_R and P_C are responsible for determining the anomaly pattern among the input image pixels by determining the difference among the pixel values. It is notable that the threshold pixels P_R and P_C are not permanent, and the pair shall be modified based on the search. The major activities of the threshold pixels P_R and P_C are determining the anomaly pattern pixels, reviewing the anomaly pixels, selecting the surrounding pixels of the anomaly pixels, and classification of the pixels. The basic architecture of the proposed AWDO-based medical image diagnosis is presented in Figure 14.3.

The mathematical models for the pixel analysis for the anomaly behavior, tracking, and the classification process are presented in this section. The parent threshold pixels P_R and P_C are supported by the other pairs of pixels at various levels, namely

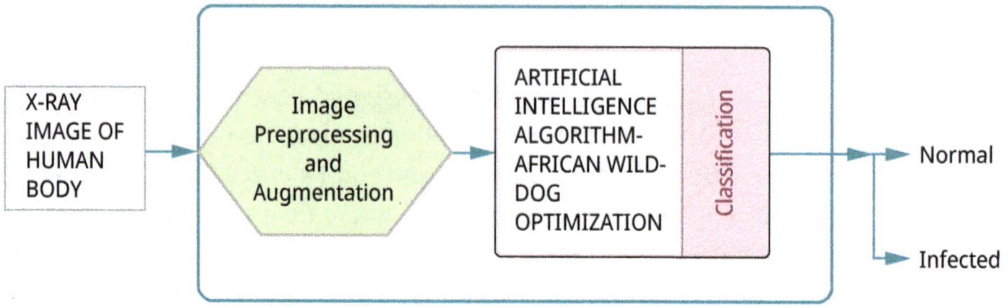

Figure 14.3: Architecture of the proposed AWDO-based medical image diagnosis.

different rows and columns. The pixel identification process, determined by the threshold pixels P_R and $P_{C,}$ is defined as follows:

$$P_r(n) = |P_{TR}(n) - \alpha \times P_{tR}(n)|\vec{P}(t) - \vec{D}(t) \tag{14.1}$$

$$P_C(n) = |P_{TC}(n) - \beta \times P_{tC}(n)|\vec{P}(t+1) - \vec{D}(t+1) \tag{14.2}$$

Equations (14.1) and (14.2) represent the target pixels in terms of rows and columns along with the vectors $\vec{\mathrm{P}}(\mathrm{t})$ and $\vec{\mathrm{D}}(\mathrm{t})$. The vectors $\vec{\mathrm{P}}(\mathrm{t})$ and $\vec{\mathrm{D}}(\mathrm{t})$ shall be determined using eqs. (14.3) and (14.4), respectively.

$$\vec{P}(t) = \vec{r_1}(t) - 2.\alpha\vec{r_2}(t) \tag{14.3}$$

$$\vec{D}(t) = \vec{r_3}(t+1) - 2.\beta\vec{r_4}(t+1) \tag{14.4}$$

During the process of iteration, the vectors α and β are chosen randomly in the range of 0 to 1. The terms r_1, r_2, $r_{3,}$ and r_4 are random vectors chosen at random in the range of pixels (0-m) rows and (0-n) columns. The threshold wild dogs (pixels) coordinates in the pixel position (*m,n*) are used to update the new position of the anomaly pixels in the input medical image. The location of the wild dogs shall be adjusted with the factors α and β in the input medical image. The new locations of the wild dogs in determining the anomaly pattern are represented in eqs. (14.5)–(14.10), respectively.

$$\vec{P}_A(t) = \left|\vec{P}(t) - \vec{D}(t)\right| \times \left|\vec{P}(t-1) - \vec{D}(t-1)\right| \times \left|\vec{P}(t+1) - \vec{D}(t+1)\right| \tag{14.5}$$

$$\vec{P}_{A1}(t) = \left|\vec{P}(t) - \vec{D}(t)\right| \times \alpha_1\beta_1 \tag{14.6}$$

$$\vec{P}_{A2}(t) = \left|\vec{P}(t-1) - \vec{D}(t-1)\right| \times \alpha_2\beta_2 \tag{14.7}$$

$$\vec{P}_{A3}(t) = \left|\vec{P}(t+1) - \vec{D}(t+1)\right| \times \alpha_3\beta_3 \tag{14.8}$$

$$\vec{P}_{A4}(t) = \left|\vec{P}(t-m) - \vec{D}(t-n)\right| \times \alpha_4\beta_4 \tag{14.9}$$

$$\vec{P}_{A5}(t) = \left|\vec{P}(t+m) - \vec{D}(t+n)\right| \times \alpha_5\beta_5 \tag{14.10}$$

The $\vec{P}_A(t)$ is the target anomaly pixel, whereas the $\vec{P}_{A1}(t)$ to $\vec{P}_{A5}(t)$ is surrounding wild dog pixels are located by the parent threshold pair of the pixels to classify the anomaly pixel. The members of the WDO support the searching process for the anomaly pattern in the input medical image. The vectors $\vec{\mathrm{P}}(\mathrm{t})$ and $\vec{\mathrm{D}}(\mathrm{t})$ are the pairs of hunting dogs that help in attacking the anomaly pixel of the input medical image. The proposed work intentionally employs the two vectors to emphasize exploration and performs the iteration process to determine the anomaly pattern. The movement of the patterns is performed using the two constants, namely α and β, respectively. The targeted pixel undergoes a classification process to determine whether the pixel is infected or normal using the conditional equations (14.11) and (14.12), respectively:

$$\text{If } \vec{P}_A(t) \geq \vec{P}_T(t) \tag{14.11}$$

$$\text{If } \vec{P}_A(t) < \vec{P}_T(t) \tag{14.12}$$

If the pixel value of the target pixel is greater than the pixel value of the threshold pixel, then it is classified as an infected pixel. Otherwise, as represented in eq. (14.12), it is classified as a normal pixel. The AWDO algorithm for performing medical image diagnosis is presented in Table 14.1.

The algorithm for the AWDO is illustrated in Table 14.1, which compares the pixel values of the neighboring pixels with the pixel value of the threshold pixel. A greater pixel value results in an infected pixel. The analysis and the comparative analysis are illustrated in the upcoming section.

14.4 Comparative Analysis

The proposed AWDO for determining anomalies in the input multi-modal medical images is analyzed against accuracy, precision, recall, and *F*1 score, and compared with the performance of state-of-the-art methodologies. The performance of the proposed work is compared with state-of-the-art algorithms like support vector machine [21], random forest [22], decision tree [23], particle swarm optimization [24], K-means clustering [25], DNN [26], and recurrent neural networks [27]. The model is trained with the BRATS dataset containing brain tumor images and is tested with images from the same dataset.

The term "accuracy" in terms of medical image diagnostics is considered a critical parameter that assists in determining further treatment. The accuracy is influenced by various parameters like the quality of the input image, tumor heterogeneity, avail-

Table 14.1: Algorithm for AWDO in medical image diagnosis.

Algorithm 1: African Wild Dog Optimization (AWDO) in Medical Image Diagnosis

Input: MRI Image (I_{mxn})
Output: Anomaly Pixel $P_A(t)$
Processes:
1: Initialize Pixel population:

$$P_{m\times n} = \frac{d}{dt}(I_{m\times n}) \tag{14.13}$$

2: Initialize Threshold Pixel:

$$P_T(t) = \frac{\int_{i=0}^{m}\int_{j=0}^{n}(I_{m\times n})_{ij}dmdn}{\sum_{i=0}^{m}\sum_{j=0}^{n}M_i + N_j} \tag{14.14}$$

3: for (i = 0; i++; i = m) && for (j = 0; j++; j = n)

4: $$P_n(t) = P(t)[i-1; i+2; j-1; j+2].flatten() \tag{14.15}$$

5: $$P_n(t) = P_n^*(t)\left[P_n^T(t)\right] = I_{m\times n}[i-1; j-1]//P_n^*(t) = \text{Neighbor pixel} \tag{14.16}$$

6: If; $P_n(t) < P_T(t)$;
7: Return: "Normal Pixel"
8: Else if $P_n(t) \geq P_T(t)$, then
9: Return, "Infected Pixel"
10: End if
11: End for
12: End processes

ability of the data, complexity of the algorithm, feature extraction process, and preprocessing algorithm, along with its effectiveness in removing noise from the input test image. The accuracy of the classification is determined in terms of a confusion matrix with true positive (TP), true negative (TN), false positive (FP), and false negative (FN), as depicted in Figure 14.4.

The mathematical representation of the accuracy in terms of the confusion matrix is defined as follows:

$$\text{Accuracy}(A) = \frac{\text{TP} + \text{TN}}{\text{TP} + \text{TN} + \text{FP} + \text{FN}} \tag{14.17}$$

Precision is the performance metric employed to evaluate the quality of the positive predictions made by the proposed classification model. Specifically, precision is the rate at which predictions were made correctly by the classification model. The mathematical representation of precision is defined as follows:

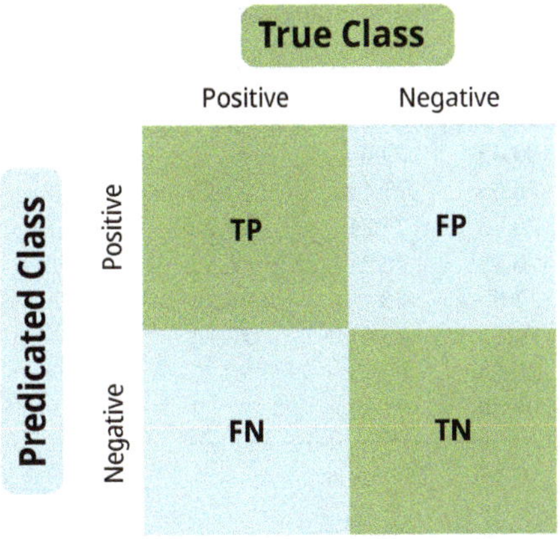

Figure 14.4: Confusion matrix for medical image prediction.

$$\text{Precision}(P) = \frac{\text{TP}}{\text{TP} + \text{FP}} \tag{14.18}$$

A high precision performance of the classification model indicates that the model is likely to be correct in terms of the classification process. Recall is the alternative parameter for precision, which could be defined as the metric used for determining the actual positive cases that were correctly identified. The term recall can also be termed as sensitivity and is mathematically represented as follows:

$$\text{Recall}(R) = \frac{\text{TP}}{\text{TP} + \text{FN}} \tag{14.19}$$

The recall parameter determines the proportion of actual positive cases that were correctly predicted. The term $F1$ score is the integrated metric of recall and precision to form a single rate of classification. The $F1$ score is usually determined to identify the balance between the precision and recall of the classification model. The mathematical representation of the $F1$ score is defined as follows:

$$F1 \text{ score} = \frac{2 \times (\text{Precision} \times \text{Recall})}{(\text{Precision} + \text{Recall})} \tag{14.20}$$

The observed values of the above-mentioned parameters by the state-of-the-art methodologies like support vector machine [21], random forest [22], decision tree [23], particle swarm optimization [24], K-means clustering [25], DNN [26], and recurrent neural networks [27] are listed in Table 14.2.

The existing and proposed classification models [28] were trained with the brain tumor datasets, and their accuracy, precision, recall, and $F1$ score are listed in Table 14.2. The graphical comparison is depicted in Figure 14.5.

Table 14.2: Comparison of performance by proposed and state-of-the-art methodologies.

Technologies	Accuracy	Precision	Recall	F1 score
Support vector machine [21]	81.15	80.45	79.46	79.95
Random forest [22]	79.11	78.65	77.15	77.89
Decision tree [23]	78.19	79.48	77.21	78.33
Particle swarm optimization [24]	77.84	76.95	76.91	76.93
K-means clustering [25]	78.48	77.8	76.18	76.98
Deep neural network [26]	80.11	79.62	80.07	79.84
Recurring neural networks [27]	81.46	80.64	80.07	80.35
Proposed AWDO	82.65	81.48	81.67	81.57

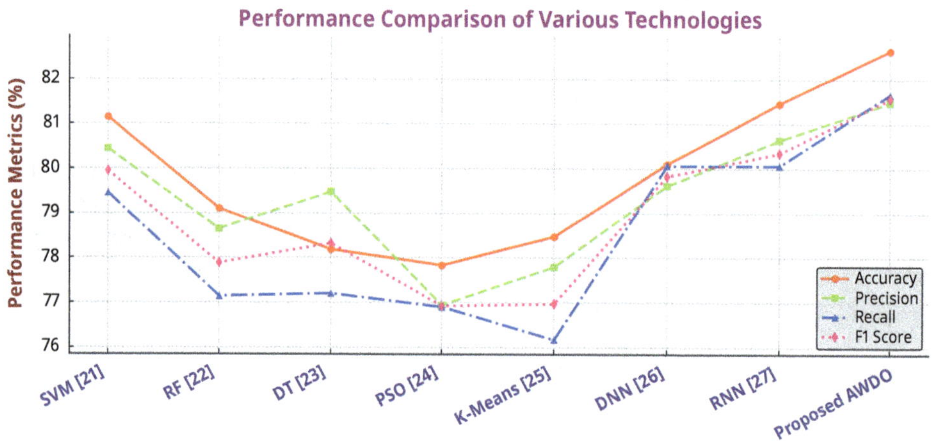

Figure 14.5: Comparison of performance by proposed and state-of-the-art methodologies.

The proposed model of the AWDO classification algorithm is known for its reduced complexity in terms of performance. In this context, the computational time [29] is measured in the proposed work by providing various types of medical MRI images. Various input images were fed to the proposed model, and the results are depicted in Table 14.3.

The proposed work consumed a maximum of 37.09 μs, which is comparatively minimal compared to the existing state-of-the-art methodologies. The performance analysis proves that the proposed AWDO work demonstrates its supremacy in terms of accuracy, precision, recall, and *F*1 score in diagnosing the medical MRI image, along with reduced computational time.

Table 14.3: Prediction of medical MRI image and its computational time.

Input image	Classification result	Computational time
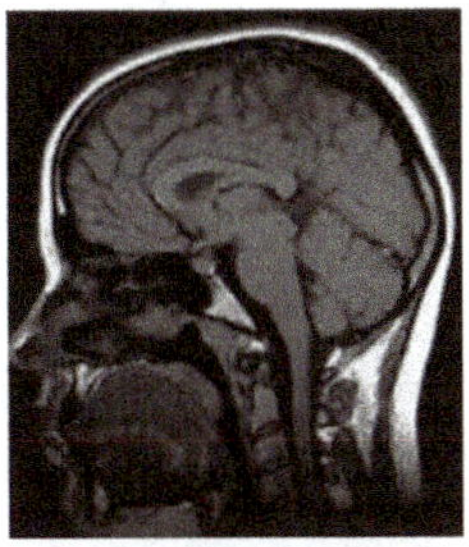	Normal	36.15 μs
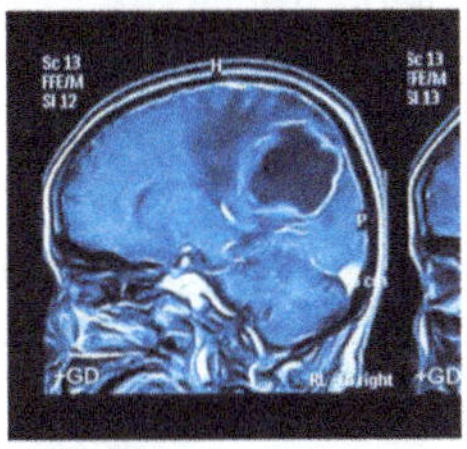	Infected cells	35.48 μs
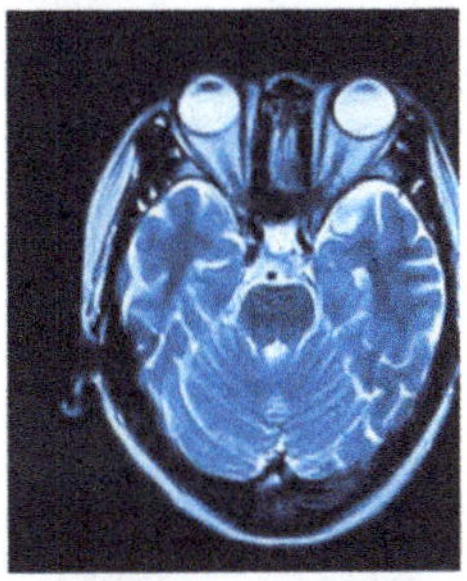	Infected cells	36.39 μs
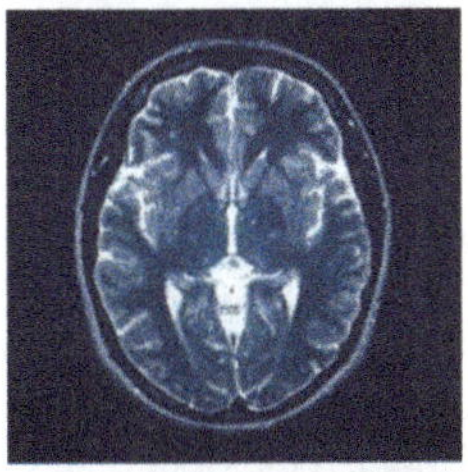	Infected cells	37.09 μs
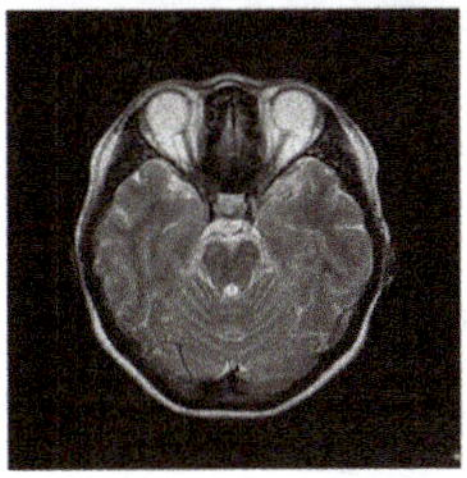	Normal	35.61 μs

14.5 Conclusion

Medical image diagnosis is a major thrust area and necessitates the incorporation of technology for accurate predictions. Various technologies have been involved in the classification of tumors in various parts of the body and cancer cells, but still experience a major setback in terms of accuracy and consume an enhanced level of computational time. This proposed work employs the AWDO model, which accepts noisy input images and consumes reduced computational time when compared with state-of-the-art methodologies. The AWDO model is designed based on the behavior of the hunting technique employed by African wild dogs. The performance of the proposed AWDO model is compared with existing methodologies, and the results prove that the proposed AWDO model stands ahead in terms of accuracy, precision, recall, and *F*1 score. In addition, the proposed AWDO model consumes reduced computational complexity. The performance of the proposed AWDO model can further be enhanced with the incorporation of fuzzy logic and rules along with the AWDO model.

References

[1] Soomro, T. A. *et al.* (2023). Image segmentation for MR brain tumor detection using machine learning: A Review. *IEEE Reviews in Biomedical Engineering*, *16*, 70–90.

[2] Chaki, J., & Woźniak, M. (2023). Brain tumor categorization and retrieval using deep brain incep res architecture based reinforcement learning network. *IEEE Access*, *11*, 130584–130600.

[3] Zhuang, Y., Liu, H., Song, E., & Hung, -C.-C. (Jan 2023). A 3D cross-modality feature interaction network with volumetric feature alignment for brain tumor and tissue segmentation. *IEEE Journal of Biomedical and Health Informatics*, *27*(1), 75–86.

[4] Kannan, S., Prabakaran, D., Dhenesh Kumar, S., & Sivaraj, S. (2023). A Deep Learning-Based Convolution Neural Networks to Forecast Wind Energy. In 2023 International Conference on Recent Trends in Electronics and Communication (ICRTEC), Mysore, India.

[5] Alagarsamy, S., Govindaraj, V., & A, S. (Nov 2023). Automated brain tumor segmentation for MR brain images using artificial bee colony combined with interval Type-II Fuzzy Technique. *IEEE Transactions on Industrial Informatics*, *19*(11), 11150–11159.

[6] Wisaeng, K. (2023). U-Net++DSM: Improved U-Net++ for brain tumor segmentation with deep supervision mechanism. *IEEE Access*, *11*, 132268–132285.

[7] Dhiman, H., Vyas, N., Pagrotra, A., & Dutt, V. (2023). Real-time traffic analysis and driver safety enhancement via IoT and computer vision with high-resolution project lamps for intelligent road illumination [Internet]. In 2023 1st International Conference on Circuits, Power and Intelligent Systems (CCPIS). IEEE. Available from: http://dx.doi.org/10.1109/CCPIS59145.2023.10291872

[8] Ramprasad, M. V. S., Rahman, M. Z. U., & Bayleyegn, M. D. (2023). SBTC-Net: Secured Brain Tumor Segmentation and Classification Using Black Widow With Genetic Optimization in IoMT. *IEEE Access*, *11*, 88193–88208.

[9] Kumar, R., Singh, S., Chauhan, S., Anand, A., & Kumar, A. (2023). ED based spectrum sensing over IRS-assisted Rayleigh-FTR fading channels. *AEU-International Journal of Electronics and Communications*, *171*, 154908.

[10] Rajendran, S. *et al.* (2023). Automated segmentation of brain tumor MRI images using deep learning. *IEEE Access*, *11*, 64758–64768.

[11] Jabbar, A., Naseem, S., Mahmood, T., Saba, T., Alamri, F. S., & Rehman, A. (2023). Brain tumor detection and multi-grade segmentation through hybrid caps-VGGNet model. *IEEE Access, 11*, 72518–72536.

[12] Mo, Y. *et al*. (June 2023). HoVer-Trans: Anatomy-Aware HoVer-Transformer for ROI-Free Breast Cancer Diagnosis in Ultrasound Images. *IEEE Transactions on Medical Imaging, 42*(6), 1696–1706.

[13] Li, Y. *et al*. (March 2024). Zero-shot medical image translation via frequency-guided diffusion models. *IEEE Transactions on Medical Imaging, 43*(3), 980–993.

[14] Wu, J., Xiao, P., Huang, H., Gou, F., Zhou, Z., & Dai, Z. (Sept 2022). An artificial intelligence multiprocessing scheme for the diagnosis of Osteosarcoma MRI Images. *IEEE Journal of Biomedical and Health Informatics, 26*(9), 4656–4667.

[15] Asiri, A. A., Soomro, T. A., Shah, A. A., Pogrebna, G., Irfan, M., & Alqahtani, S. (2024). Optimized brain tumor detection: A dual-module approach for MRI image enhancement and tumor classification. *IEEE Access, 12*, 42868–42887.

[16] Shamshad, N. *et al*. (2024). Enhancing brain tumor classification by a comprehensive study on transfer learning techniques and model efficiency using MRI Datasets. *IEEE Access, 12*, 100407–100418.

[17] Palkar, A., Dias, C. C., Chadaga, K., & Sampathila, N. (2024). Empowering glioma prognosis with transparent machine learning and interpretative insights using explainable AI. *IEEE Access, 12*, 31697–31718.

[18] Asif, A. *et al*. (2024). Machine learning based diagnostic paradigm in viral and non-viral hepatocellular carcinoma. *IEEE Access, 12*, 37557–37571.

[19] Prakash, P. S., Rao, P. K., Babu, E. S., Khan, S. B., Almusharraf, A., & Quasim, M. T. (2024). Decoupled SculptorGAN Framework for 3D Reconstruction and Enhanced Segmentation of Kidney Tumors in CT Images. *IEEE Access, 12*, 62189–62198.

[20] Siddique, A. A., Raza, A., Alshehri, M. S., Alasbali, N., & Abbasi, S. F. (2024). Optimizing tumor classification through transfer learning and particle swarm optimization-driven feature extraction. *IEEE Access, 12*, 85929–85939.

[21] Alqhtani, S. M. *et al*. (2024). Improved brain tumor segmentation and classification in brain MRI With FCM-SVM: A diagnostic approach. *IEEE Access, 12*, 61312–61335.

[22] Raets, C. *et al*. (2024). Optimizing rectal cancer patient care: Dworak TRG prediction via Bayesian evolutionary Fourier-domain random subspace forest. *IEEE Transactions on Instrumentation and Measurement, 73*, 1–10.

[23] Batool, A., & Byun, Y.-C. (2024). Toward improving breast cancer classification using an adaptive voting ensemble learning algorithm. *IEEE Access, 12*, 12869–12882.

[24] Sun, Y., Wen, S., Shi, S., & Chen, Y. (March 2024). Individual adaptive regulation strategy inspired by artificial fish swarm algorithm for tumor targeting. *IEEE Transactions on Molecular, Biological, and Multi-Scale Communications, 10*(1), 132–143.

[25] Faizi, M. I., & Adnan, S. M. (2024). Improved segmentation model for melanoma lesion detection using normalized cross-correlation-based k-means clustering. *IEEE Access, 12*, 20753–20766.

[26] Almufareh, M. F., Imran, M., Khan, A., Humayun, M., & Asim, M. (2024). Automated brain tumor segmentation and classification in MRI Using YOLO-Based Deep Learning. *IEEE Access, 12*, 16189–16207.

[27] Vasantbhai Patel, V., Yadav, A. R., Jain, P., & Cenkeramaddi, L. R. (2024). A systematic kidney tumour segmentation and classification framework using adaptive and attentive-based deep learning networks with improved crayfish optimization algorithm. *IEEE Access, 12*, 85635–85660.

[28] Prabakaran, D., & Ramachandran, S. (09 June 2022). Secure Channel for Financial Transactions in Cloud Environment Using Blockchain Technology. *PREPRINT (Version 1).*

[29] Aishwarya, N., Praveena, N. G., Priyanka, S. *et al*. (2023). Smart farming for detection and identification of tomato plant diseases using light weight deep neural network. *Multimedia Tools and Applications, 82*, 18799–18810.

S. Saranya

15 Artificial Intelligence-Based Control Strategies for COVID-19 That Target Different Age Groups

Abstract: This chapter describes artificial intelligence (AI) tools in the mathematical model of COVID-19 disease with medication and various methods to control its spread. The age-structured populations' general MC Kendrick equations are examined to create the model equations utilizing AI approaches to improve the precision and utility of a prey–predator system that pairs one predator with three age groups of prey. AI is useful for accounting for intricate interactions between predators and prey as well as for parameter estimation. We study the equilibria's existence, positiveness, boundedness, and stability in relation to the inherent carrying capacity of the prey. The primary outcome of this study is the population's distribution across three age groups and how to prevent disease transmission from predators by using both local and Hopf bifurcation methods. AI models the effects of various public health initiatives like lockdown and social distancing measures to control the spread of the virus. Lastly, the model's output included numerical examples of Rossler-type AI tools in Maple software.

Keywords: Point of equilibrium, age group targets, Hopf bifurcation, mathematical model of stability switch

15.1 Introduction

Many of the ideas and principles of the science being discussed are best communicated through mathematical models [17]. With the use of artificial intelligence (AI) approaches, predator–prey relations and rivalry are frequently seen as the two primary obstacles in population models based on mathematics [13, 18]. The behavior of predators and prey can be simulated mathematically to create more accurate predation models. Understanding and modeling the impact of their phenotypic traits require studying how predators and prey respond. Since these models rely heavily on empirical knowledge, they must be compared with the test results gathered in the virtual environment with the individuals present.

The COVID-19 pandemic was first identified in China and rapidly spread to the remainder of the world, leading the World Health Organization to proclaim a worldwide public health emergency. The third respiratory disease outbreak linked to the corona virus in the last 20 years is COVID-19, following Middle East respiratory syndrome and severe acute respiratory syndrome. COVID-19 has devastated the global healthcare system, resulting in thousands of deaths when it first surfaced in Wuhan,

https://doi.org/10.1515/9783111436180-015

China. In the fight against the COVID-19 pandemic, it is critical to identify and screen affected persons as soon as possible because failure to do so could result in the spreading of virus to other people [20]. The actual COVID-19 data is a sequence of observations organized over time, and time-series prediction techniques are well-established in the statistics domain and include structure-based approaches, meta-predictors, and machine learning-based approaches [16].

Data is a fundamental prerequisite for building a framework that the AI community needs to handle this issue. The lack of appropriate data about patients infected with COVID-19 makes it difficult to create identification and analysis models [11]. Due to reluctance among patients to seek medical attention in order to avoid potential confinement, very little information is known about the COVID-19 test-positive patients. Furthermore, some infected individuals do not exhibit any COVID-19 symptoms, making it difficult to pinpoint the exact number of positive cases in large populations [14]. An infected individual becomes more dangerous for others the longer it takes to diagnose them. For this reason, the analysis depends heavily on the availability of vast amounts of data.

Numerous publications examining the impacts of COVID-19 from a statistical, mathematical, and computational standpoint have been released. Susceptible–infectious–recovered is one of the computational frameworks for examining the dynamics of COVID-19 virus. It calculates the spread of an epidemic by taking advantage of a time-dependent system of differential equations [11]. A thorough analysis of the research on predator–prey interactions conducted over the past century reveals how crucial this environment is to understand animal biology's evolutionary and adaptive processes [18]. Providing for human needs and luxuries continuously depletes the natural world's ecological balance [3, 7]. The similar underlying methodology based on the aging group population has been utilized in mathematical models to describe the dynamics of diseases [19, 21]. Cushing inquired about numerous formulas for age-structured populations that were obtained from the Mc Kendrick model [15].

After highlighting the three distinct age groups of prey, we can say that:

i) The age range of the first prey population is from birth to 14 years.
ii) Age range of the second prey population is 15–39 years.
iii) The third age group of prey is those over 40.

Due to their high degree of immunity and implementation of corrective measures, the first and second prey populations will experience a lower incidence of sickness. For the third prey population, which is 40 years of age and over, getting enough nutrients is essential for lowering the danger and severity of viral infections as well as strengthening the defense mechanism. The constant τ is interpreted as a gain in immunity, either through complementary medicine or government-provided vaccinations. In this study, we examine an AI model that simulates the dependence of temporal control measures (facial mask, elbow cough, hand washing, social distancing, and self-quarantine), based on the theory that people infected with COVID-19 can also carelessly transfer the illness [2].

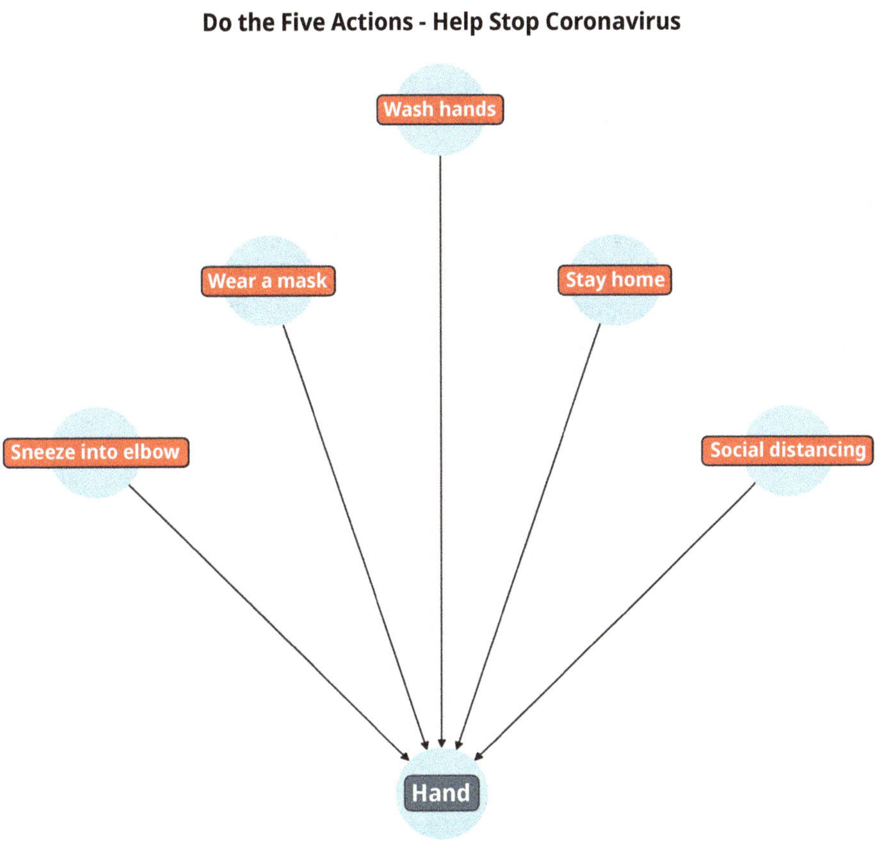

Figure 15.1: Control measures for corona virus.

Our primary objectives are: We investigate the optimal performance of AI techniques under the supposition that the control measures (social distancing, face masks, hand washing, elbow coughing, and self-quarantine) limit the infection and the spread of illness. We refer to the continual τ as a natural medicine or vaccination to increase the immune level of prey population, which is people under 40 years old. Distinct prey models are being created, representing distinct physical interactions between prey and predator species, derived from Lotka and Volterra's groundbreaking work [12, 11]. Here, we are utilizing the carrying capacity for each of the three preys and the control mechanisms for the third prey [3]. We examine the Hopf bifurcation's direction and stability as it arises at the boundary equilibrium point [9, 14].

This study retrieves the system and solves the guidelines for the system's local stability [8]. Maintaining the balance of sickness in the prey–predator system is a crucial aspect of the ecosystem for natural coexistence and stability [5, 6]. The results of our investigation into how three preys interact with a single predator model under varying incidences and disease-related mortality rates including vaccination and natural medicine are highlighted [2, 16]. Lastly, we display the numerical solution findings for every model at different parameter values. Symbolic computation and algebra, machine learning integration, natural language processing, predictive analysis, and other AI technologies are used in Maple 15 programming.

15.2 Technique for Choosing Values for Parameters

Three prey population subpopulations, shown by three prey populations, are created by the model from the whole population as $N_1(t)$, $N_2(t)$, $N_3(t)$ at time t, and one predator population $N_4(t)$. Examine the 4D model system as follows:

$$\frac{dN_1}{dt} = aN_1\left(1-\frac{N_1}{k_1}\right) - \alpha\ N_1N_4$$

$$\frac{dN_2}{dt} = bN_2\left(1-\frac{N_2}{k_2}\right) - \beta\ N_2N_4$$

$$\frac{dN_3}{dt} = cN_3\left(1-\frac{N_3}{k_3}\right) - \gamma\ N_3N_4 - eN_3 + U\tau N_3$$

$$\frac{dN_4}{dt} = \eta\ N_1N_4 + \delta\ N_2N_4 + \mu\ N_3N_4 - fN_4 \tag{15.1}$$

with initial conditions

$$N_l(t) \geq 0, N_2(t) \geq 0,\ N_3(t) \geq 0, N_4(t) \geq 0 \tag{15.2}$$

The first, second, third, and predator are denoted by the parameters $N_1(t)$, $N_2(t)$, $N_3(t)$, and $N_4(t)$, respectively, and may be found in this case. Here is a description of the model parameters:

The first, second, and third prey's respective specific growth rates are **a, b, c.**

The corresponding carrying capabilities of the first, second, and third preys are k_1, k_2, k_3.

The predator interacts with three preys (N_1, N_2, N_3) in the following ways: **α**, **β**, and **γ**.

e is the third prey population's natural death. **U** represents the control shown in Figure 15.1. τ denotes the increased immunity with the intake of natural medicines.

The conversion factors for predators from the three prey populations are **η**, **δ**, and **μ**.

The natural death of the predator is *f*.

This model includes certain assumptions, consisting of:

1. In the first, second, and third prey populations, interactions occur through predators.
2. Strong immune growth from infancy to 39 years allows the first and second prey populations to recover after interactions with predators.

3. Three populations of prey possessing carrying capacities and intrinsic growth rates of a, b, and c and $k_1,\ k_2, k_3$ were fitted using the Leslie–Gower model, and no interaction was found between them.
4. The virus only affects one-third of the prey population, as immunity develops and dies naturally. Also, we raise the immunity growth by AI through parameter estimation, such as constants $U\tau$, where U is the control measure (see Figure 4.1) and τ is the immunization or alternative medicine:
   τ = cardamom, ginger, turmeric, cumin seeds, and pepper
5. After implementing control measures like $U\tau$, the afflicted prey recovers in a third prey population as a result of higher immune growth.
6. Predators are weakened in primary and secondary prey populations, but not in tertiary prey populations.

15.3 Boundedness and Positivity of the Specified System

15.3.1 Positivity

Theorem 15.1: The given equation system (1) is never negative. Therefore, every potential outcome for system (15.1) is positive.

Proof.
Given that system (15.1)'s right-hand side is locally Lipschitzian on C and entirely continuous, condition (15.2) is distinct on $[0,\ \xi)$, where $0 < \xi \leq +\infty$ as in [16]. With initial conditions (15.2) applied to system (15.1), we have

$$N_1(t) = N_1(0)\exp\left[\int\left\{a\left(1-\frac{N_1(t)}{k_1}\right) - \alpha N_4(t)\right\}dt\right] > 0$$

$$N_2(\mathrm{t}) = N_2(0)\exp\left[\int\left\{b\left(1-\frac{N_2(t)}{k_2}\right) - \beta N_4(t)\right\}dt\right] > 0$$

$$N_3(t) = N_3(0)\exp\left[\int\left\{c\left(1-\frac{N_3(t)}{k_3}\right) - \gamma N_4(t) + U\tau\right\}dt\right] > 0$$

$$N_4(t) = N_4(0)\exp\left[\int\{\eta N_1(t) + \delta N_2(t) + \mu N_3(t) - f\}dt\right] > 0$$

which brings the proof to an end.

15.3.2 Boundedness

Theorem 15.2: The domain contains bounds for the trajectories of system (15.1).
$R^4_+ = \{(N_l, N_2, N_3,\ N_4) \in R^4_+ : N_l \geq 0,\ N_2 \geq 0,\ N_3 \geq 0,\ N_4 \geq o\}$.

Proof.
Define the function $\Delta = N_l + N_2 + N_3 + N_4$ Take its temporal derivative with respect to solution (15.1):

$$\frac{dN}{dt} = \frac{d\,N_l}{dt} + \frac{d\,N_2}{dt} + \frac{d\,N_3}{dt} + \frac{d\,N_4}{dt}$$

Since the solutions starting in $R_+{}^4$ continue in the non-negative quadrant and the parameters are non-negative, $\frac{dN}{dt} + \rho N = aN_1\left(1 - \frac{N_1}{k_1}\right) + bN_2\left(1 - \frac{N_2}{k_2}\right) + cN_3\left(1 - \frac{N_3}{k_3}\right) - eN_3 + U\tau N_3 - fN_4 + \rho N_1 + \rho N_2 + \rho N_3 + \rho N_4$

$= (\rho + a)N_l + (\rho + b)N_2 + \frac{(\rho + c + U\tau)}{e} N_3 + (\rho - f)N_4 - \frac{aN_1^2}{k_1} - \frac{bN_2^2}{k_2} - \frac{cN_3^2}{k_3}$

where ρ is an increasing constant for $\rho > f$ and given $\epsilon > 0$ exist to the extent that t on $t\ \geq\ t_0$

$$\frac{dN}{dt} + \rho N \leq m + \epsilon, m = \min\left\{(\rho + a), (\rho + b), \frac{(\rho + c + U\tau)}{e}, (\rho - f)\right\}$$

Lemma 15.3 Give ϕ an absolutely continuous function that satisfies the inequality in differential equations $\frac{d\phi(t)}{dt} + \alpha_1\phi(t) \leq \alpha_2, t \geq 0$, where$(\alpha_1, \alpha_2) \in R^3_+, \alpha_1 \neq 0$ then$\forall t \geq T \geq o$, $\phi(t) \leq \frac{\alpha_2}{\alpha_1} - \left(\frac{\alpha_2}{\alpha_1}\phi(T)\right)e^{-\alpha_1(t-T)}$.

From the above lemma [1], we get $N(t) \leq N(t_0)e^{-\rho(t-t_0)} + \left(\frac{m+\varepsilon}{\rho}\right)\left(1 - e^{-\rho(t-t_0)}\right)$, where $t \to \infty$ then letting $\varepsilon \to o$, $sup\ N(t) \leq \frac{m}{\rho}$

There are bounds to system (15.1) on the initial circumstances.

15.4 Qualitative Analysis of the Given System

15.4.1 Equilibrium Points

For the dynamic model (15.1), the equilibrium points may be found using the steady-state equations $\frac{d\,N_l}{dt} = \frac{d\,N_2}{dt} = \frac{d\,N_3}{dt} = \frac{d\,N_4}{dt}$. There are 15 equilibrium points in the system. The equilibrium points that are nontrivial, axial, and trivial are obtained using algebraic computation as follows.

Table 15.1: Stability evaluation of equilibrium points.

Equilibrium point	Explanation	Stable/unstable
$E1\{N_1 = o, N_2 = 0, N_3 = 0, N_4 = 0\}$	Trivial	Saddle point
$E3\{N_1 = 0, N_2 = k_2, N_3 = o, N_4 = 0\}$	Prey 1, prey 3, and predator free	Saddle point
$E4\{N_1 = 0, N_2 = k_2, N_3 = \theta_1, N_4 = 0\}$	Prey 1 and predator free	Saddle point
$E5\{N_1 = k_1, N_2 = 0, N_3 = o, N_4 = 0\}$	Prey 2, prey 3, and predator free	Saddle point
$E6\{N_1 = k_1, N_2 = 0, N_3 = \theta_1, N_4 = 0\}$	Prey 2 and predator free	Saddle point
$E7\{N_1 = k_1, N_2 = k_2, N_3 = 0, N_4 = 0\}$	Prey 3 and predator free	Stable
$E8\{N_1 = k_1, N_2 = k_2, N_3 = \theta_1, N_4 = 0\}$	Predator free	Stable
$E9\{N_1 = 0, N_2 = o, N_3 = \theta_2, N_4 = \theta_3\}$	Prey 1 and prey 2 free	Stable
$E10\{N_1 = 0, N_2 = \theta_4, N_3 = o, N_4 = \theta_5\}$	Prey 1 and prey3 free	Stable
$E12\{N_1 = \theta_6, N_2 = 0, N_3 = 0, N_4 = \theta_7\}$	Prey 2 and prey 3 free	Stable
$E15\{N_1 = N_1^*, N_2 = N_2^* N_3 = N_3^*, N_4 = N_4^*\}$	Coexistence	Stable

The analysis of equilibrium points in Table 15.1 were $\theta_1 = \frac{(c+U\tau-e)}{c}$, $\theta_2 = \frac{f}{\mu}$,

$$\theta_3 = -\frac{-U\mu\tau k_3 - c\mu k_3 + e\mu k_3 + cf)}{\gamma\mu k_3},\ \theta_4 = \frac{f}{\delta},\ \theta_5 = -\frac{b(-\delta k_2 + f)}{\beta\delta k_2},\ \theta_6 = \frac{f}{\eta},\ \theta_7 = -\frac{a(-\eta k_1 + f)}{\eta a k_1}.$$

15.4.2 Analysis of Stability and Existence of Equilibrium Points

The existence and stability behavior of system (15.1) at the equilibrium locations are examined in this section [10]:

- The trivial equilibrium point $E1$ is a saddle point.
- The axial equilibrium points $E3$ and $E5$ are saddle points.
- The planar equilibrium points $E4$ and $E6$ are saddle points.
- The planar equilibrium of $E2$, $E11$, $E13$, and $E14$ is unstable.
- The planar equilibrium of $E7$, $E8$, $E9$, $E10$, and $E12$ are asymptotically stable locally. The equilibrium points $E7$, $E8$, $E9$, $E10$, $E12$, and the internal equilibrium point $E15$ will be interpreted.
- At equilibrium, the Jacobian matrix of system (15.1) will interpret the equilibrium points $E7$, $E8$, $E9$, $E10$, $E12$, and the interior equilibrium point $E15$.

System (15.1)'s equilibrium point Jacobian matrix $E = (N_1(t), N_2(t), N_3(t), N_4(t))$ is given by

$$J = [m_{11}\ 0\ 0 - N_1\alpha\ 0\ m_{22}\ 0 - N_2\beta\ o\ 0\ m_{33} - N_3\gamma\ N_4\eta\ N_4\delta\ N_4\mu\ m_{44}]$$

where $m_{11} = a - 2\frac{a\,N_1}{k_1} - N_4\alpha$, $m_{22} = b - 2\frac{b\,N_2}{k_2} - N_4\beta$, $m_{33} = c - 2\frac{c\,N_3}{k_3} - N_4\gamma - e + U\tau$, $m_{44} = N_1\eta + N_2\delta + N_3\mu - f$.

Due to the nature of the eigenvalues, eq. (15.1) is stable if all eigenvalues are negative for the real roots of the Jacobian characteristic equation above, and have negative real parts for the complex roots. Otherwise, the system will become unstable.

Theorem 15.4: The given system (l) is a saddle at an equilibrium point

$$E1\ (N_1 = 0, N_2 = 0, N_3 = o, N_4 = 0)$$

Proof.
The Jacobian matrix $E1 = [a\,0\;0\;0\;0\;b\;0\;o\;0\;0\,c + U\tau - e\,0\;0\;0\;0 - f]$
Now the eigenvalues are $\lambda_1 = a > 0,\;\; \lambda_2 = b > o,\;\; \lambda_3 = U\tau + c - e < 0$ with the condition $U\tau + c < e,\; \lambda_4 = -f < 0$.
Hence, the equilibrium point is a saddle point.

Theorem 15.5: The given system (l) is a saddle point at an equilibrium point $E3\,(N_1 = o, N_2 = k_2, N_3 = 0,\;\; N_4 = o)$.

Proof.
The Jacobian matrix $E3 = [a\,0\,0\,0\,0 - b\,0\; - k_2\beta\,0\,0\,c + U\tau - e\,0\,0\,0\,0\,\delta k_2 - f]$.
Now the eigenvalues are $\lambda_1 = a > 0,\;\; \lambda_2 = -b < o,\;\; \lambda_3 = U\tau + c - e < 0,\; \lambda_4 = \delta k_2 - f < 0$ with the condition $U\tau + c < e,\;\; \delta k_2 < f$. Hence, the equilibrium point is a saddle point.

Theorem 15.6: The given system (l) is a saddle point at an equilibrium point $E4\;(N_1 = 0, N_2 = k_2,\;\; N_3 = \theta_1,\;\; N_4 = 0)$.

Proof.
The Jacobian matrix

$$E4 = \left[a\,0\,0\,0\,0 - b\,0 - k_2\,\beta\,0\,0 - U\tau - c + e\,\frac{-\gamma k_3(c + U\tau - e)}{c}\,0\,0\,0\,\delta k_2 + \frac{\mu k_3(c + U\tau - e)}{c} - f\right]$$

Now the eigenvalues are $\lambda_1 = a > 0, \lambda_2 = -b < 0,\; \lambda_3 = -U\tau - c + e < o, \lambda_4 = \frac{U\mu\tau k_3 + c\delta k_2 + c\mu k_3 - e\mu k_3 - cf)}{c} < 0$ with the condition $e < U\tau + c,\;\; U\mu\tau k_3 + c\delta k_2 + c\mu k_3 < e\mu k_3 + cf$. Hence, the equilibrium point is a saddle point.

Theorem 15.7: The given system (l) is a saddle point at an equilibrium point $\mathrm{E5} = (N_1 = 0, N_2 = k_2, N_3 = 0,\;\; N_4 = 0)$

Proof.
The Jacobian matrix $\mathrm{E5} = [-a\;0\;0 - k_1 a\;0\;b\;0\;0\;0\;0\;U\tau + c - e\;0\;0\;0\;0\;\eta k_1 - f]$.
Now the eigenvalues are $\lambda_1 = -a < 0, \lambda_2 = b > 0,\;\; \lambda_3 = U\tau + c - e < 0,\; \lambda_4 = \eta k_1 - f < 0$ with the condition $U\tau + c < e, \eta k_1 < f$. Hence, the equilibrium point is a saddle point.

Theorem 15.8: The given system (l) is a saddle point at an equilibrium point. $E6\ (N_1 = k_1, N_2 = 0, N_3 = \theta_1,\ N_4 = 0)$.

Proof.
The Jacobian matrix.

$$E6 = \left[-a00 - k_1 a0b0000 - U\tau - c + e \frac{-\gamma k_3(c + U\tau - e)}{c} 000\eta k_1 + \frac{\mu k_3(c + U\tau - e)}{c} - f \right]$$

Now the eigenvalues are $\lambda_1 = -a < 0,\ \lambda_2 = b > o,\ \lambda_3 = -U\tau - c + e < 0,\ \lambda_4 = \frac{U\mu\tau k_3 + c\eta k_1 + c\mu k_3 - e\mu k_3 - cf}{c} < 0$ with the condition $e < U\tau + c,\ U\mu\tau k_3 + c\eta k_1 + c\mu k_3 < e\mu k_3 + cf$. Hence, the equilibrium point is a saddle point.

Theorem 15.9: The given system (l) is stable at an equilibrium point $E7(N_1 = k_1,\ N_2 = k_2,\ N_3 = 0,\ N_4 = 0)$ provided that $U\tau + c < e \ and\ \delta k_2 + \eta k_1 < f$.

Proof.
The Jacobian matrix $E7 = [-a00\ -k_1 a0\ -b0\ -k_2\beta 00\ c + U\tau - e\,0000\ -\delta k_2 + \eta k_1 - f]$
Now the eigenvalues are $\lambda_1 = -a < 0,\ \lambda_2 = -b < 0,\ \lambda_3 = U\tau + c - e < 0,\ \lambda_4 = \delta k_2 + \eta k_1 - f < 0$; consequently, the equilibrium point is only stable if $U\tau + c < e$ and $\delta k_2 + \eta k_1 < f$ and $\lambda_1,\ \lambda_2$ have negative equal parts.

Theorem 15.10: At an equilibrium point $E8(N_1 = k_1,\ N_2 = k_2,\ N_3 = \theta_1,\ N_4 = 0)$, system (l) is locally asymptotically stable, provided that $U\tau + c > e$ and $c\delta k_2 + \eta k_1 + U\mu\tau k_3 + c\mu k_3 < e\mu k_3 + cf$.

Proof.
The Jacobian matrix

$$E8 = \left[-a00 - k_1 a0 - b0 - k_2\beta 00 - U\tau - c + e \frac{-\gamma k_3(c + U\tau - e)}{c} 000\delta k_2 + \eta k_1 + \frac{\mu k_3(c + U\tau - e)}{c} - f \right]$$

Now the eigenvalues are $\lambda_1 = -\mathrm{a} < \mathrm{o},\quad \lambda_2 = -\mathrm{b} < 0,\quad \lambda_3 = -\mathrm{U}\tau - \mathrm{c} + \mathrm{e} < 0$, $\lambda_4 = -\frac{-U\mu\tau k_3 - c\eta k_1 - c\delta k_2 - c\mu k_3 + e\mu k_3 + cf}{c} < 0$ hence the above equilibrium point is locally asymptotically stable if $\lambda_1,\ \lambda_2$ have negative equal parts.
and $\lambda_3,\ \lambda_4$ *satisfies* $\mathrm{U}\tau + \mathrm{c} > \mathrm{e},\ c\delta k_2 + \eta k_1 + U\mu\tau k_3 + c\mu k_3 < e\mu k_3 + cf$.

Theorem 15.11: The given system (l) is locally asymptotically stable at an equilibrium point $E9(N_1 = 0,\ N_2 = 0,\ N_3 = \theta_2,\ N_4 = \theta_3)$, provided that $U\alpha\mu\tau k_3 + \alpha c\mu k_3 > \gamma\mu a k_3 + \alpha e\mu k_3 + acf$ and $U\beta\mu\tau k_3 + \beta c\mu k_3 > \gamma\mu b k_3 + \beta e\mu k_3 + \beta cf$.

Proof.
The Jacobian matrix $E9 = \left[\theta_a\, 0\,0\,0\,0\,\theta_b\, 0\,0\,0\,0\,\theta_c\, \frac{-\gamma f}{\mu}\, \theta_d\, \theta_e - \frac{-U\mu\tau k_3 - c\mu k_3 + e\mu k_3 + cf}{\gamma k_3}\, 0\right]$

where $\theta_a = a + \frac{\alpha(-U\mu\tau k_3 - c\mu k_3 + e\mu k_3 + cf)}{\gamma\mu k_3}$, $\theta_b = b + \frac{\beta(-U\mu\tau k_3 - c\mu k_3 + e\mu k_3 + cf)}{\mu\gamma k_3}$,

$\theta_c = U\tau + c - e + \frac{-U\mu\tau k_3 - c\mu k_3 + e\mu k_3 + cf}{\mu k_3} - \frac{2cf}{\mu k_3}$,

$\theta_d = -\frac{\eta(-U\mu\tau k_3 - c\mu k_3 + e\mu k_3 + cf)}{\gamma\mu k_3}$, $\theta_e = \frac{\delta(-U\mu\tau k_3 - c\mu k_3 + e\mu k_3 + cf)}{\gamma\mu k_3}$.

Now the eigenvalues are

$$\lambda_1 = \frac{-U\alpha\mu\tau k_3 + \gamma\mu a k_3 - \alpha c\mu k_3 + \alpha e\mu k_3 + \alpha cf}{\gamma\mu k_3}, \lambda_2 \frac{-U\beta\mu\tau k_3 + \gamma\mu b k_3 - \beta c\mu k_3 + \beta e\mu k_3 + \beta cf}{\gamma\mu k_3},$$

$$\lambda_3 = \frac{1}{2}\frac{-cf + i\sqrt{\theta_f}}{\mu k_3}, \lambda_4 = -\frac{1}{2}\frac{cf + i\sqrt{\theta_f}}{\mu k_3},$$

where $\theta_f = 4Ufk_3{}^2\mu^2\tau + 4cfk_3{}^2\mu^2 - 4efk_3{}^2\mu^2 - 4cf^2\mu k_3 - f^2c^2$

The equilibrium point mentioned above is therefore locally asymptotically stable if $U\alpha\mu\tau k_3 + \alpha c\mu k_3 > \gamma\mu a k_3 + \alpha e\mu k_3 + \alpha cf$, $U\beta\mu\tau k_3 + \beta c\mu k_3 > \gamma\mu b k_3 + \beta e\mu k_3 + \beta cf$ and λ_3, λ_4 have negative real parts.

Theorem 15.12: The given system (l) is locally asymptotically stable at an equilibrium point

$$E10(N_1 = 0,\ N_2 = \theta_4,\ N_3 = 0,\ N_4 = \theta_5) \text{ provided that } a\beta\delta k_2 + abf < ab\delta k_2 \,\&$$

$$U\beta\delta\tau k_2 + \beta\delta c k_2 + bf\gamma < b\delta\gamma k_2 + \beta\delta e k_2.$$

Proof.
The Jacobian matrix

$$E10 = \left[a + \frac{ab(-\delta k_2 + f)}{\beta\delta k_2}\ 0\,0\,0\,0\ \theta_g\ 0\ \frac{-\beta f}{\delta}\ 0\,0\ U\tau + c - e + \frac{b\mu(-\delta k_2 + f)}{\beta\delta k_2}\ 0 - \frac{b\eta(-\delta k_2 + f)}{\beta\delta k_2}\right.$$
$$\left. - \frac{b(-\delta k_2 + f)}{\beta k_2} - \frac{b\mu(-\delta k_2 + f)}{\beta\delta k_2}\ 0\right]$$

where $\theta_g = \mathrm{b} - \frac{-2bf}{\delta k_2} + \frac{b(-\delta k_2 + f)}{\beta\delta k_2}$. Now the eigenvalues are $\lambda_1 \frac{a\beta\delta k_2 - ab\delta k_2 + abf}{\beta\delta k_2}$,

$$\lambda_2 = \frac{U\beta\delta\tau k_2 - b\delta\gamma k_2 + \beta\delta ck_2 - \beta\delta ek_2 + bf\gamma}{\beta\delta k_2},\ \lambda_3 = \frac{1}{2}\frac{-bf + i\sqrt{4bfk_2{}^2\delta^2 - 4b\delta k_2 f^2 - f^2b^2}}{\delta k_2},$$

$$\lambda_4 = -\frac{1}{2}\frac{bf + i\sqrt{4bfk_2{}^2\delta^2 - 4b\delta k_2 f^2 - f^2b^2}}{\delta k_2}.$$

Accordingly, the equilibrium point mentioned above is locally asymptotically stable if $a\beta\delta k_2 + abf < ab\delta k_2, U\beta\delta\tau k_2 + \beta\delta ck_2 + bf\gamma < b\delta\gamma k_2 + \beta\delta ek_2$
and $\lambda_3,\ \lambda_4$ have negative real parts.

Theorem 15.13: The given system (l) is locally asymptotically stable at an equilibrium point

$E12(N_1 = \theta_6,\ N_2 = 0,\ N_3 = 0,\quad N_4 = \theta_7)$ provided that $a\beta\eta k_1 > ab\eta k_1 + a\beta f$,

$Ua\eta\tau k_1 + a\eta ck_1 + af\gamma < a\eta\gamma k_1 + a\eta ek_1$.

Proof.

$$E12 = \left[a - \frac{2af}{\eta k_1} + \frac{a(-\eta k_1 + f)}{\eta k_1} 0 0 \frac{-\alpha f}{\eta} 0 b + \frac{a\beta(-\eta k_1 + f)}{\alpha\eta k_1} 0000 U\tau + c - e + \frac{a\gamma(-\eta k_1 + f)}{\eta\alpha k_1} 0 - \frac{a(-\eta k_1 + f)}{\alpha k_1}\right.$$
$$\left. - \frac{a(-\eta k_1 + f)}{\alpha\eta k_1} - \frac{a\mu(-\eta k_1 + f)}{\alpha\eta k_1} 0\right].$$

Now the eigenvalues are

$$\lambda_1 = \frac{-a\beta\eta k_1 + ab\eta k_1 + a\beta f}{a\eta k_1},\ \lambda_2 = \frac{Ua\eta\tau k_1 - a\eta\gamma k_1 + a\eta ck_1 - a\eta ek_1 + af\gamma}{a\eta k_1},$$

$$\lambda_3 = \frac{1}{2}\frac{-af + i\sqrt{4afk_1{}^2\eta^2 - 4a\eta k_1 f^2 - f^2a^2}}{\eta k_1},$$

$$\lambda_4 = -\frac{1}{2}\frac{af + i\sqrt{4fak_1{}^2\eta^2 - 4a\eta k_1 f^2 - f^2a^2}}{\eta k_1}.$$

Hence, the above equilibrium point is locally asymptotically stable if $a\beta\eta k_1 > ab\eta k_1 + a\beta f$,
$Ua\eta\tau k_1 + a\eta ck_1 + af\gamma < a\eta\gamma k_1 + a\eta ek_1$

and $\lambda_3,\ \lambda_4$ have negative real parts.

Theorem 15.14: The interior equilibrium point $\{N_1 = N_1^*, N_2 = N_2^*,\ N_3 = N_3^*,\ N_4 = N_4^*\}$ is locally asymptotically stable.

Proof.
The variation of the Jacobian matrix is $J_{E15} = [m_{11}\,0\,0 - N_1^*\alpha\,0\,m_{22}\,0 - N_2^*\beta\,0\,0\,m_{33} - N_3^*\gamma$ $N_4^*\eta\,N_4^*\delta\,N_4^*\mu\,m_{44}]$

where

$$m_{11} = a - 2\frac{aN_1^*}{k_1} - N_4^*\alpha,\; m_{22} = b - 2\frac{bN_2^*}{k_2} - N_4^*\beta,\; m_{33} = c - 2\frac{N_3^*c}{k_3} - N_4^*\gamma - e + U\tau,$$
$$m_{44} = N_1^*\eta + N_2^* + N_3^*\mu - f$$

The characteristic equation is $\Lambda(\lambda) = \lambda^4 + M_3\lambda^3 + M_2\lambda^2 + M_1\lambda^1 + M_0$

where $M3 = 4I - (m11 + m22 + m33 + m44) = 4I - X$

$$M2 = 6I^2 - (m11 + m22 + m33 + m44)3I + m22m33 + m22m44 + m33m44 + m11m22 + m11m33 + m11m44 - N_3^*N_4^*\mu\gamma + N_1^*N_4^*\eta\alpha$$
$$= 6I^2 - 3XI + (Y + j_1)$$

$$M1 = 4I^3 - (m11 + m22 + m33 + m44)3I^2 + (2m22m33 + 2m22m44 + 2m33m44 + 2m22m11 + 2m33m11 + 2m44m11 + 2N_3^*N_4^*\gamma\mu + 2N_1^*N_4^*\alpha\eta + N_2^*N_4^*\delta\beta)I - m22m33m44 - m22m33m11 - m22m44m11 - m33m44m11 - N_3^*N_4^*\gamma\mu m22 - N_3^*N_4^*\gamma\mu m11 - N_2^*N_4^*\delta\beta m11 - N_1^*N_4^*\alpha\eta m22 - N_1^*N_4^*\alpha\eta m33$$
$$= 4I^3 - 3XI^2 + (2(Y + j_1) + j_2)I - Z$$

$$M0 = I^4 - (m11 + m22 + m33 + m44)I^3 + (m11m22 + m11m33 + m11m44 + m22m33 + m33m44 + m22m44 - N_3^*N_4^*\gamma\mu - N_2^*N_4^*\delta\beta - N_1^*N_4^*\alpha\eta)I^2 - (m11m22m33 + m11m22m44 + m11m33m44 + m22m33m44 + m11N_3^*N_4^*\gamma\mu + N_3^*N_4^*\gamma\mu m22 + N_2^*N_4^*\delta\beta m33 + N_2^*N_4^*\delta\beta m11 + N_1^*N_4^*\alpha\eta m22 + N_1^*N_4^*\alpha\eta m33)I + m11m22m33m44 - m11\,N_3^*N_4^*\gamma\mu m22 + N_2^*N_4^*\delta\beta m11m33 + N_1^*N_4^*\alpha\eta m33m22$$
$$= I^4 - XI^3 + (Y - j_1 + j_2)I^2 - (Z + j_3) + W$$

By Routh Hurwitz's criterion, all the eigenvalues of J_{E15} have negative real parts if
(i) $M_3 > 0$,
(ii) $M_3M_2 > M_1$,

where $M_3M_2 = [4I - X][6I^2 - 3XI + (Y + j_1)]$
$$= 24I^3 - 18XI^2 + (4Y + j_1) + 3X^2)I - (Y + j_1)X$$
$$> 4I^3 - 3XI^2 + (2(Y + j_1) + j_2)I - Z = M_1$$

Hence, we proved $M_3M_2 > M_1$. Similarly, we proved (iii) $M_3M_2M_1 > M_1{}^2 + M_3^2\,M_0$.

Consequently, around the non-negative equilibrium point, the given system of the non-linear differential equation (15.1) is locally asymptotically stable. $\{N_1 = N_1^*, N_2 = N_2^*, N_3 = N_3^*,\ N_4 = N_4^*\}$

if the conditions mentioned in the theorem hold.

15.5 Hopf bifurcation

The Hopf bifurcation around the interior equilibrium point $E15$ was studied in this section. The bifurcation parameter N_3 is identified by the basic parameter c, which represents the prey's growth rate. The Jacobian matrix J_{E15} experiences Hopf bifurcation, wherein two of its eigenvalues are solely imaginary, while the remaining eigenvalues containing non-positive real components are $\mathrm{Re}[d\lambda/dc]|_{c=c_0} \neq 0[4]$. Assume that the interior equilibrium point's characteristic equation $E15$ is as follows:

$$\lambda^4 + A_3\lambda^3 + A_2\lambda^2 + A_1\lambda^1 + A_0 = 0 \tag{15.3}$$

The characteristic polynomial coefficients (15.3) meet the following requirement for just imaginary eigenvalues $A3A2A1 - A_3^2 A0 - A_1^2 = 0$.

Assume that $\pm i\omega$ represents two purely imaginary eigenvalues that match c0. From the characteristic equation (15.3),

$$\left[4\lambda^3 + 3A_3\lambda^2 + 2A_2\lambda + A_1\right]\frac{d\lambda}{dc} + \left(\lambda^3\frac{dA}{dc} + \lambda^2\frac{dA}{dc} + \lambda\frac{dA}{dc} + \frac{dA}{dc}\right) = 0$$

Hence

$$\frac{d\lambda}{dc} = -\left(\frac{\lambda^3\ \frac{dA}{dc} + \lambda^2\ \frac{dA}{dc} + \lambda\ \frac{dA}{dc} + \frac{dA}{dc}}{4\lambda^3 + 3A_3\lambda^2 + 2A_2\lambda + A_1}\right) \tag{15.4}$$

When we change $i\omega$ in eq. (15.4), we get

$$\frac{d\lambda}{dc}|_{i\omega} = -\left(\frac{-i\omega^3\ \frac{dA_3}{dc} - \frac{\omega^2 dA_2}{dc} + \frac{i\omega dA_1}{dc} + \frac{dA_0}{dc}}{-4i\omega^3 - 3A_3\omega^2 + 2A_2 i\omega + A_1}\right)$$

Hence

$$\mathrm{Re}\left(\tfrac{d\lambda}{dc}|_{i\omega}\right) = -\left(\frac{\left[A - 3A_3\omega^2\right]\left[\frac{dA_0}{dc} - \omega^2\frac{dA_2}{dc}\right] + \left[2A_2\omega - 4\omega^3\right]\left[\omega^3\frac{dA_3}{dc} - \omega\frac{dA_1}{dc}\right]}{\left[A_1 - 3A_3\omega^2\right]^2 + \left[2A_2\omega - 4\omega^3\right]^2}\right)$$

Theorem 15.15: If we assume that "*c*" is the bifurcation parameter, then system (1) maintains the given Hopf bifurcation $\left(\left[A_1 - 3A_3\omega^2\right] \left[\frac{dA_0}{dc} - \omega^2 \frac{dA_2}{dc}\right] + \left[2A_2\omega - 4\omega^3\right] \left[\omega^3 \frac{dA_3}{dc} - \omega \frac{dA_1}{dc}\right] \right) \neq 0.$

15.6 Numerical Solutions

The system for numerically solving the nonlinear differential equation (15.1) is discussed. This section contains our current techniques of AI using Maple to analyze the solutions under stability analysis:

1. We hold the system's parameters first as ρ_1= (a = 1.3, b = 1.05, c = 1.450, α = 2.236, β = 1.05, k_1=1.25, k_2= 0.23, k_3 = 0.005, U = 0.545, δ = 1.236, η = 0.212, μ = 2.024, τ = 1.07, γ = 2.1, e = 2.05, f = 2.11) at the population (N_1, N_2, N_3, N_4) = (2.35, 3.46, 2.06, 3.13).

   We now employ the system's $\rho 1$ parameter, as previously stated. The original requirement is thus satisfied with $(N_1, N_2, N_3, N_4) = (0.040, 0.04, 0, 0), (0.020, 0.02, 0, 0)$, and the contact rate of first and second prey interactions is shown in Figure 15.2.

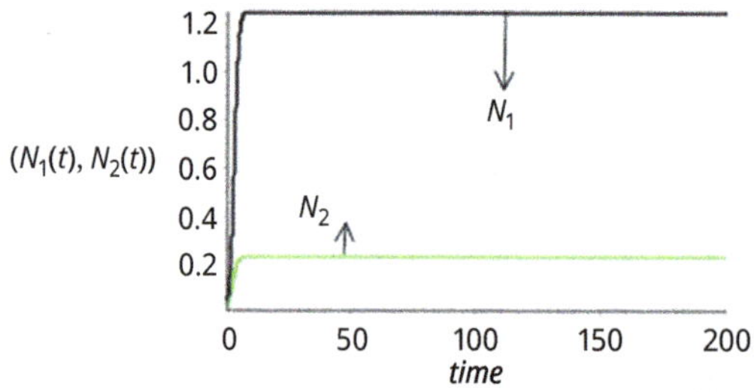

Figure 15.2: Interaction of prey 1 and prey 2.

2. Now we accept the parameter of the system $\rho 1$ as mentioned above. Then the initial condition satisfies $(N_1 N_2\ N_3\ N_4) = ((0.040, 0.04, 0.01, 0), (0.020, 0.02, 0.05, 0), (0.04, 0.04, 0.01, 0))$, and the contact rate of three prey interactions is shown in Figure 15.3.

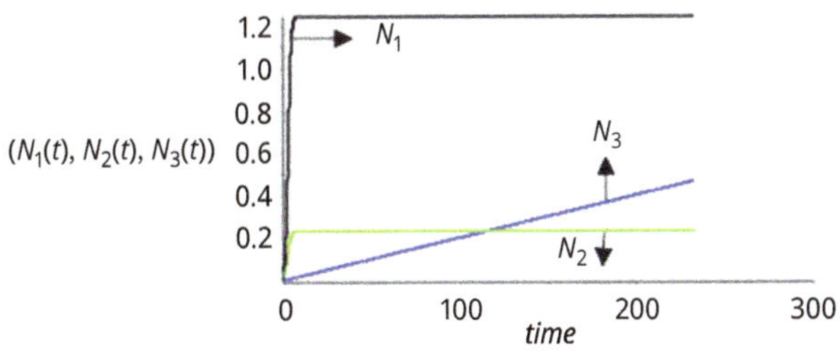

Figure 15.3: Interaction of three preys.

3. As previously stated, we are now gathering the system's parameter $\rho 1$. Then the initial condition satisfies with $(N_1, N_2,\ N_3,\ N_4) = ((0, 0, 0.1, 1), (0, 0, 0.6, 0.2))$, and the contact rate of third prey–predator interaction is shown in Figure 15.4.

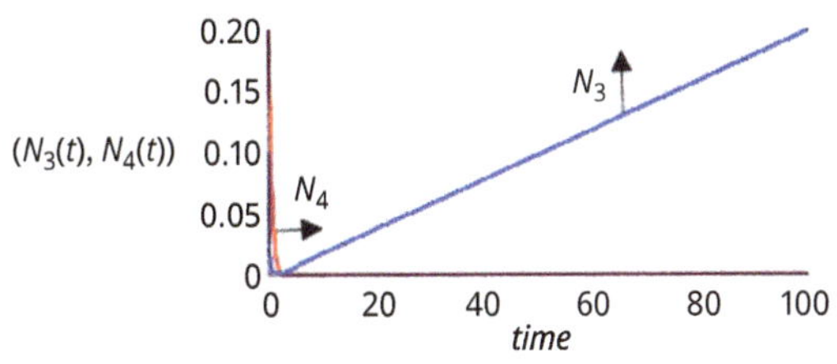

Figure 15.4: Interaction of prey 3 and predator.

4. As previously stated, we are now gathering the system's parameter $\rho 1$. Then the initial condition satisfies with $(N_1, N_2,\ N_3,\ N_4) = ((0, 0.04, 0, 0.040), (0, 0.003, 0, 0.002))$, and the contact rate of the second prey–predator interaction is shown in Figure 15.5.

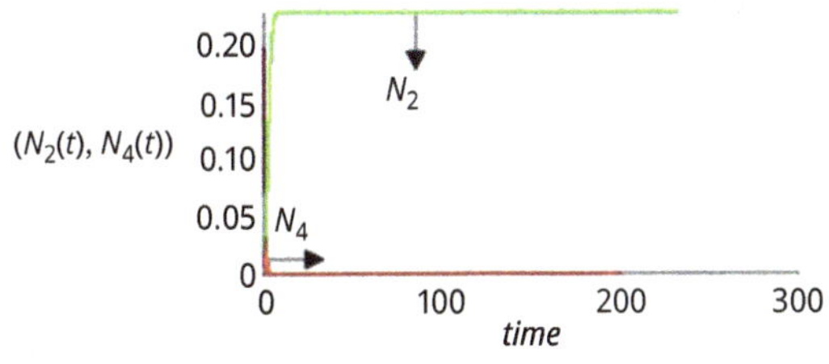

Figure 15.5: Interaction of prey 2 and predator.

5. As previously stated, we are now gathering the system's parameter $\rho 1$. Then the initial condition satisfies with $(N_1, N_2,\ N_3,\ N_4) = ((0.40, 0, 0, 0.4), (0.2, 0, 0, 0.1))$, and the contact rate of the first prey and predator interaction is shown in Figure 15.6.

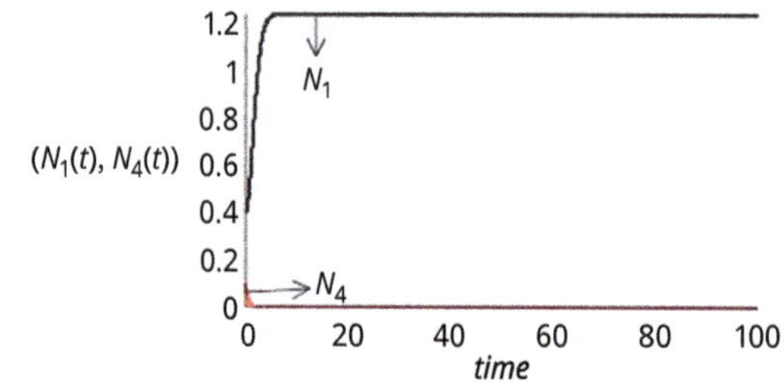

Figure 15.6: Interaction of prey 1 and predator.

6. As previously stated, we are now gathering the system's parameter $\rho 1$. Then the initial condition satisfies

   $(N_1,N_2,N_3,N_4)=((0.40,0.40,1.0,0.2),(1.20,0.2,0.1,1.0),(0.2,0.1,0.3,1.20),(0.2,1.2,0.4,1.0))$.

   The interaction of three preys and predator is shown in Figure 15.7.

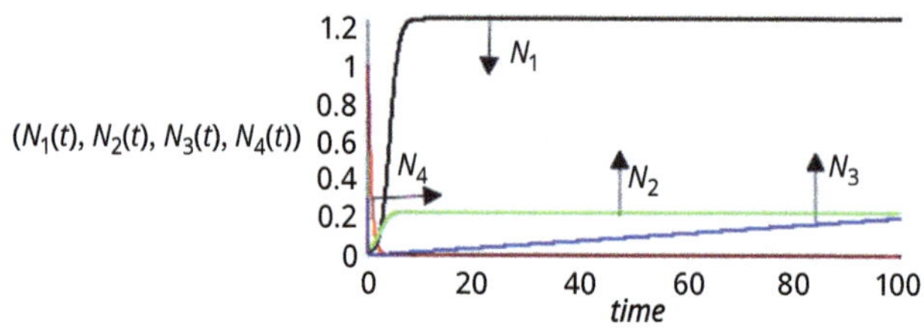

Figure 15.7: Interaction of three preys and predator.

15.7 Discussions and Conclusions

There are some quite intriguing dynamics in an eco-epidemiology model with three prey and one predator as the sickness spreads to the third prey group. In this case, we imagined that all three preys develop logistically, with the exception of the third, which is incapable of procreating. Therefore, with the use of Maple and AI approaches, all of our significant analytical discoveries are quantitatively validated. The time-delayed system experiences a Hopf bifurcation at approximately E^* at $c = c^*$, taking time delay c as the bifurcation parameter.

Finally, take some control measure $U\tau$, which plays an important role in controlling population stability. This model will analyze the stability, which is shown in Table 15.1:

- From Figure 15.2, we observed that the first prey N_1 belongs to the child age increases and becomes constant, and the second prey N_2 belongs to adult age and becomes constant due to lower immunity growth compared to the first prey.

- From Figure 15.3, we find that the first prey N_1 and second prey N_2 becomes constant, but the third prey N_3 increases slowly because of infection by the predator N_4.
- From Figure 15.4, we have shown that the third prey N_3 increases because of the control measure $U\tau$ acting as a key role but the predator N_4 decreases.
- From Figure 15.5, we find that the second prey N_2 goes to a constant, while the predator N_4 decreases.
- From Figure 15.6, we observe that the first prey N_1 increases and then becomes constant, while the predator N_4 decreases.
- From Figure 15.7, we find that the first prey N_1 and the second prey N_2 become a constant, and the third prey N_3 increases because of the parameter value of the control measure $U\tau = 0.58315$ so as to avoid the disease from spreading, while the predator N_4 diminishes instantly.

In our real-life situations, we compare this model by following the constant mentioned in $U\tau$ as a preventive way to avoid the disease from spreading. Therefore, to avoid the spread of infection and disease, the predator spread in a third prey population could be reduced by applying the prevention and control measures described in Figure 15.1, as τ control is recommended.

References

[1] Aziz-Alaoui, M. A., & Okiye, M. D. (2003 Oct 1). Boundedness and global stability for a predator-prey model with modified Leslie-Gower and Holling-type II schemes. *Applied Mathematics Letters*, *16*(7), 1069–1075.

[2] Gaff, H., & Schaefer, E. (2009 Jun 1). Optimal control applied to vaccination and treatment strategies for various epidemiological models. *Mathematical Biosciences & Engineering*, *6*(3), 469–492.

[3] Ganguli, C., Kar, T. K., & Mondal, P. K. (2017 Jul 12). Optimal harvesting of a prey–predator model with variable carrying capacity. *International Journal of Biomathematics*, *10*(05), 1750069.

[4] Ghasemabadi, A., & Rahmani Doust, M. H. (2021 May 1). Investigating the dynamics of lotka –Volterra model with disease in the prey and predator species. *International Journal of Nonlinear Analysis and Applications*, *12*(1), 633–648.

[5] Batta, P., Ahuja, S., & Kumar, A. (2023). A hybrid framework for secure data transfer for enhancing the blockchain security. In 2023 Seventh International Conference on Image Information Processing (ICIIP), pp. 645–650. IEEE.

[6] Hugo, A., & Simanjilo, E. (2019). Analysis of an eco-epidemiological model under optimal control measures for infected prey. *Applications and Applied Mathematics: An International Journal (AAM)*, *14*(1), 8.

[7] Kar, T. K. (2006 Jan 1). Modelling and analysis of a harvested prey–predator system incorporating a prey refuge. *Journal of Computational and Applied Mathematics*, *185*(1), 19–33.

[8] Kazeem, O. K., Makinde, O. D., & Takaidza, I. (2013 Mar 15). Impact of optimal control on the treatment of HIV/AIDS and screening of unaware infectives. *Applied Mathematical Modelling*, *37*(6), 3802–3820.

[9] Dhiman, H., Vyas, N., Pagrotra, A., & Dutt, V. (2023). Real-Time Traffic Analysis and Driver Safety Enhancement via IoT and Computer Vision with High-Resolution Project Lamps for Intelligent Road Illumination [Internet]. In 2023 1st International Conference on Circuits, Power and Intelligent Systems (CCPIS). IEEE. Available from: http://dx.doi.org/10.1109/CCPIS59145.2023.10291872

[10] Sharma, S., & Samanta, G. P. (2015 Sep). Analysis of a two prey one predator system with disease in the first prey population. *International Journal of Dynamics and Control, 3*, 210–224.

[11] Saranya, S., & Vijaya, S. (2023). Covid-19 model with SI for Beddington DeAngelis prey-predator species. *International Journal of Ecology and Development, 38*(1), 13–25.

[12] Simon, J. S., & Rabago, J. F. (2018 May 31). Optimal Control for a Predator-Prey Model with Disease in the Prey Population. *Malaysian Journal of Mathematical Sciences, 12*(2), 269–285.

[13] Upadhyay, R. K., & Iyengar, S. R. (2013 Jul 23). Introduction to mathematical modeling and chaotic dynamics. CRC Press.

[14] Saranya, S. (2023 Aug 11). Optimal control measures of Beddington DeAngelis response of lotka Volterra model. *Int. Journal of Applied Mathematics and Statistics*, 62.

[15] Werner, E. E., & Gilliam, F. J. (1984). The ontogenetic niche and species interactions in size-structured populations. Annual review of ecology and systematics. *Annual Reviews, 15*, 393–425.

[16] Saranya, S., Vijaya, S., & Rekha, E. (2023 Sep). Susceptible-Infected-Treatment model with disease infected in prey population recovered by treatment. *Novel Research Aspects in Medicine and Medical Science, 6*, 92.

[17] Bezabih, A. F., Edessa, G. K., & Rao, K. P. (2021). Epidemiological modelling and analysis of COVID-19 pandemic with treatment. *Applied Mathematical Modelling, 6*(1), 1.

[18] Legreneur, P., Laurin, M., & Bels, V. (2012 Feb). Predator-prey interaction paradigm: A new tool for artificial intelligence. *Adaptive Behavior, 20*(1), 3–9.

[19] Musa, S. S., Qureshi, S., Zhao, S., Yusuf, A., Mustapha, U. T., & He, D. (2021 Jan 1). Mathematical modeling of covid-19 epidemic with effect of awareness programs. *Infectious Disease Modelling, 6*, 448–460.

[20] Hamadneh, N. N., Tahir, M., & Khan, W. A. (2021 Jan 18). Using artificial network with prey predator algorithm for prediction of the COVID19: The case of Brazil and Mexico. *Mathematics, 9*(2), 180.

[21] Ackleh, A. S., Salceanu, P., & Veprauskas, A. (Aug 3). A nullcline approach to global stability in discrete-time predator-prey models. *Journal of Difference Equations and Applications, 27*(8), 1120–1133.

Neha, Shailja, and Umesh Kumar Lilhore

16 Model Optimization in Deep Learning: Theory and Application

Abstract: Model optimization is the most important component in the deep learning field, impacting performance and model training. In this chapter, techniques of optimization and their theoretical and practical implementations will be discussed. We will focus on the optimization techniques used in deep learning, which improve the accuracy of the model, convergence speed, and efficiency. Optimization techniques such as stochastic gradient descent and its variants, including momentum, RMSProp, and Adam, are explored in depth. Different types of optimizers will be used for updating the learning rate to minimize the objective function. Some advanced optimization methods, such as adaptive learning and gradient clipping, have been discussed in depth. A comparative analysis of variants of gradient descent and various other optimizers will be discussed, highlighting the important features.

Keywords: Optimization algorithm, stochastic gradient descent, momentum, RMSProp, Adam optimizer

16.1 Basic theory

Deep learning (DL) is a powerful machine learning (ML) technique that can learn through data patterns and enhance functions by repeatedly examining algorithms. In DL, optimization plays a significant role in enhancing the performance of artificial intelligence (AI) models. Model optimization provides a foundation for DL by introducing an iterative method of adjusting model features (such as weights and biases) to minimize the error rate and enhance the accuracy of the model. It generates a high-dimensional parameter space that results in the most accurate predictions on given problems related to classification, regression, or the generation of new data. The most important features of optimization are hyperparameter tuning, in which the rate of learning, size of batch, and the count of layers are used to achieve optimal performance [1].

Techniques of regularization like L2 regularization are used to avoid overfitting and also help in improving the ability of the model to specify the unseen data. Some more methods used to minimize the size and computational requirements of the model are pruning and quantization. In DL, pruning is used to build a neural network model that is small in size and more effective [2]. The objective of pruning is to remove the unwanted neurons and connections so that we can get a cost-effective model that takes less training time. Quantization methods are used to shrink the

https://doi.org/10.1515/9783111436180-016

weights of neurons to one-fourth of the original weight to improve the performance of the model. The main objective of optimizers in DL is to develop models whose accuracy is high as well as more effective and scalable. This chapter tries to bridge the gap between theoretical aspects and practical implementations, providing a thorough understanding of various optimization techniques and their relevance to DL [3].

16.2 Fundamentals of optimization

Optimization techniques are most widely used in the field of ML as well as DL. These are used to learn about the parameters of the data taken as input. Hence, optimization techniques provide the successful execution of results to real-world problems. Various researches have been directed to find out the best algorithms for the problems to give the best solution, as there is no particular method that can solve various types of problems [3]. In this chapter, various optimization algorithms will be discussed like all the variants of gradient descent (GD) such as stochastic GD (SGD) and mini-batch GD (MBGD). Both differ based on the size of data used for calculating the learning rate and also for training DL models. The biggest challenge faced by the optimizer is how to select the varying learning rate [2]. Many optimizers will be discussed to overcome the challenges. The key elements of optimization algorithms are summarized in Table 16.1.

Table 16.1: Elements of optimization algorithms.

Keyword	*Description*
Gradient descent	Optimizer for finding the minimum of a function
Learning rate	A hyperparameter that controls how much to adjust the weights of our network with respect to the loss gradient
Batch size	The number of training samples used for an iteration
Epoch	One iteration through the whole training dataset
Overfitting	After training on data, when the model fits too well, it results in overfitting
Loss function	A function that measures how well the model's predictions match the actual data
Backpropagation	The gradient of the function is calculated with respect to weights
Hyperparameters	Parameters that are set before the learning process starts and control the process of training

16.2.1 Elements of optimization algorithms

16.2.1.1 Hyperparameters

Hyperparameters are the most significant component used in the optimization of DL and ML models. As parameters used in the model are identified during training, hyperparameters are identified before the process of training starts. Some common approaches used for selecting hyperparameters are grid search, random search, and Bayesian optimization. Hyperparameters need to be chosen carefully for the best results [4].

16.2.1.2 Backpropagation

Backpropagation means backward propagation of errors. It is a method that is used to minimize the error in a neural network. Backpropagation is the most vital algorithm used in neural network optimization, which enables them to acquire from data by modifying weights and biases. It helps in minimizing the overall error. Some of the basic key concepts used in backpropagation are forward pass, loss function, and backward pass [5].

16.2.1.3 Loss Function

In the field of DL, a loss function measures the difference between the output predicted by the model and the original data. It is also known as an objective function or cost function. There are various optimizers used to minimize the loss function for improving the performance of the model. Some of the commonly used loss functions are mean squared error, mean absolute error, and so on [6].

16.2.1.4 Overfitting

Overfitting means that while training, the model learns from so much data, including noise, resulting in learning some patterns that are not relevant with respect to the data. Because of this, the model becomes too complex and fits so well that it captures not only the correct patterns but also random fluctuations and outliers. It leads to high accuracy. To reduce overfitting in the model, we can improve the quality of data by removing unimportant features, extracting meaningful patterns, and mitigating the risk of fitting the noise or features that are not relevant [7].

16.3 Optimizers in deep learning

16.3.1 Gradient-Based Optimization Methods

GD is measured as the most important concept of optimization in DL. This method uses mathematical concepts to update the value and achieve the local minima, as shown in Figure 16.1. When the model is trained by GD in DL, a step size has to be selected, that is, alpha. The most common problem faced by everyone is choosing the best value of alpha. If the value of alpha is too small, then the performance of the optimizer is very slow, and if it is too big, the optimizer fails to converge. To find the optimal alpha value, we need to find the derivative of the objective function with respect to alpha, as well [8]:

$$y\_new = y - alpha^{*} f(x) \tag{16.1}$$

Here in eq. (16.1), y is the current value, alpha represents the learning rate, and $f(x)$ is the derivative of the function at y. The above equation is used to calculate gradient. In this, alpha is used to denote the step size, which means how far to travel with each gradient for every iteration [9].

Following is the step-by-step working of GD:

1. At first, we will begin with the starting coefficients.
2. Then calculate the cost of coefficients and find out which has the lesser cost compared to the current one.
3. We move towards the lower weight and modify the coefficient's value.
4. Repeat this process iteratively.
5. The process will be repeated until the local minimum point is reached. A local minimum is a point beyond which it cannot proceed.

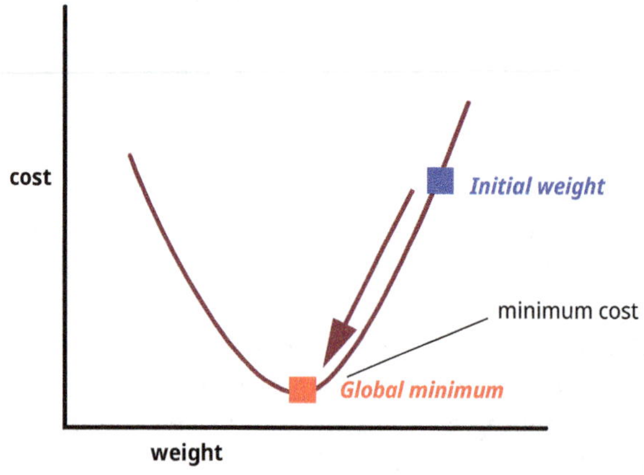

Figure 16.1: Gradient descent.

The GD algorithm works best for convex functions but not for non-convex functions. To overcome the problem of GD when dealing with large datasets, there are many ways to address this using various types of optimization algorithms.

Generally, there are three types of GD:

1. SGD
2. BGD (batch GD)
3. MBGD

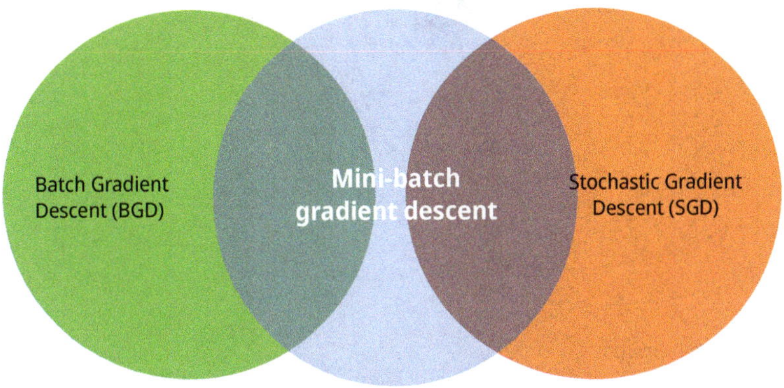

Figure 16.2: Variants of gradient descent.

Every type of GD is differentiated based on the size of data used to calculate the gradient of the loss or minimum function, the accuracy of the coefficient updates, and the time taken for updating the coefficients (Figure 16.2) [1]. For a detailed comparison of BGD, MBGD, and SGD, refer to Table 16.2.

16.3.1.1 Stochastic Gradient Descent

The SGD optimizer is an important optimization algorithm in ML as well as in the DL field, especially for big datasets. It is an iterative method that modifies the values by considering steps proportional to the gradient of the loss function. SGD calculates the gradient based on an individual random data point or a small part of the dataset at each iteration, as shown in Figure 16.3. This randomness allows for the selection of data batches, reducing the need for large datasets and significant computational time [10].

The SGD optimizer is ideal for DL with massive data and significant computational time [10]. So, we can say that in DL, mostly the SGD optimizer is preferred when the size of the data is too large, and it provides efficient computational time. The update rule is similar to GD as follows:

$$y_{\text{new}} = y - \text{alpha}^{*} f(x) \tag{16.2}$$

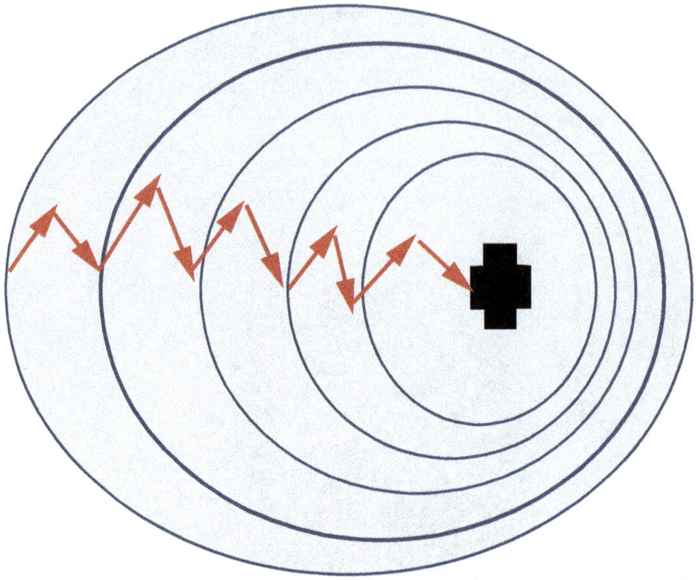

Figure 16.3: Stochastic gradient descent.

16.3.1.2 Batch Gradient Descent (BGD)

Batch gradient descent (BGD) is also called vanilla gradient descent, which is used to calculate the gradient of the cost function for the complete dataset [9, 10]. In this method, the gradient is calculated for the complete dataset, as shown in Figure 16.4, to make just one update of parameters, as follows:

$$y = y - \eta^{*} \nabla Z(y) \tag{16.3}$$

where

- y: Model features
- η: Learning rate
- $\nabla Z(y)$: Gradient of the cost function J with respect to θ.

16.3.1.3 Mini-Batch Gradient Descent

This is the another most widely used variant of gradient descent known as mini-batch. In this method, the breakdown of a large dataset is done into small batches, as shown in Figure 16.5, and after every batch, the value of the model's parameters is updated, ensuring global minima in fewer steps [11, 12].

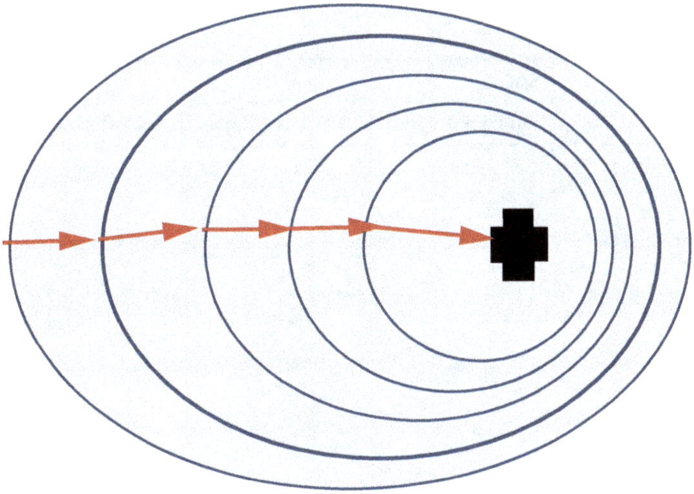

Figure 16.4: Batch gradient descent.

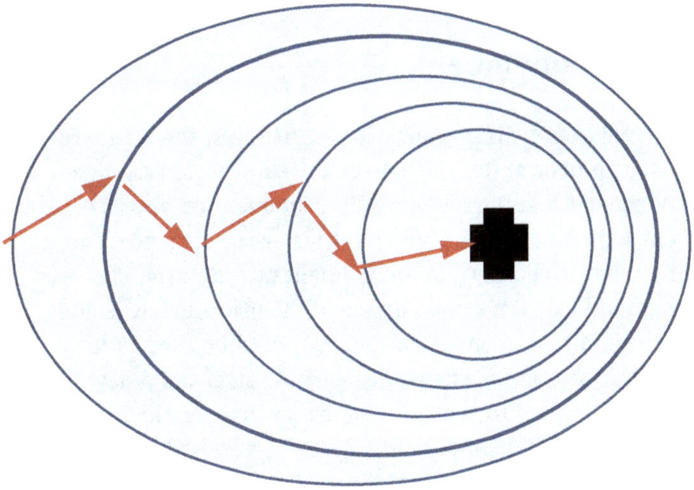

Figure 16.5: Mini-batch gradient descent.

The advantage of a mini-batch optimizer is the use of less memory and low variance in the model. MBGD inherits features from both SGD and vanilla gradient descent (BGD). This can help in reducing noise in the update of model parameters and improve convergence [13, 14].

Table 16.2: Comparison of gradient descent methods.

Feature	BGD	SGD	MBGD
Data used	**Entire dataset**	**Single data point**	**Small subset (batch) of data**
Gradient calculation	Calculated on entire dataset	Calculated on a single data point	Calculated on a batch of data
Convergence	Slower, but more stable	Faster, but less stable	Faster than BGD, more stable than SGD
Computational cost	High	Low	Medium
Memory usage	High	Low	Medium
Risk of local minima	Lower	Higher	Medium
Typical use cases	Smaller datasets and convex problems	Large datasets and online learning	The most common balance is between BGD and SGD

16.3.2 Momentum-Based Optimizer

As we have already discussed about GD optimization algorithms, the standard SGD optimizer requires a greater number of iterations to achieve lower minima due to the noisy process. Computational speed suffers as a result. A momentum approach is applied to solve this problem, aiding in the loss function's faster convergence. This algorithm modifies weights based on oscillations in the gradient direction [14, 15].

It can, however, go more quickly if a small portion of the prior update is added. It is necessary to remember that a high momentum phase should be used to lower the learning rate. This method is based on the last value, learning rate, and a new parameter is added as gamma, which is used to control this history update. The update rule for a momentum-based optimizer will be like $y = y - v$, where v is the history element [16, 17], as follows:

$$v = \text{gamma}^* v + \text{alpha}^* f'(a) \tag{16.4}$$

$$y_{\_new} = a - v \tag{16.5}$$

16.3.3 Nesterov accelerated gradient (NAG)

NAG is one of the most widely used optimization techniques, which is based on the classical momentum approach in GD. While the momentum method focuses on the historical values for the update of parameters, the NAG technique focuses on the "look-ahead" mechanism. In this mechanism, before calculating the gradient, the algorithm predicts the future positions of the model parameters based on the momentum [18] as follows:

$$vt = \gamma vt - 1 + \eta \nabla \theta J(\theta - \gamma vt - 1) \tag{16.6}$$

$$y_t = \text{gamma}^* y_t - 1 + \text{alpha} \nabla\ \theta J\ (\theta - \text{gamma}^* y_{t-1}) \tag{16.7}$$

Here y_t is the momentum, alpha is the learning rate, and $\theta J(\theta – \text{gamma}^* y\_(t–1)$ = gradient of the objective function.

Now, we can say that we are able to modify the error function with the help of history and next values, which results in speeding up the convergence by using the NAG technique [18].

16.3.4 Adagrad (Adaptive Gradient Descent)

As in the above algorithms, the same learning rate was updated for all the parameters; the adaptive GD algorithm changes the training parameters to use different learning rates for each iteration based on differences in training parameters. The Adagrad optimizer directs to provide an adaptive learning rate by declining the learning rate proportional to the modified gradient history. The only negative of this technique is that the learning rate falls violently, ultimately approaching zero. The advantage of using Adagrad is that it eliminates the need to modify the learning rate manually [19].

Calculate the squared gradient by using eq. (16.8) and update the parameters as follows:

$$G = G + f'(y)^2 \tag{16.8}$$

Update parameters calculated by using the following equation:

$$y\_\text{new} = \frac{a - \text{alpha}}{\sqrt{G + \text{epsilon}}} {}^* f'(y) \tag{16.9}$$

In this approach, the learning rate is not modified manually. The Adagrad approach is used more compared to other GD variants. Sometimes the learning rate becomes too slow, which causes the model to be unable to gain more knowledge and results in less accuracy of the model [20].

16.3.5 RMSProp

Root mean square propagation, or RMSProp, is an optimizer algorithm similar to the Adagrad approach. Although it has not been published, the concept of this approach is well-known in the community. It focuses on solving the problem of fluctuating gradients. In other optimizers, the value of the gradient keeps changing; sometimes it is very small, and sometimes it is very large. This problem cannot be solved with an individual learning rate. This optimizer is a progression of the Adagrad optimizer, as it reduces the repetitive decrease in the learning rate [21]. RMSProp can be measured by the following equations:

$$E[g\wedge 2)\_t = 0.9 * E[g\wedge 2)\_\{t-1\} + 0.1 * g\_t\wedge 2 \tag{16.10}$$

$$a\_\text{new} = a - \text{alpha}/\,\text{sqrt}(E[g\wedge 2)\_t + \text{epsilon}) * g\_t \tag{16.11}$$

16.3.6 Adaptive moment estimation (Adam)

Adaptive moment estimation optimizer is an optimization algorithm used in the field of DL. It is an advancement of the SGD optimization algorithm and is used to modify the weights during the process of training in a neural network. This is also known as the Adam optimizer. This combines the important features of Adagrad and RMSProp. The Adam optimizer calculates the learning rate for individual data considering the history of gradients and their second moment [20, 21]. By calculating the mean (first moment) and uncentred variance (second moment), this optimizer gets an adaptive learning rate, resulting in speeding up convergence and better performance of the neural network [22]. As presented in Table 16.3, each optimization algorithm has distinct advantages and drawbacks depending on the problem context.

Table 16.3: Comparison of different optimization algorithms, highlighting features, advantages, and disadvantages.

Optimizer	Key features	Advantages	Drawbacks
SGD (stochastic gradient descent)	Weights are updated with the help of gradients	Simple and easy to implement	Can be slow to converge and may get stuck in local minima; sensitive to the learning rate
Momentum	Add the last update to the current	Accelerates convergence	May overshoot the minimum if the momentum is too high

Table 16.3 (continued)

Optimizer	Key features	Advantages	Drawbacks
Nesterov accelerated gradient (NAG)	Look-ahead approach is used for calculating the gradient at the anticipated future position	Improves upon momentum by anticipating updates, which can lead to faster convergence	Similar issues arise with momentum, as it is sensitive to learning rate and momentum parameters
Adagrad	Adjusts the learning rate based on the frequency of parameter updates; accumulates squared gradients	Suitable for sparse data, it performs well on problems with sparse features	Learning rate decreases over time, potentially becoming too small for further training
RMSProp	The moving average of squared gradients is used to scale the learning rate	Prevents rapid decrease in learning rate and stabilizes training	Requires careful tuning of the decay rate; it may perform poorly on some non-stationary tasks
Adam (adaptive moment estimation)	Combines the benefits of Adagrad and RMSProp, maintaining a per-parameter learning rate and momentum	Adaptive learning rates and momentum help in faster convergence and handling sparse gradients	May not generalize well under some conditions; sensitive to hyperparameter tuning

16.4 Applications of Optimization in Various Fields

16.4.1 Engineering and Operations Research

16.4.1.1 Structural Optimization

In engineering design, structural optimization is used to enhance the performance of structures like bridges and buildings. Topology optimization is a process that focuses on optimal material layout, finding the best distribution of material to maximize structural efficiency. Structural optimization is used in various fields to improve performance and resource utilization, including aerospace, automotive, and civil engineering [22].

16.4.1.2 Supply Chain Optimization

In supply chain network design, optimization plays an important role in determining the best locations for warehouses, cost reduction, and inventory management. Complex problems can be resolved with the use of mathematical programming and heuristic techniques, resulting in reliable answers with practical applications. Better resource use, more efficient operations, and heightened market competition are the outcomes of this optimization [23].

16.4.1.3 Production Planning

Optimizers are used to design production schedules that are effective. This involves determining start and end times, assigning jobs to machines, and considering factors like machine capacity and due dates. The aim is to minimize production time, costs, or other relevant metrics [24].

16.4.2 Machine Learning

16.4.2.1 Neural Network Training

Millions of parameters are usually used to train a neural network. The main objective of the optimizers (such as GD) is to adjust these parameters to minimize the error between the network's predictions and the actual data. **GD** uses an iterative approach in ML and DL, where it is used to minimize error functions [25].

16.4.2.2 Hyperparameter Tuning

ML models have settings called hyperparameters, and finding the optimal hyperparameter values is also a challenge that significantly impacts performance. Bayesian optimization is a powerful technique to find optimal values. Therefore, optimizers can automatically find the best combination of hyperparameters to achieve the desired outcome [26].

16.4.3 Finance

16.4.3.1 Portfolio Optimization and Risk management

Optimizers are also helpful in creating investment portfolios by optimizing asset allocation based on statistical measures of risk and return. Financial institutions use optimizers to assess and manage risks. Techniques like value at risk help quantify potential losses, and optimizers can be used to find strategies to mitigate these risks [27].

16.4.4 Other Fields

16.4.4.1 Healthcare

By identifying substances with specific characteristics, optimizers are used to create novel pharmaceuticals. One popular method used by optimizers to forecast a molecule's affinity for a target protein is called molecular docking [28, 29].

16.4.4.2 Robotics

In robotics, the models use optimizers to plan effective routes, avoid obstacles, and carry out jobs as efficiently as possible. They are employed in many different applications, including autonomous cars and industrial robots.

16.5 Challenges in optimization techniques

As optimization techniques are used by various fields to make models more efficient and effective, they also come with several challenges. These challenges can range from the complexity of data, changes in constraints, dynamic environments, and many other problems listed below:

i. **Scalability**: When the size of the problem increases, many optimization problems become difficult computationally. This becomes particularly difficult for combinatorial optimization problems, where the number of possible solutions grows exponentially.
ii. **Non-convexity**: Finding the global optimum in non-convex problems is challenging because traditional optimization methods can get stuck in local optima, as **non-convexity** is a situation where a single function has more than one local minimum and there is no guarantee of a unique optimal solution [8].
iii. **High Dimensionality**: Optimization problems with a large number of variables can be an issue due to the effect of dimensionality. The search space increases exponentially with the number of dimensions, making it harder to identify the optimal solution.
iv. **Data Quality and Availability**: The main factor in the efficiency of optimization techniques is the availability and accuracy of data. Incomplete, incorrect, or noisy data can result in wrong results.
v. **Algorithm Selection and Computational Resources**: The performance of the model also depends on algorithm selection. Therefore, choosing the right optimization algorithm is also a challenge. Along with this, proper computational power

and sufficient time are required by some optimization algorithms, which is sometimes not possible in real-life applications.

vi. **Interpretability and Implementation**: Even if an optimal solution is found, it must be interpretable and implementable in practice. A complex algorithm includes complex calculations, formulas, and variables, which make them difficult to understand. Therefore, even if a solution is mathematically optimal, it may face problems if it is not understandable or actionable by the client [30].

16.6 Conclusion

In this chapter, we focus on the fundamentals of optimization and various types of optimizers. The optimization techniques play a vital role in training DL models. This chapter focuses on updating the learning rate in different ways to minimize the loss function using optimizers such as variants of DG, momentum, Adam, RMSProp, Adagrad, and NAG. There are many aspects and features that can be considered, such as the nature of the problem, the size of the data for training, and hyperparameter sensitivity for selecting the most suitable optimizer for performing the task. This also focuses on the comparison of different optimizers, highlighting the features of every optimizer.

There are many challenges faced by optimizers, such as algorithm selection, data quality, and scalability. Optimizers are used in various fields like healthcare, engineering, finance, machine learning, and robotics. The research is still ongoing to improve optimization algorithms, which will be helpful in increasing the performance of DL models.

References

[1] Wang, M., Zhang, G., Liu, T., & Wang, W. (2023 Apr). Determination of elastoplastic properties of 2024 aluminum alloy using deep learning and instrumented nanoindentation experiment. *Acta Mechanica Solida Sinica*, *36*(2), 327–39.

[2] Cheng, B., Yao, Y., Qu, X., Zhou, Z., Wei, J., Liang, E., Zhang, C., Kang, H., & Wang, H. (2024 Oct 1). Multi-objective parameter optimization of large-scale offshore wind Turbine's tower based on data-driven model with deep learning and machine learning methods. *Energy*, *305*, 132257.

[3] Raponi, A., & Marchisio, D. (2024 Jun 1). Deep learning for kinetics parameters identification: A novel approach for multi-variate optimization. *Chemical Engineering Journal*, *489*, 151149.

[4] Kaveh, A. (2024). Applications of Artificial neural networks and machine learning in Civil Engineering. *Studies in Computational Intelligence*, 1168.

[5] Pandit, P., Abdusalamov, R., Itskov, M., & Rege, A. (2024 Jan 17). Deep reinforcement learning for microstructural optimisation of silica aerogels. *Scientific Reports*, *14*(1), 1511.

[6] Azevedo, B. F., Rocha, A. M., & Pereira, A. I. (2024 Jan). Hybrid approaches to optimization and machine learning methods: A systematic literature review. *Machine Learning*, *24*, 1–43.

[7] Mamoudan, M. M., Mohammadnazari, Z., Ostadi, A., & Esfahbodi, A. (2024). Food products pricing theory with application of machine learning and game theory approach. *International Journal of Production Research*, *62*(15), 5489-5509.
[8] Lilhore, U. K., Simaiya, S., Kaur, A., Prasad, D., Khurana, M., Verma, D. K., & Hassan, A. (2021 Dec). Impact of deep learning and machine learning in industry 4.0: Impact of deep learning. In Cyber-Physical, IoT, and Autonomous Systems in Industry 4.0 (pp. 179–197, Vol. 23). CRC Press.
[9] Simaiya, S., Lilhore, U. K., Sharma, Y. K., Rao, K. B., Maheswara Rao, V. V., Baliyan, A., Bijalwan, A., & Alroobaea, R. (2024 Jan 16). A hybrid cloud load balancing and host utilization prediction method using deep learning and optimization techniques. *Scientific Reports*, *14*(1), 1337.
[10] Lilhore, U. K., & Simaiya, S. (2024 Aug 5). 10 Quantum machine learning in natural language processing: Opportunities and challenges. *Quantum Machine Learning: Quantum Algorithms and Neural Networks*, 199.
[11] Chauhan, A. S., Lilhore, U. K., Gupta, A. K., Manoharan, P., Garg, R. R., Hajjej, F., Keshta, I., & Raahemifar, K. (2023 Apr 17). Comparative analysis of supervised machine and deep learning algorithms for kyphosis disease detection. *Applied Sciences*, *13*(8), 5012.
[12] Lilhore, U. K., Simaiya, S., Dalal, S., & Damaševičius, R. (2024). A smart waste classification model using hybrid CNN-LSTM with transfer learning for sustainable environment. *Multimedia Tools and Applications*, *83*(10), 29505-29529.
[13] Ruder, S. An overview of gradient descent optimization algorithms. arXiv preprint arXiv:1609.04747; 2016.
[14] Zaheer, R., & Shaziya, H. (2019). A study of the optimization algorithms in deep learning. In 2019 Third International Conference on Inventive Systems and Control (ICISC) (pp. 536–539). IEEE, Coimbatore, India. doi: 10.1109/ICISC44355.2019.9036442.
[15] Goodfellow, I., Bengio, Y., & Courville, A. (2016). Deep learning. MIT Press, London.
[16] Rasouli, A., & Jeon, M. (2019). Overview of artificial intelligence techniques in smart cities. In Jeon, M. (ed.). Emotional AI and Human-AI Interaction: Theory, Framework, and Applications (pp. 17–29). Woodhead Publishing. doi: 10.1016/B978-0-08-101041-9.00002-8.
[17] Darken, C., Chang, J., & Moody, J. (1992). Learning rate schedules for faster stochastic gradient search. In Neural Networks for Signal Processing II., Proceedings of the 1992 IEEE-SP Workshop (pp. 3–12). IEEE.
[18] Bottou, L. (2009). Curiously fast convergence of some stochastic gradient descent algorithms. In Proceedings of the Symposium on Learning and Data Science (Vol. 8). Paris.
[19] Efficient DL Mini-Batch Gradient Descent Explained [Internet]. StatusNeo. Available from: https://statusneo.com/efficientdl-mini-batch-gradient-descent-explained/.
[20] Dhiman, H., Vyas, N., Pagrotra, A., & Dutt, V. (2023). Real-Time Traffic Analysis and Driver Safety Enhancement via IoT and Computer Vision with High-Resolution Project Lamps for Intelligent Road Illumination [Internet]. In 2023 1st International Conference on Circuits, Power and Intelligent Systems (CCPIS). IEEE. Available from: http://dx.doi.org/10.1109/CCPIS59145.2023.10291872
[21] Nesterov, Y. (1983). A method for unconstrained convex minimization problem with the rate of convergence o(1/k2). *Dokl Akad Nauk SSSR (Sov Math Dokl)*, *269*, 543–547.
[22] Ruder, S. Gradient descent optimization algorithms [Internet]. www.ruder.io. Available from: https://ruder.io/optimizing-gradient-descent.
[23] Sutskever, I., Martens, J., Dahl, G., & Hinton, G. On the importance of initialization and momentum in deep learning. Proceedings of the 30th International Conference on Machine Learning (ICML); 2013.
[24] Kingma, D. P., & Ba, J. Adam: A method for stochastic optimization. International Conference on Learning Representations (ICLR); 2015.
[25] Shahriari, B., Swersky, K., Wang, Z., Hoffman, M. D., & De Freitas, N. (2015). Taking the human out of the loop: A review of Bayesian optimization. *Proceedings of the IEEE*, *104*(1), 148–175.

[26] Jorion, P. (2007). Value at risk: The new benchmark for managing financial risk. McGraw-Hill.
[27] Lilhore, U. K., & Simaiya, S. (2022 Apr). Role of Swarm Intelligence and Neural Network in Intelligent Traffic Management. *Multimedia Computing Systems and Virtual Reality*, *5*, 27–43.
[28] Lilhore, U. K., Dalal, S., Faujdar, N., Simaiya, S., Dahiya, M., Tomar, S., & Hashmi, A. (2024 Jan). Unveiling the prevalence and risk factors of early stage postpartum depression: A hybrid deep learning approach. *Multimedia Tools and Applications*, *25*, 1–35.
[29] Dalal, S., Lilhore, U. K., Faujdar, N., Simaiya, S., Agrawal, A., Rani, U., & Mohan, A. (2024 Jul). Enhancing thyroid disease prediction with improved XGBoost model and bias management techniques. *Multimedia Tools and Applications*, *1*, 1–32.
[30] Kukkar, A., Lilhore, U. K., Frnda, J., Sandhu, J. K., Das, R. P., Goyal, N., Kumar, A., Muduli, K., & Rezac, F. (2023 Jan 1). ProRE: An ACO-based programmer recommendation model to precisely manage software bugs. *Journal of King Saud University-Computer and Information Sciences*, *35*(1), 483–98.

Roselin Kiruba R., L. Sharmila, M. Misba, J. Kavitha,
and J. Jude Moses Anto Devakanth

17 Quantitative Analysis for LMS Using Mathematical Modeling by Artificial Intelligence

Abstract: Learning management systems (LMSs) driven by artificial intelligence (AI) are causing a transformative change in the educational sector. This study aims to explore their impact on the teaching and learning method inside higher education. The study deploys a mixed method of qualitative and quantitative approaches to understand the current applications and AI-LMS advantages. This research uses qualitative analysis to discover key themes in student performance, learning problems, and interventions. Quantitative analysis will be conducted using the Paradis corpus consisting of naturalistic language samples from 25 children learning English, using descriptive, category, correlation, and inferential statistics. Machine learning (ML) algorithms will be adopted to classify and predict student performance. The study's novelty lies in the combination of quantitative and qualitative methods to provide evidence-based suggestions for addressing challenges detected in student performance. The study suggests deploying advanced algorithms for ML and qualitative interviews to improve the accuracy of predicting models for student achievement. This might include engaging an extensive range of participants from varying origins and educational contexts.

Keywords: Learning management systems, artificial intelligence, teaching, learning, higher education, student performance

17.1 Introduction

Student individualization and the process of differentiation of instructions and recommendations based on the needs of a student have become achievable due to learning management systems (LMSs) that incorporate artificial intelligence (AI). With these tools, activities are accomplished mechanically, which allows to focus on individual interactions with students and the ways of enhancing oratory skills and other pedagogical skills among teachers [27, 28]. Another advantage of the application of AI algorithms is that it offers relevant information on the growth and learning of students who assist teachers in changing their strategies. Further, the advanced AI LMSs offer enhanced access and flexibility to enable learning to support a variety of learner options, including impaired learners, and to offer any time and any place learning opportunities based on the lives of today's learners. The context of education has also gone through a tremendous transformation in the last 100 years, and it is defined by the expansion of technol-

https://doi.org/10.1515/9783111436180-017

ogy [11]. LMSs that have been integrated with AI are gradually building themselves as sources that are revolutionizing the process of learning and teaching, especially in higher learning institutions [13, 26]. This is one of the most important breakthroughs. These systems, which are powered by AI, are an innovation in that they feature self-contained methods that can be tailored to the needs of the learner as a person [16]. The following greatly transforms the traditional methods of teaching and allows teachers to delve deeper into individual instructional methods [6]. When it comes to current educational environments, there are major problems involved in both comprehending and resolving the wide range of elements that influence the academic achievement of students [12]. There is a need for a complete analysis that combines quantitative evaluations from accessible datasets with qualitative perspectives collected from the current literature. Although previous research provides insights into these elements, there is still a need for such an analysis. Quantitative and qualitative methods were used to fill a research gap in the fields of education technology and pedagogy. It sought a comprehensive knowledge of how AI-driven LMSs affect higher education teaching, learning, and achievement among students [15, 18]. This study integrated quantitative prognosis with qualitative perspectives to improve decision-making and give expert opinions over efficient approaches, indicators, and policies. It also proposed adaptive measurement and strategies to create positive, acceptable, and productive learning contexts by identifying important concepts, issues, and approaches [29, 30].

To address the research gap and get a clear view of AI-driven LMSs in higher learning and suggest practical solutions and policy, the study employed both quantitative and qualitative research approaches. Past studies have reviewed the impact of AI-LMS on the learning approach of students individually. Nevertheless, there is an overall lack of integrative research in which the qualitative and quantitative assessments are included. As a result, there are few studies that blend the quantitative outcomes from the datasets with the qualitative findings from the current literature, thereby restraining the growth of a holistic appreciation of the multiple effects of AI-LMS in the higher learning process. Therefore, there is a need to conduct research that strategically integrates different approaches to get an overall and comprehensive understanding of the effects of AI-based LMSs on students' performance, faculty instruction, and the learning environment.

17.2 Literature Review

17.2.1 Application of AI in Learning Management Systems and Higher Education

AI has become an essential component of daily life across several industries, including the field of learning. Nuzil et. al [20] Investigated the use of AI in LMS in higher level study. They analyzed 356 articles from 2010 to 2022. Among them, a total of 33 were selected for the final analysis. The data suggested that Moodle is the most often utilized LMS for AI processes in learning. AI is mostly used for assessing student management through methods like decision trees, support vector machine, neural networks, K-means, naive Bayes, and random forest. The report emphasized the increasing significance of AI in streamlining educational processes [25]. Al-Ansi and Fatmawati [4] examined the incorporation of information and communication technology (ICT) into higher education at Thamar University in Yemen. A total of 264 participants, including staff members and learners, completed questionnaires. The findings indicated that ICT components such as the Internet, structures, and LMS had a beneficial influence, while the educational institutions had an adverse effect. The report also identified a dearth of backing from institutions and government officials for e-learning throughout the COVID-19 epidemic, indicating the need for more financing as well as instruction for those involved in education. The COVID-19 epidemic has highlighted the significance of AI in LMS. Ali et al. [5] conducted a study with a large number of children from two common region institutions in Pakistan, stating that AI plays a crucial role in LMS, where natural language processing (NLP) and reasoning are closely interconnected. NLP greatly enhances the utilization of LMS by students, while the connection between AI and LMS plays an important role in logical thinking. An AI-powered LMS improves learning experiences and offers systematic answers for intricate inquiries.

17.2.2 AI and its impact on educational innovation and online learning

It is important to mention that AI technologies have become key influencing factors in the sphere of research, innovations, and growth in the methods of education and instruction. It is particularly useful for designing a learner model for optimizing education in E-Learning Management System Technology (ELMST) that can convert conventional classroom teaching into digital media. ELMST enhances teaching methods by incorporating illustrations, movements, and other forms of display and media content used in teaching. Data mining categorizes learner behavior with clusters, and AI helps in the discovery of pupils based on learning abilities [19]. Supervising innovation in education is important in a world that has gone global and one that is characterized by new technology, and this fosters dynamic learning environments. However, this task involves many aspects such as student characteristics, available projects, research findings, and environmental trends. Innovation in the field of management implies the search and introduction of new ideas, their implementation, and evaluation in order to improve the efficiency of education. Nevertheless, there are always challenges encountered when trying to put into practice the activities that have been proposed. The incorporation of AI systems into educational systems has helped to revolutionize traditional learning systems to provide better learning, improved efficiency, and intelligent development. The following challenges have however been realized: infrastructure development, personnel training, data privacy issues, and culture change. These challenges have been effectively solved by adopting a participatory approach with all stakeholders, and at the same time, the gains of the AI learning process have been optimized as found by Ökördi and Molnár [21]. The growth of the education sector and the new business models grounded on available data have led to the shift to data-driven pedagogies. However, the use of AI and learning analytics (LA) by organizations to personalize education is still limited. The motivations and obstacles that impact data-driven pedagogical and curricular pathways within EdTech companies are the subject of this study. The traditional objective of education and the innovative concept of knowledge sharing and instruction are opposed. An imperfect understanding of data, ambiguity, and a lack of data sovereignty impede the development and implementation of pertinent solutions. The discourse surrounding AI-based learning systems is nevertheless propelled by the need for customization and adaptation [2].

17.2.3 AI applications in educational administration

Ouyang et al. [22] examined the functions, consequences, and significance of AI solutions in online higher education studies. Beginning with 434 articles from the database up to the 2020, 32 articles were found suitable for analysis. In the review, different types of AI were used in performance prediction, satisfaction prediction, recommendation of resources, and assessment of learning, as well as measures to boost the education level. First, it is important to understand that despite the cross-industry integration of conventional AI technologies, the review showed that genetic algorithms and deep learning are used quite sparingly in this domain. The integration of AI applications revealed several benefits, such as increased online participation, improvement of academic performance, and accuracy of predictions due to the existence of AI algorithms. According to the review, there is a need to incorporate theories in education and learning in AI-enabled online learning environments in order to come up with concrete and measurable effects of AI in higher education via the Internet. This also called for the use of advanced AI technologies to handle real-time data analysis and stressed the need for the possibilities of AI in this educational environment as more beneficial [24]. Ahmad et al. [1] focused on the opportunities of using AI in the administration of institutions and in academic work within education. Educators, responsible for imparting knowledge, were often burdened with administrative tasks. Academic support provided by AI to instructors included LA, virtual reality, grading/assessments, and admissions. By eliminating administrative duties, these tools enabled educators to focus more on learning and counseling students. AI facilitated various administrative tasks, lightened the load on instructors, and significantly enhanced student learning. However, the study needed a quantitative assessment to determine its generalizability and acceptance. Technological advancements such as big data, machine learning (ML), and AI have given the comprehension of unique requirements by individuals, thereby creating prospects for customization within the education industry. Goksel et al. [31] consolidated the literature on AI for education personalization and examined key themes regarding how AI-driven approaches altered the current education system. An exhaustive examination of scholarly works from China, India, and the USA revealed that AI effectively catered to particular learning needs, routines, and capabilities by directing students along optimized learning trajectories. AI additionally enhanced educational content by tailoring it to suit the specific requirements of each learner and identifying potential challenges that may arise. This improved the learning experience by optimizing the instruction and learning environment. However, obstacles such as concerns regarding data privacy, the availability of digital resources, and financial constraints have impeded the implementation of AI in everyday life.

The growth of AI in education (AIEd) has been swift, yet its adoption poses persistent challenges for numerous educators. The landscape of AIEd research is dominated by STEM and computer science subjects, with empirical studies predominantly employing quantitative methodologies across a systematic review of 2,984 articles. The review delineated four primary categories of findings within AIEd literature: AI tutoring, personalization, evaluation, and prediction. However, it underscored the need for a more expansive exploration into ethical and pedagogical approaches within higher education. Alam and Mohanty [3] aimed to foster a more comprehensive understanding of the complexities surrounding AIEd, urging a deeper examination of its ethical implications and pedagogical applications in higher education.

RQ1: What factors significantly influence academic performance in learning English as a second language?

RQ2: How effective are different predictive algorithms in forecasting academic success, and what disparities exist between their predictive capabilities?

RQ3: What distinctive patterns of academic achievement exist across different areas of study, and how do various variables impact grades, offering insights into common student performance?

17.3 Materials and methods

The quantitative analysis will make use of the dataset from the CHILDES website, which has samples of 25 children with different first languages. Classification and prediction approaches, such as ML algorithms (e.g., decision trees and logistic regression), will be used to draw predictive insights and to make assertions that pertain to student performance. Recent literature evaluations will be used as a source for qualitative analysis, which will uncover major themes, difficulties, and viable countermeasures.

This research aims to synthesize quantitative forecasts with qualitative suggestions by comparing the results from the dataset and the literature.

17.3.1 Research paradigm

This work employs a mixed-method research design to examine the effect of AI-based LMS on students' performance. The dataset used is the UCI dataset, which samples the learning of English in 25 children. In order to predict the necessity of further educational help, the naive Bayes classifier is applied to student characteristics. The study also reviews the literature to identify main themes, difficulties, and possible approaches to AI-based LMS. A thematic analysis is therefore conducted on AI-based technologies, social isolation as a result of absence, and the need for interventions. The study objectives are to combine numerical findings from the dataset with ideas from the literature review to link students' interaction with the AI-enabled LMS, the impact of high absenteeism on academic performance, and the possibilities of AI-based solutions to overcome academic difficulties:

- The naive Bayes classifier is given as follows:

$$P(x) = \frac{P(C_k)\Pi_{i=1}^{n} P(x_i|C_k)}{P(x)} \tag{17.1}$$

where $P(x)$ is the posterior probability, the probability of class C_k, $P(C_k)$ is prior probability of class C, $P(x_i|C_k)$ is the likelihood, the probability of observing feature x_i given the class C_k, $P(x)$ is the evidence or normalizing constant, x is the feature vector $x = (x_1, x_2, \ldots, x_n)$, n is the number of features in the feature vector x, and m is the number of possible cases.

Naive Bayes classifiers are probabilistic classifiers based on Bayes' theorem with the assumption of feature independence. There are several types of naive Bayes classifiers, each suited to different types of data.

1. Gaussian naive Bayes is shown as follows:

$$P(C_k) = \frac{1}{\sqrt{2\pi\sigma_k^2}} \exp\left(-\frac{(x_i - \mu_k)^2}{2\sigma_k^2}\right) \tag{17.2}$$

where X_i is the feature value, C_k is the class, μ_k and σ_k^2 are the mean and variance of the feature x_i for class C_k.

2. Multinomial naive Bayes

This is used for discrete data, typically for the text classification, where features show the counts of frequencies using the following equation:

$$P(C_k) = \frac{\text{Count}(x_i, C_k) + 1}{\sum_{j=1}^{n} n \,\text{Count}(x_i, C_k) + n} \tag{17.3}$$

where $\text{Count}(x_i, C_k)$ is the number of times the feature x_i appears in documents of class C_k, and n is the total number of possible features.

3. Bernoulli naive Bayes
This is used for binary or Boolean features, where features are binary indicators as follows:

$$P(C_k) = p_{ki}^{x_i} \cdot (1 - p_{ki})^{1-x_i} \tag{17.4}$$

where p_{ki} is the probability of feature x_i is 1 for the given class C_k, x_i is 0 or 1, indicating the absence or presence of the feature.

4. Complement naive Bayes
This is a variant of the multinomial naive Bayes, designed to handle imbalanced data more effectively, often used in text classification using the following equation:

$$P(C_k) = \frac{\text{Count}(x_i, \text{not}\, C_k) + 1}{\sum_{j=1}^{n} \text{Count}(x_i, \text{not}\, C_k) + n} \tag{17.5}$$

where $\text{Count}(x_i, \text{not}\, C_k)$ is the count of feature x_i in documents not belonging to class C_k.

17.3.2 Data collection

The Paradis corpus comprises a compilation of natural linguistic samples obtained from 25 children. The transcripts are transcribed using English syntax, with anonymous identities. The data was gathered in 2002 in Edmonton, Canada, via discussions with a student research assistant. The research assistant formulated interview inquiries; however, children had the autonomy to offer their subjects of discussion. The dataset may be accessed on the CHILDES website, including the data collected during the inaugural phase in 2002. The whole longitudinal corpus is accessible [14], as shown in Figure 17.1.

17.3.3 Data analysis

The method entails the preparation of a dataset that contains characteristics such as the amount of time spent in studying, the level of education of the parents, the support of the family, Internet access, and the health state of the individual. The need for further educational help is the variable that will be targeted. A preprocessing step is performed on the dataset, which involves encoding categorical variables and dividing them into training and testing sets. The training dataset is used to train the naive

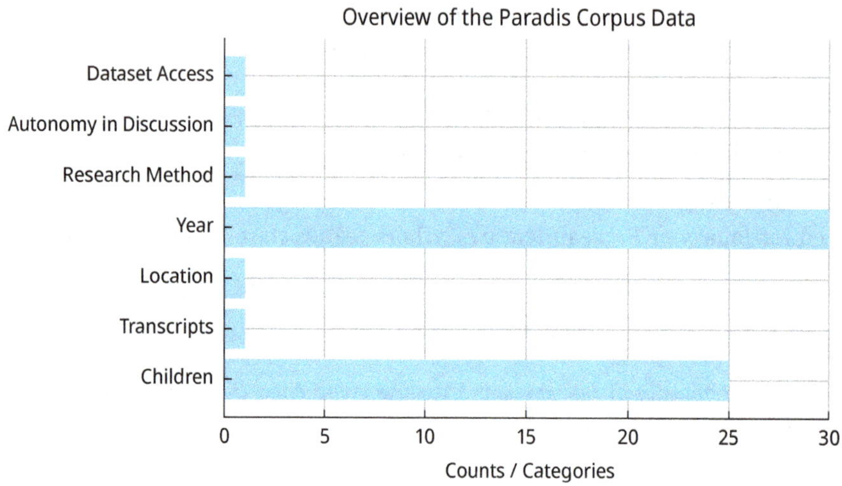

Figure 17.1: Important aspects of the Paradis corpus data.

Bayes classifier, and the model is then used to make predictions about the possibility that students will need further financial assistance for their schooling. Several measures, including accuracy, precision, and recall, are used to evaluate the model. The categorical variables are defined as follows:

$$X_{\text{encoded}} = \text{OneHotEncode}\left(X_{\text{categorical}}\right) \tag{17.6}$$

where $X_{\text{categorical}}$ is the original categorical features and X_{encoded} is transformed.

17.3.3.1 Quantitative analysis

For the analysis, the datasets known as "guide_to_files.csv," "student-mat.csv," and "student-por.csv" were used to gather and preprocess data from 25 students. Several datasets were combined, categorical variables were encoded, and numerical characteristics were scaled. During the feature selection process, it was necessary to determine pertinent characteristics such as demographics and study habits. It was determined by the correlation analysis in which features should be selected.

In Figure 17.2, the X-axis denotes the percentage of students, while the Y-axis signifies the different languages (Romanian, Korean, Ukrainian, Japanese, Cantonese, Spanish, Farsi, Mandarin, and Arabic). In the process of predicting academic failure based on student characteristics, the analysis produced encouraging findings. These results provided educators and policymakers with actionable insights that can be used to help students who are not performing up to their potential.

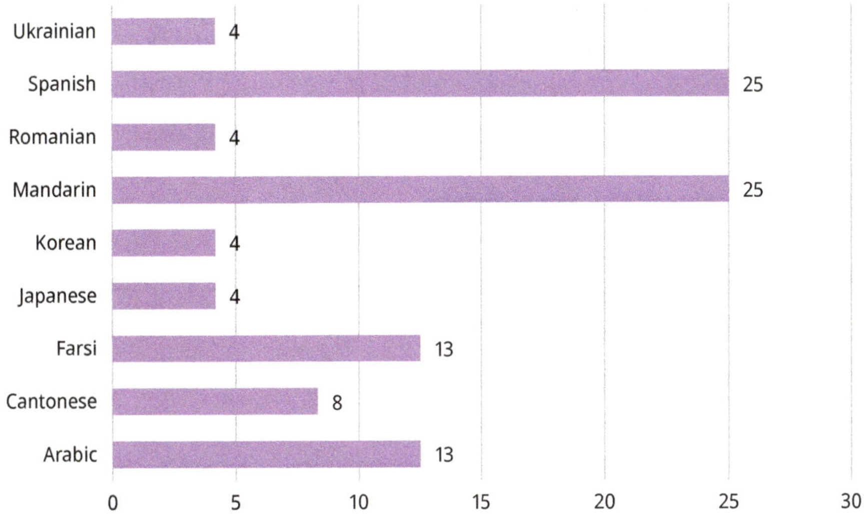

Figure 17.2: A comparison among Romanian, Korean, Ukrainian, Japanese, Cantonese, Spanish, Farsi, Mandarin, and Arabic using a bar plot.

The scaling numerical features are calculated using the following equation:

$$X_{\text{scaled}} = \text{Scale}(X_{\text{numerical}}) \tag{17.7}$$

where: Scale is the min-max scaling or standardization and $X_{\text{numerical}}$ is the numerical value of x.

Feature selection using correlation analysis is given as follows:

$$X_{\text{selected}} = \text{Select}(X, \text{Correlation}(X, Y) > \text{threshold})$$

where X represents features and Y represents the target variable.

17.3.1.2 Qualitative analysis

The target of the subject examination displayed in Table 17.1 is to explore the impact that computer-based intelligence-controlled learning the executive frameworks have on scholastic execution, the chance of detachment due to nonattendances, and the requirement for mediation for students who are frequently absent. The relationship between students' involvement with AI-driven technologies integrated into LMS and its impact on academic performance is investigated using a thematic analysis. The topical examination places unique accentuation on the effect that rising absences have on learning outcomes. Increased absence rates among students, as stated by Kivinen [16], Gutierrez and Sequeda [11], and Atherton [8], may indicate a lack of engagement with

AI-powered learning aids or difficulty comprehending course material, both of which ultimately result in lower academic achievement, as shown in Figure 17.3.

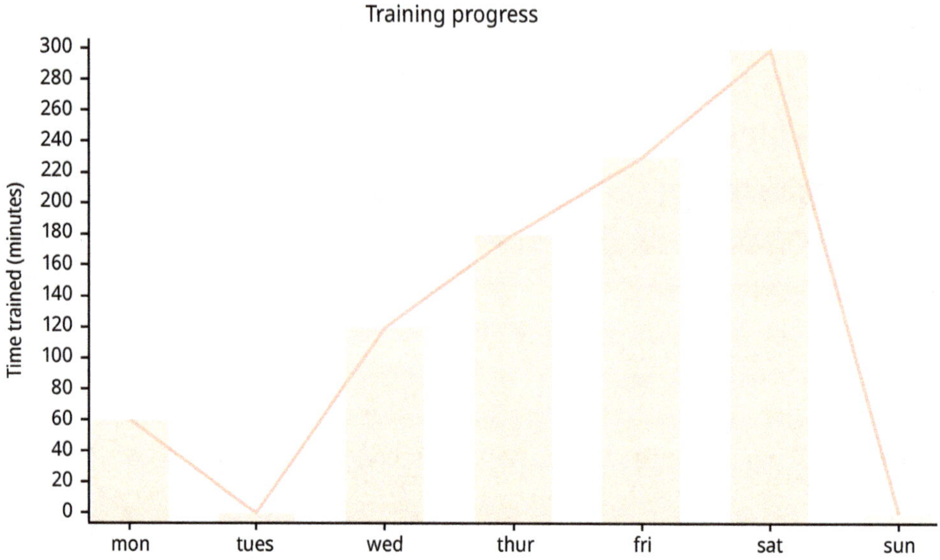

Figure 17.3: LMS performance of AI.

Table 17.1: Thematic analysis.

Theme	Findings
Impact on academic performance	An increase in absence may be an indication of disengagement or difficulties in learning, which may be the result of confusing course content or inadequate interaction with AI-powered learning tools in LMS [8, 11, 16].
Potential disconnection	Students can lose out on AI-driven adaptive learning features or personalized material inside the LMS if they skip lessons, which might affect their intellectual performance [9, 13, 24].
Need for intervention	Students who have frequent absences may be in danger of academic underperformance, which allows for the implementation of interventions such as personalized learning plans or targeted assistance via suggestions produced by artificial intelligence inside LMS driven by AI [7, 17, 21, 23]. These interventions are designed to assist students in catching up and improving their grades.

The presentation highlights the possible detachment that students may sense when they miss courses while using LMS. This disengagement is directly related to the loss of access to adaptive learning features enabled by AI, as well as personalized materials that are accessible inside the platform. As stated by Gligorea et al. [13], Burbules et al. [10], and Tan [24], this loss has the potential to considerably hamper the intellectual development of pupils and limit their academic advancement. In addition, the findings of the investigation highlight the critical need for prompt assistance for kids who have relatively high rates of absence. By referring to the works in [7, 17, 21, 23], this chapter suggests the implementation of personalized learning plans or targeted help via the use of AI-generated ideas inside the LMS. The use of these interventions is to provide essential assistance to students who are at risk, to help them make up for lost ground, and to improve their academic performance. In conclusion, the analysis sheds light on the intricate relationship that exists between students' participation in the LMS and the use of AI-powered tools, the impact that increased absences have on students' educational experiences, and the possibility that AI-driven interventions could help students overcome academic challenges while simultaneously enhancing their academic advancement.

17.4 Results

The research seeks to optimize educational practices and interventions to boost academic achievement. The naive Bayes algorithm demonstrated favorable outcomes in forecasting student academic achievement. Employing ML algorithms such as naive Bayes may aid educators and institutions in detecting students who may need supplementary assistance, allowing proactive interventions and tailored coaching to increase overall academic achievement.

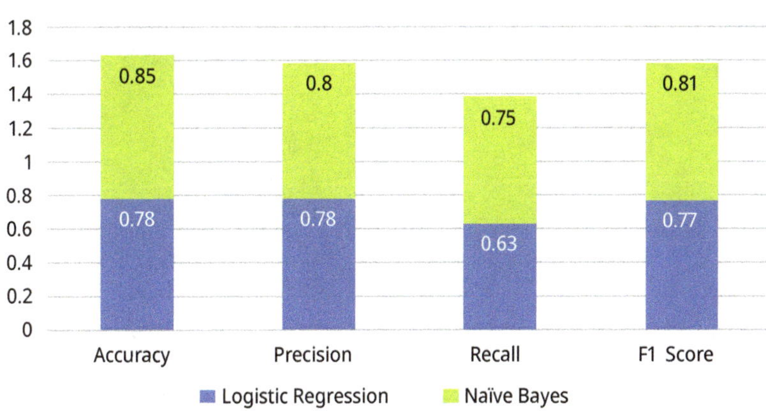

Figure 17.4: Comparative performance of logistic regression and naive Bayes classifiers using mathematical modeling.

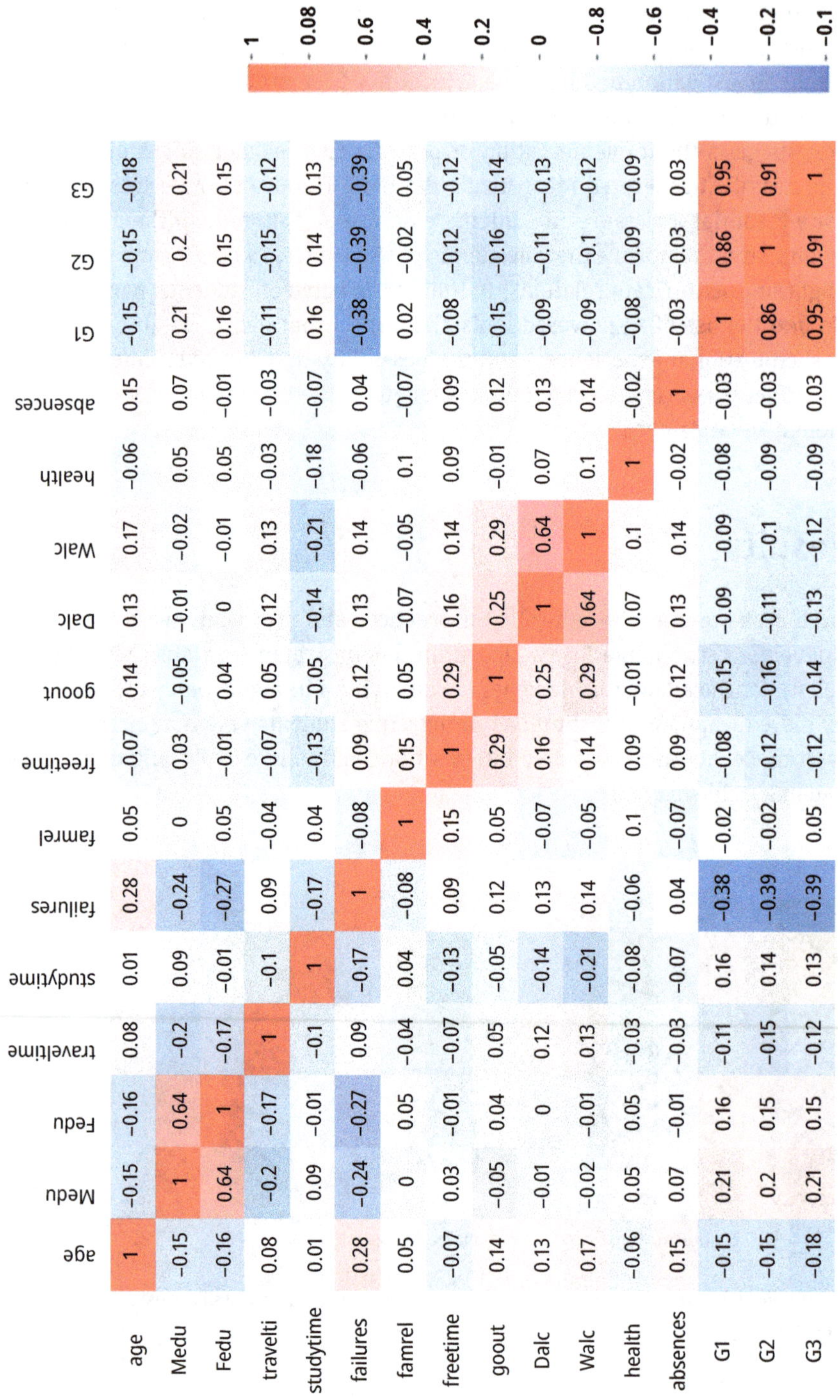

Figure 17.5: Correlation analysis using mathematical modeling.

Figure 17.4 indicates that naive Bayes performs better than logistic regression across many assessment criteria, such as accuracy, recall, and *F*1 score. This suggests that naive Bayes may be a more appropriate model for forecasting student academic success. Nevertheless, the algorithm's decision for actual implementation should be guided by further study and use-case requirements.

The naive Bayes model is a forecasting tool that identifies crucial characteristics that have a major impact on academic success. These factors include prior academic performance, allocated study duration, and the educational background of one's family. Despite the model's satisfactory accuracy, it may be enhanced by the exploration of further characteristics or the refinement of current ones. Struggling pupils may receive help via the use of early intervention measures, such as tutoring or counseling. Additional development and validation of the model may improve its resilience and capacity. The model's predictions may be used to provide individualized recommendations, such as customizing study schedules or offering supplementary materials depending on the projected performance levels.

The analysis demonstrates a robust positive link between academic performance, parental education, alcohol consumption, study time, health, and age. From Figure 17.5, the substantial correlation between the three final grades (G1, G2, and G3) indicates that a student's achievement in one period is highly indicative of their performance in the following periods. In the context of AI-powered LMSs in higher education, the correlation matrix demonstrates a modest link between increased absences and poorer marks in all three periods. This is concerning since it indicates that the correlation is moderate.

17.5 Discussion

This study explores the determinants of scholastic accomplishment in language courses by the utilizing a blended strategy procedure. The review assesses the viability of several calculations in determining scholastic accomplishment and recognizes significant disparities. The review seeks to fathom the collaboration between several factors that impact understudy execution and their capabilities in shaping scholarly achievements across various fields. In addition, it aims to answer research questions, evaluate the significance of individual traits, and offer insights into common patterns in student achievement across a variety of academic fields. The relationship network uncovers areas of strength for a relationship between liquor consumption and scholastic performance, indicating that increased liquor consumption on both weekdays and weekends is associated with poorer academic performance throughout all three evaluation periods. This can possibly bring about a deterioration in scholastic performance, health problems, and the improvement of bad social habits. Drinking too much alcohol can make it harder to focus, get motivated, and stick with studying for a long time, which can affect how well you do in school across a variety of assessment times. The health implications include numerous health concerns, such as ailments, frequent absences, and other health-related difficulties. Negative social behavior is one example of a behavioral consequence, and it prevents students from participating in productive academic activities and results in lower academic achievement. To address this issue, several techniques might include advocating for a more health-conscious approach to everyday life, increasing information about the adverse impacts of liquor on scholastic achievement, and establishing emotionally supportive networks to assist students maintain an amicable balance between their academic and personal responsibilities.

This study shows that simulated intelligence-fueled LMS further develops student performance. The monitoring of absences, instructional strategies, and student engagement might all benefit from AI. Learning may be customized by insights generated by AI. Working on convenient help and resource allocation is possible. The report suggests looking into AI, student participation, ongoing progress, and ethics. The advanced gap and equivalent admittance to AI-fueled innovations are likewise worried. In order to enhance student experiences and academic achievement, this research emphasizes the importance of carefully integrating AI into education.

17.5.1 Limitations

The study is subject to several restrictions, such as data bias, the selection of features, computational limits, and characteristics that are peculiar to the setting. The conclusions are derived from a dataset that may not accurately reflect the characteristics of larger student groups, and the study may have been constrained by the factors that were included. The efficacy of ML models, such as naive Bayes, might be constrained by the particular method used, and the analysis may be deficient in context-specific data.

17.6 Conclusion

With a particular emphasis on the use of ML algorithms and qualitative analysis, the research investigates the influence that AI-driven LMSs have on higher education. To provide predicted insights on student performance, the quantitative analysis makes use of the UCI dataset. On the other hand, the qualitative analysis highlights themes, issues, and countermeasures in the contemporary educational environment. The purpose of the study is to bridge the gap between data-driven insights and contextual understandings, which will allow for the creation of evidence-based policies and interventions to address the complexity of student accomplishment. In the future, the use of sophisticated ML algorithms and a wide range of qualitative interviews offers the potential to improve prediction models and enhance qualitative viewpoints, therefore contributing to the development of educational environments that are both inclusive and effective. To define the transformative effects of AI-powered LMSs, the study highlights the significance of a harmonious blend of quantitative analysis, predictive modeling, and qualitative insights. This will pave the way for informed interventions, strategies, and educational policies that cater to the diverse needs of students and ensure that educational environments are equitable and effective.

17.7 Future works

Future research should use more diverse datasets, advanced feature engineering techniques, thorough algorithmic investigations, contextual analysis, and longitudinal studies to improve model performance and applicability. Algorithmic exploration involves sophisticated ML algorithms, while contextual analysis uses qualitative data or surveys to understand students' experiences with AI-powered learning systems. Longitudinal studies can uncover changing trends and provide insights into the long-lasting effects of AI-driven learning environments on academic achievement.

References

[1] Ahmad, S. F., Alam, M. M., Rahmat, M. K., Mubarik, M. S., & Hyder, S. I. (2022). Academic and administrative role of artificial intelligence in education. *Sustainability*, *14*(3), 1101. Jan 19.

[2] Alam, A., & Mohanty, A. (2022 Dec 16). Foundation for the future of higher education or 'misplaced optimism'? Being human in the age of artificial intelligence. In InInternational Conference on Innovations in Intelligent Computing and Communications (pp. 17–29). Springer International Publishing, Cham.

[3] Alam, A., & Mohanty, A. (2022 Nov 18). Business models, business strategies, and innovations in EdTech companies: Integration of learning analytics and artificial intelligence in higher education. In In2022 IEEE 6th Conference on Information and Communication Technology (CICT) (pp. 1–6). IEEE.

[4] Al-Ansi, A. M., & Fatmawati, I. (2023). Integration of ICT in higher education during Covid-19 pandemic: A case study. *International Journal of Learning and Change*, *15*(4), 430–442.

[5] Ali, M., Yousaf, M., & Behlol, G. (2023 Apr 30). Artificial Intelligence in Learning Management System: A Case Study of the Students of Mass Communication. *Voyage Journal of Educational Studies*, *3*(2), 92–114.

[6] Alshahrani, A. (2023 Aug 18). Revolutionizing Blended Learning: Exploring Current Trends and Future Research Directions in the Era of ChatGPT. In 2023 7th International Conference on Business and Information Management (ICBIM) (pp. 41–47). IEEE.

[7] Anderson, N., McGowan, A., Adhikari, J., Cutting, D., Galway, L., & Collins, M. Does identifying and addressing academic difficulties early on contribute to enhanced student success and higher retention rates for a distance learning course? In InInternational Conference on Education and New Developments 2023 Jun 26 (pp. 125–128). Science Press.

[8] Atherton, P. (2023 Feb 28). Goal-setting and Problem-solving in the Tech-enhanced Classroom: A Teaching and Learning Reboot. Routledge.

[9] Dhiman, H., Vyas, N., Pagrotra, A., & Dutt, V. (Internet 2023). Real-time traffic analysis and driver safety enhancement via IoT and computer vision with high-resolution project lamps for intelligent road illumination. In 1st International Conference on Circuits, Power and Intelligent Systems (CCPIS), IEEE, 2023. Available from. http://dx.doi.org/10.1109/CCPIS59145.2023.10291872.

[10] Burbules, N. C., Fan, G., & Repp, P. (2020 Jun 1). Five trends of education and technology in a sustainable future. *Geography and Sustainability*, *1*(2), 93–97.

[11] Gutierrez, C., & Sequeda, J. F. (2020 Oct 19). Knowledge graphs: A tutorial on the history of knowledge Graph's Main ideas. In Proceedings of the 29th ACM international conference on information & knowledge management (pp. 3509–3510)

[12] George, B., & Wooden, O. (2023 Aug 29). Managing the strategic transformation of higher education through artificial intelligence. *Administrative Sciences*, *13*(9), 196.

[13] Gligorea, I., Cioca, M., Oancea, R., Gorski, A. T., Gorski, H., & Tudorache, P. (2023 Dec 6). Adaptive learning using artificial intelligence in e-learning: A literature review. *Education Sciences*, *13*(12), 1216.

[14] Golberg, H., Paradis, J., & Crago, M. Lexical acquisition over time in minority children learning English as a L. *Applied Psycholinguistics*.

[15] Kabudi, T., Pappas, I., & Olsen, D. H. (2021 Jan 1). AI-enabled adaptive learning systems: A systematic mapping of the literature. *Computers and Education: Artificial Intelligence*, *2*, 100017.

[16] Kivinen, L. K. AI-driven chatbot as a support tool for developers during the onboarding process.

[17] Kustitskaya, T. A., Kytmanov, A. A., & Noskov, M. V. (2022 Mar 1). Early student-at-risk detection by current learning performance and learning behavior indicators. *Cybernetics and Information Technologies*, *22*(1), 117–133.

[18] Rajmohan, R., Kumar, T. A., Julie, E. G., Robinson, Y. H., Vimal, S., Kadry, S., & Crespo, R. G. (2022 May). G-Sep: A deep learning algorithm for detection of long-term sepsis using bidirectional gated recurrent unit. *International Journal of Uncertainty, Fuzziness and Knowledge-Based Systems*, *30*(Supp01).

[19] Manhiça, R., Santos, A., & Cravino, J. The use of artificial intelligence in learning management systems in the context of higher education: Systematic literature review. In In2022 17th Iberian Conference on Information Systems and Technologies (CISTI) 2022 Jun 22 (pp. 1–6). IEEE.

[20] Nuzli, M., Nadeak, B., Mini, M., Jejen, A., & Hardiyanti, N. R. (2023 Nov 18). Management innovation in education: application case studies technology learning based intelligence artificial in higher education. *International Journal of Educational Research Excellence (IJERE)*, *2*(2), 493–500.

[21] Ökördi, R., & Molnár, G. (2022 Aug 18). Computer-based intervention closes learning gap in maths accumulated in remote learning. *Journal of Intelligence*, *10*(3), 58.

[22] Ouyang, F., Zheng, L., & Jiao, P. (2022). Artificial intelligence in online higher education: A systematic review of empirical research from 2011 to 2020. *Education and Information Technologies*, *27*(6), 7893–7925. Jul.

[23] Tamada, M. M., Giusti, R., & Netto, J. F. (2022 Feb 5). Predicting students at risk of dropout in technical course using LMS logs. *Electronics*, *11*(3), 468.

[24] Tan, S. (2023 Feb). Harnessing artificial intelligence for innovation in education. In Learning intelligence: Innovative and digital transformative learning strategies: Cultural and social engineering perspectives (pp. 335–363, Vol. 10). Springer Nature Singapore, Singapore.

[25] Turnbull, D., Chugh, R., & Luck, J. (2020 Jun). Learning management systems, an overview. *Encyclopedia of Education and Information Technologies*, *14*, 1052–1058.

[26] UCI Machine Learning Repository. (n.d.). Student Performance Data Set. Retrieved from (Accessed June 5, 2023 at, https://data.world/uci/student-performance)

[27] Sharmila, L., & Sakthi, U. (2020 Jun). An artificial immune system-based algorithm for abnormal pattern in medical domain. *The Journal of Supercomputing*, *76*, 4272–4286.

[28] Sharmila, L., & Sakthi, U. (2018 Sep). Chronological pattern exploration algorithm for gene expression data clustering and classification. *Wireless Personal Communications*, *102*, 1503–1519.
[29] Roselinkiruba, R. (2023 Sep). Reversible data hiding using optimization, interpolation and binary image encryption techniques. *Multimedia Tools and Applications*, *82*(23), 35757–35780.
[30] Roselinkiruba, R., & Bhuvaneshwari, G. (2024 Feb). Feature extraction based pixel segmentation techniques data hiding and data encryption. *Multimedia Tools and Applications*, *83*(7), 19259–19276.
[31] Goksel, N., & Bozkurt, A. (2019). Artificial Intelligence in Education. In Advances in Educational Technologies and Instructional Design (pp. 224–236). IGI Global. https://doi.org/10.4018/978-1-5225-8431-5.ch014.

S. S. Uma, K. Suresh Kumar, P. Suresh, and P. Dhivya

18 Optimizing Neural Network Training by Addressing Key Challenges and Advanced Techniques

Abstract: A neural network can be successfully trained when several key challenges and solutions are addressed effectively. This chapter outlines the critical aspects of optimization and theory that contribute to successful training. It begins by examining issues like gradient explosion and vanishing, which can impede learning by causing gradients to become excessively large or small. It also discusses the undesirable spectrum of weight matrices that can affect training dynamics. Practical solutions include careful initialization of weights to prevent these problems and normalization methods such as batch normalization, which help stabilize training. The chapter then reviews various optimization methods, including stochastic gradient descent (SGD), adaptive gradient techniques like Adam, and distributed methods that enhance training efficiency. Theoretical results on these algorithms provide insights into their convergence properties and practical performance. Finally, the chapter explores broader research topics such as bad local minima, which can trap the model in suboptimal solutions, mode connectivity, which examines how different solutions relate to each other, the lottery ticket hypothesis, which identifies sparse subnetworks capable of achieving high performance, and infinite-width analysis, which studies network behavior as layer widths grow infinitely. Understanding these factors and ongoing research helps in designing neural networks that are robust, efficient, and capable of effective learning.

Keywords: Gradient explosion and vanishing, weight matrix spectrum, batch normalization, stochastic gradient descent (SGD), bad local minima, lottery ticket hypothesis

18.1 Introduction

The artificial intelligence community defines deep learning as using multilayered neural networks, known as "depth," to find and understand complex data patterns. Hierarchical structures help deep learning algorithms extract complex features from unstructured data. Thus, they excel in natural language processing, speech and image recognition, and autonomous driving. These models utilize techniques such as gradient descent optimization to adapt network parameters and minimize errors in order to learn from large datasets. Although deep learning models are highly efficient, they require significant resources, such as computational power and data acquisition. Two additional scenarios that frequently arise are overfitting and interpretability concerns. Nevertheless, deep learning continues to significantly contribute to the advancement of machines' abilities and the

https://doi.org/10.1515/9783111436180-018

achievement of significant advancements in various domains [1]. A neural network can be successfully trained when it is provided with a sufficiently large and high-quality dataset, appropriate model architecture, and effective optimization strategies. Adequate data ensures that the network has enough examples to learn from and generalize well, while a well-chosen architecture allows the model to capture the complexities of the data without overfitting. Effective optimization algorithms, coupled with proper hyperparameter tuning, enable the network to converge to a good solution efficiently. Proper initialization of weights and the use of regularization techniques help maintain stability during training and prevent overfitting. A suitable loss function provides meaningful feedback for the network to learn from, and sufficient computational resources are crucial to handle the demands of training deep models. When these conditions are met, the network can learn effectively, achieve high performance on both training and unseen data, and perform robustly in practical applications.

18.2 Why Optimization in Deep Learning?

Deep learning can be viewed as an optimization framework in many ways. The primary objective is to ascertain the optimal values for a neural network's parameters, including its weights and biases, with the aim of effectively minimizing a specific loss function. Optimization plays a vital role in deep learning for various reasons, as stated below. Typically, the loss function evaluates the extent to which the model's predictions align with the observed data, and optimization algorithms aid in minimizing the loss function. Enhancing efficiency results in a shorter training period, allowing the model to pick up knowledge faster and more efficiently. The accuracy of the model as well as its generalizability are both improved through the utilization of effective optimization techniques. Finding the optimal combination of parameters, which includes weights and biases, is the key to achieving greater accuracy in the model's predictions on newly observed data. This can be accomplished by utilizing the model. It is not uncommon for deep learning models to exhibit loss surfaces that are complex and contain a large number of local minima. Effective optimization makes it simpler to navigate this intricate landscape in search of a minimum that is satisfactory, if not optimal, which ultimately results in improved model performance. Through the utilization of optimization strategies, deep learning models are able to handle larger models and a greater quantity of data with greater efficiency. The ability of optimization to increase the efficiency of computational resources can result in a number of benefits, including a reduction in the amount of time required for training and the capacity to use models on a large scale. With the assistance of more advanced optimization techniques, it is possible to train models that are more stable. These techniques reduce the number of instances in which problems such as exploding or disappearing gradients occur.

18.3 Decomposition of Theory

18.3.1 Generalization, Optimization, and Representation

The goal of deep learning is to reduce the magnitude of this loss function to the greatest extent that is realistically possible. A loss function is a tool that is utilized to determine the degree of disparity that exists between the desired target values and the outputs that are predicted by a model. When performing regression tasks, the loss function will typically use a metric like mean squared error (MSE), and when performing classification tasks, it will use cross-entropy.

The function is then used to create predictions using newly collected test data once it has been fine-tuned to perfection. Because the set of functions that was chosen has a limited ability to precisely represent the true underlying function, a representation error may arise. The root cause of optimization error is the impossibility of precisely identifying the ideal parameters that minimize the loss function. A gauge of the model's performance on fresh data is the generalization error. It illustrates how effectively the function that was acquired during training can be used in real-world scenarios. Each kind of error clarifies a distinct facet of learning and advances our knowledge of the overall effectiveness of the model [2].

18.3.2 Decomposition of Optimization Issues

Deep learning requires disassembling optimization problems into their component parts in order to find solutions. In this method, issues that may cause the learning algorithm to fail to converge or perform poorly are identified and resolved.

18.3.3 Gradient Explosion and Vanishing

Gradient explosion is a major issue during deep neural network training. This occurs when the loss function gradients with respect to model parameters are very large. This problem is common in deep neural networks, especially when gradients propagate through many layers.

As gradients are backpropagated, they can grow exponentially if they are already large at earlier layers, leading to instability in training. Gradient explosion primarily leads to unpredictable and excessively significant modifications to the network weights. This can destabilize the training process, resulting in substantial fluctuations or divergence of the model parameters. In severe cases, this instability can lead to numerical issues such as NaN (not a number) values in the model's weights, effectively halting the training process. To address gradient explosion, several techniques are employed: gradient clipping limits the maximum value of gradients to prevent them from exceeding a predefined threshold; proper weight initialization ensures that weights are scaled appropriately to maintain stable gradients; and normalization methods, such as batch normalization, help in stabilizing the distribution of activations and gradients throughout the network [3]. By managing gradient explosion effectively, one can ensure more stable and reliable training of deep neural networks.

Deep neural networks, in which gradients are dispersed across a large number of layers, are prone to experiencing this issue quite frequently. As these gradients are backpropagated, they can shrink exponentially due to the repeated application of the chain rule, especially when using activation functions that squash outputs into a narrow range, such as sigmoid or tanh. The result is that the gradients diminish to near zero, making it difficult for the network to perform significant weight updates and thus hindering the learning process. Consequently, the model may experience very slow convergence or fail to capture complex patterns in the data, leading to suboptimal performance. Solutions to gradient vanishing include using activation functions like ReLU (rectified linear unit) that do not compress outputs into a small range, employing proper weight initialization techniques to maintain stable gradient magnitudes, and leveraging architectures with residual connections, such as ResNets, which provide alternative paths for gradients to flow more effectively through the network. By addressing gradient vanishing, one can improve the stability and efficiency of training deep neural networks.

18.4 Gradient Descent – Basic Analysis

This basic optimization algorithm reduces the loss function to its minimum value. The next step is to adjust the model's parameters to minimize loss immediately after computing the gradient (or derivative) of the loss function for each parameter. Iteratively approaching the loss function's minimum yields the best results. The basic idea is that gradient descent is an optimization algorithm that minimizes the loss function $L(\theta)$, where θ represents the model's parameters. It helps reduce the error or difference between the predicted output and the actual output by iteratively adjusting the model parameters [4]. The goal is to gradually reduce loss by adjusting parameters against the loss function's gradient. At its core, the gradient descent update rule is

$$\sum_{t=1}^{n} \theta_{t+1} = \sum_{t=1}^{n} \theta_t - \sum_{t=1}^{n} r(n).\eta \nabla L(\theta_t) \tag{18.1}$$

where

- The symbol θt is used to represent the parameters at the iteration level.
- The learning rate (η) is a small positive scalar that determines the size of the step in the opposite direction of the gradient.

- The gradient of the loss function L with respect to parameters θ at iteration t is denoted as $\nabla L(\theta t)$.
- $r(n)$ represents the number of parameter inclusions at various t values.

18.4.1 Variants of Gradient Descent

18.4.1.1 Stochastic Gradient Descent (SGD)

Stochastic gradient descent (SGD) calculates gradients and updates model parameters. Mini-batches, which are random subsets of training data, achieve this. This is a mini-batch. This causes gradient estimation interference, which can help avoid local minimum points and accelerate convergence [5]. Let w be the parameters (weights) of a model. Let $L(w;\, x_i, y_i)$ be the example of a single training sample, and η be the learning rate. Consider the loss function of the gradient with respect to the parameters, and let it be represented by the symbol ∇L.

According to the update rule for SGD,

$$\sum_{t=1}^{n} wt + 1 = \sum_{t=1}^{n} wt - \sum_{t=1}^{n} \eta \nabla L(wt)^{*} \sum_{i=1}^{50} (x_i, y_i) \tag{18.2}$$

In order to determine the gradient of the loss function L with respect to the parameters $\iiint L(w;\, x_i, y_i)$, it is necessary to compute the gradient for each training sample (x_i, y_i). The expression that represents the mean square error of the loss function is presented in the following manner:

$$\sum_{t=1}^{n} L(wt)^{*} \sum_{i=1}^{r} (x_i, y_i) = 21 \sum_{i=1:t=1}^{n} (y_i - w_t x_i)^2 \tag{18.3}$$

The gradient of this loss function with respect to w is given as

$$\nabla Lw \sum_{i=1}^{n} (x_i, y_i) = - \sum_{i=1}^{n} [(y_i - wTx_i)x_i] \tag{18.4}$$

The SGD update rule is given as

$$\sum_{t=1}^{n} w_{t+1} = \sum_{t=1}^{n} w_t + \eta \sum_{i=1:t=1}^{n} \left[(y_i - {w_t}^T x_i)x_i\right] \tag{18.5}$$

In this rule, the gradient is included in the weight (wt) because we are traveling in the direction that will cause the least amount of loss. The gradient's opposite was used by the general update rule.

18.4.1.2 Batch Gradient Descent

Batch gradient descent is a machine learning algorithm that, rather than updating model parameters by using individual examples or small subsets of data, updates them by taking into account the entire dataset. This is in the place of the traditional method of updating model parameters. The purpose of this approach is to bring about a reduction in a loss function. The computation of the gradient is defined as the process of determining the average of the gradients that are found across all of the training samples. The update rule is defined as wt + 1 = wt – $\eta\iiint L$(wt), where wt represents the current weight, η represents the learning rate that determines the size of the update steps, and $\iiint L$(wt) represents the average gradient calculated from the entire dataset. Because it requires processing all examples in each iteration, batch gradient descent is a computationally heavy and slow algorithm, particularly when applied to large datasets.

However, it is able to provide accurate and reliable gradient estimates in other situations. Having said that, two of its most highly regarded characteristics are its ability to easily integrate updates and its ability to efficiently converge. It is therefore effective when applied to problems that involve dataset sizes and computational resources that are manageable.

This following equation is used to determine the gradient of the loss function $L(w)$ for a given dataset that contains N training examples (x_i, y_i), where i can range from 1 to N:

$$\nabla L(W) = \frac{1}{N}\sum_{i=1}^{N} \nabla L(w; x_i, y_i) \tag{18.6}$$

When the direction of the calculated gradient is reversed, the parameters w are updated to reflect the new values:

$$\sum_{t=1}^{n} w_{t+1} = \sum_{t=1}^{n} [w_t - \eta \nabla L(w_t)] \tag{18.7}$$

The parameter vector $w_t + 1$ represents the updated values of the parameters. At iteration t, the parameter vector is symbolized by the symbol θ, while the size of the update step is determined by the learning rate, which is denoted by the symbol η. For the purpose of evaluating the effectiveness of a model on a dataset that contains N samples, the MSE loss function is utilized [6, 7]:

$$L(w) = \frac{1}{2N}\sum_{i=1}^{N} \left(y_i - w^T x_i\right)^2 \tag{18.8}$$

Utilizing this gradient results in a transformation of the update rule for the batch gradient descent algorithm:

$$w_{t+1} = w_t + \eta \left(\frac{1}{N}\sum_{i=1}^{N} (y_i - w_t^T x_i) x_i\right) \tag{18.9}$$

18.4.1.3 Mini-Batch Gradient Descent

Mini-batch gradient descent is an optimization algorithm that aims to balance SGD and batch gradient descent [8]. Mini-batch gradient descent illustrates this. Mini-batch gradient descent uses a random-selected mini-batch to compute the gradient. Unlike the traditional method of using the entire dataset or one training example, this method balances stable convergence and efficient computation by combining the best of the two approaches. To compute the gradient of the loss function with respect to the model parameters, all instances from a mini-batch must be considered and used:

$$\nabla L(w) = \frac{1}{B}\sum_{i=1}^{B} \nabla L(w; x_i; y_i) \tag{18.10}$$

It is necessary to move in the opposite direction of the computed mini-batch gradient in order to update the parameters of the mathematical model:

$$w_{t+1} = w_t - \eta \nabla L(w_t;\ \text{mini} - \text{batch}) \tag{18.11}$$

where B represents the mini-batch size, and (x_i, y_i) are the examples within the mini-batch.

The mini-batch gradient descent technique is a method that is frequently utilized for the purpose of training machine learning models on a large scale. It is favored due to the fact that it is effective and has the capability of developing a balanced convergence. The computational requirements of batch gradient descent and the faster, albeit more erratic, updates of SGD are both met by this method, which provides a pragmatic middle ground between the two.

18.4.2 Advanced Optimization Techniques

18.4.2.1 Momentum

Momentum works by accumulating the gradients of past iterations to adjust the parameter updates in the current iteration. This helps the optimization process to continue moving in the direction of consistent gradients, reducing oscillations and speeding up convergence, especially in areas with a lot of noise or small gradients.

18.4.2.2 Adaptive Moment Estimation

Adam, which stands for adaptive moment estimation, is a technique that combines the benefits of momentum and RMSProp – two additional SGD enhancements. The parameter θ is used to represent the weight of the model, which is set to zero. Within the first moment vector $m0$ and the second moment vector $v0$, it is necessary to incorporate zero vectors that have the same dimension as θ. For the first moment estimate, the decay rate is assumed to be $\beta 1$, which is typically 0.9. On the other hand, for the second moment estimate, the decay rate is set to $\beta 2$, which is typically 0.999. One way to represent the learning rate is the symbol α. The algorithm updates the parameter θ using the following process.

When performing optimization using deep learning, the process of updating the first moment estimate involves calculating an exponentially weighted moving average of the gradient, and is expressed as follows:

$$\sum_{i=1}^{n} m_i = \sum_{t=1}^{n} (\beta_1 \cdot m_{t-1} + x(1 - \beta_1)) \tag{18.12}$$

where m_{t-1} is the estimate of the first moment (mean of gradients).

The "second moment estimate" in optimization refers to the calculation of the variance in order to enhance the stability of gradient-based learning. This is particularly important for adaptive algorithms and is given as

$$v_t = \beta_2 . v_{t-1} + (1 - \beta_2) g_t^2 \tag{18.13}$$

In this context, the value of v_t denotes the estimation of the second moment, which is the variance of gradients from the uncentered position.

In order to reduce the impact of bias in adaptive optimization algorithms, the bias-corrected first moment estimate makes adjustments to the average of both gradients:

$$m_t = \frac{m_t}{1 - \beta_1^t} \tag{18.14}$$

This corrects the bias in m_t because it starts at zero.

The second moment, corrected for bias estimate, improves the stability of optimization algorithms like Adam by mitigating the bias in adaptive learning rates:

$$v_t = \frac{v_t}{1 - \beta_2^t} \tag{18.15}$$

When it comes to machine learning, the parameter update refers to the process of modifying model parameters while the model is being trained in order to reduce loss and significantly improve performance:

$$\theta = \theta - \alpha . \frac{m_t}{\sqrt{v_t} + \epsilon} \tag{18.16}$$

Adam is renowned for its capacity to effectively solve a wide variety of problems by utilizing the benefits of both momentum and adaptive learning rates. This ability has earned Adam considerable acclaim.

18.4.2.3 Adaptive Gradient Algorithm

The adaptive gradient algorithm, or AdaGrad, is a technique for optimization that modifies the learning rate for each parameter individually, taking into account the historical gradient information. AdaGrad is also known as AdaGrad. In situations where the data being worked with is not densely populated or when dealing with features that have varying degrees of importance, this approach is particularly useful [10].

Let us initialize the parameter θ (the weights of your model). Initialize the gradient accumulation vector G_0 to zero vectors of the same dimension as θ. Let the learning rate be α (often around 0.01). Let the small constant be ϵ (to avoid division by zero, typically 10 − 810^{−8}10 − 8).

Gradient Computation:
For each parameter θ, compute the gradient g_t at time step t.

Update Rules
The algorithm updates parameters θ using the following steps:

Update the Accumulated Squared Gradients

$$G_t = G_{t-1} + g_t^2 \tag{18.17}$$

At time step t, the squared gradients have added up to G_t.

Update Parameters

$$\boldsymbol{\theta} = \boldsymbol{\theta} - \frac{\boldsymbol{a g_t}}{\sqrt{\boldsymbol{G_t} + \boldsymbol{\varepsilon}}} \tag{18.18}$$

The learning rate is represented by α in this context, and ϵ is a small value used to avoid division by zero. The main advantage of AdaGrad is that it can change the learning rate on the fly, which makes it great at handling sparse features and gradients of different sizes. AdaGrad adjusts the learning rate of each parameter by considering the past gradients, allowing it to effectively handle sparse data and features that are updated at different frequencies. Nevertheless, the rapid decrease in learning rates can occasionally result in suboptimal performance if the learning rate diminishes too rapidly.

RMSProp
Root mean square propagation, more commonly referred to as RMSProp, is an optimization algorithm that is applicable to adaptive learning rates. This method was developed with the express purpose of overcoming certain limitations that are associated with conventional gradient descent methods such as AdaGrad. By calculating the moving average of squared gradients, RMSProp and AdaGrad both attempt to normalize the gradient that is currently being applied. This is in contrast to AdaGrad, which uses

an excessively low learning rate for parameters that have large historical gradients. The purpose of this moving average is to promote a more equitable and steady convergence, as well as to protect against the impact of learning rate adjustments [11, 12]. Due to the fact that RMSProp modifies the learning rate for each parameter, it is suitable for dealing with gradients of varying magnitudes. By utilizing a moving average of squared gradients, RMSProp is able to alleviate the problem of rapid decay in learning rate that AdaGrad is experiencing.

In order to update each parameter θ, RMSProp takes into consideration its gradient g_t and a moving average v_t of the squared gradients obtained from the previous parameter. It is necessary to update the moving average v_t by employing a decay rate ρ, which is typically around 0.9, and a small constant ϵ, such as 10 − 8, in order to avoid division by zero.

The parameter update rule is given as follows:

Update the Moving Average of Squared Gradients:

$$v_t = \rho . v_{t-1} + (1 - \rho) g_t^2 \tag{18.19}$$

where v_t is the moving average of the squared gradients

$$\theta = \theta - \frac{\alpha . g_t}{\sqrt{v_t} + \varepsilon} \tag{18.20}$$

Specifically, the symbol α is used to denote the rate of learning, while the symbol ϵ is a small constant that is employed to avoid division by zero. Through the process of adjusting the learning rate for each parameter and smoothing the gradient information, RMSProp significantly improves the stability and efficiency of the training process. This is especially helpful in circumstances in which the magnitudes of the gradients are different. When it comes to training deep neural networks and addressing problems that are characterized by sparse gradients or significant variance, RMSProp is particularly advantageous. The gradient descent method is improved by advanced optimization algorithms such as Momentum, Adam, and RMSProp. These algorithms incorporate adaptive learning rates and smoothing mechanisms, which ultimately leads to improved convergence and training performance.

18.4.4 Global Optimization of Neural Networks

Global optimization of neural networks involves determining the optimal arrangement of network parameters in order to minimize a loss function while avoiding local minima and saddle points. This is accomplished through the process of global optimization. Even though it might not seem significant at first glance, neural networks frequently exhibit non-convex loss landscapes that contain a large number of saddle points and local minima. The goal of training is to optimize this function by either maximizing or minimizing it. The configuration space encompasses all conceivable combinations of weights and biases that are linked to the neural network.

Local Minima: These are points where the loss function is lower than the nearby points but may not be the lowest overall.

Global Minima: These are the points where the loss function reaches its absolute lowest value.

18.4.5 Strategies for Global Optimization

18.4.5.1 Gradient-Free Optimization

Gradient-free optimization refers to a category of optimization techniques that do not rely on gradient information to find optimal solutions. When working with objective functions that do not have differentiability, contain noise, or require expensive evaluation, these methods are especially useful because of their versatility. They work by exploring the solution space through mechanisms like random sampling, evolutionary strategies, or heuristic search, rather than calculating and following gradients. Gradient-free methods are often employed in complex, high-dimensional problems where traditional gradient-based methods may struggle or be impractical [13].

Gradient-free optimization involves finding the best solution to an optimization problem without using gradient information. The Nelder–Mead algorithm optimizes a function $f(x)$ without requiring gradient information. It uses a simplex method to iteratively refine a set of candidate solutions.

1. **Initialization:** Start with an initial simplex, which is a geometric figure with $n+1$ vertices in n-dimensional space. For example, in two dimensions, the simplex is a triangle with three vertices.
2. **Simplex Transformation:** The algorithm iteratively performs a series of transformations on the simplex:
   - **Reflection:** Perform a reflection of the vertex with the highest negative value through the centroid of the remaining vertices.
   - **Expansion:** If the point of reflection surpasses the optimal point, proceed to expand further in that particular direction.
   - **Contraction:** If the point of reflection does not produce any improvement, then the simplex should be contracted in the direction of the centroid.
   - **Shrinkage:** If contraction does not improve, shrink the simplex toward the best vertex.

The simplex transformation steps can be described by the following update equations:

- **Reflection Step:**

$$x_{\text{ref}} = x_{\text{centroid}} + \alpha(x_{\text{worst}} - x_{\text{centroid}}) \tag{18.21}$$

where x_{ref} is the reflected vertex, x_{centroid} is the centroid of the simplex excluding the worst vertex, and α is a reflection coefficient.

- **Expansion Step:**

$$x_{\text{exp}} = x_{\text{centroid}} + \gamma(x_{\text{ref}} - x_{\text{centroid}}) \tag{18.22}$$

where x_{exp} is the expanded vertex and γ is the expansion coefficient.

- **Contraction Step:**

$$x_{\text{contract}} = x_{\text{centroid}} + \rho(x_{\text{worst}} - x_{\text{centroid}}) \tag{18.23}$$

where x_{contract} is the contracted vertex and ρ is the contraction coefficient.

The algorithm continues to iterate through these steps until convergence criteria are met, such as a small enough change in function values or vertex positions. This approach is useful in various fields, including optimization problems in machine learning, engineering design, and parameter tuning, where gradient information may not be readily available or reliable.

18.4.5.2 Bayesian Optimization

The Bayesian optimization technique makes use of probabilistic models in order to optimize objective functions that are difficult to understand, expensive, or noisy. In most cases, a Gaussian process (GP) is utilized by the system as the probabilistic model to assess the behavior of the underlying function and to direct the search for optimal solutions. This is followed by the utilization of an acquisition function to ascertain the subsequent sampling location, ensuring that both new and promising regions are investigated. This process can be broken down into two primary components. Bayesian optimization is a method that iteratively updates the model with new data and improves its predictions to find the best solutions. This method requires a lower amount of testing in comparison to other approaches that are more conventional. This method is useful in situations where function evaluation can be expensive, such as when optimizing engineering designs or fine-tuning hyperparameters for machine learning, for example [14].

Bayesian optimization is a technique used to optimize an objective function $f(x)$ that is costly to evaluate. It achieves this by utilizing a probabilistic model. A GP models the objective function as follows:

$$f(x) = \mathcal{GP}(m(x),\ K(x,x')) \tag{18.24}$$

The mean value, which is typically zero, is represented by the function $m(x)$, and the covariance or kernel function is represented by the function $k(i, j)$. Through the use of the acquisition function, the optimal location for the subsequent sampling point can be determined. Common choices for this are given as

$$\text{Expected improvement, } \mathrm{EI}(x) = E[\max(f^* - f(x), 0)] \tag{18.25}$$

The variable f^* represents the highest recorded value. Then the

$$\text{Probability of improvement, } \mathrm{PI}(x) = \Phi\left(\frac{\mu(x) - f^* - \xi}{\sigma(x)}\right) \tag{18.26}$$

A parameter is represented by the symbol ξ, while the symbol Φ represents the cumulative distribution function:

$$\sum_{x=0}^{a} \mathrm{UCB}(x) = \sum_{x=0}^{a} (\mu(x) + \kappa\sigma(x)) \tag{18.27}$$

where UCB is the upper confidence bound.

Bayesian optimization efficiently finds the optimum by balancing exploration and exploitation, using a probabilistic model to guide the search process.

18.4.5.3 Simulated Annealing

The stochastic optimization technique that is commonly referred to as “simulated annealing” takes its cues from the metallurgical annealing process. By gradually heating a material and then cooling it once more, it is possible to determine the state of the material that has the least amount of energy. In the field of optimization, this approach involves systematically exploring the range of possible solutions by repeatedly creating new potential solutions and deciding whether to accept them based on a probability that is influenced by the current temperature and the disparity in objective function values. The temperature schedule, usually following a geometric or exponential decrease, controls the equilibrium between exploration and exploitation. Simulated annealing is a highly effective method for discovering solutions that are close to being optimal in search spaces that are both large and complex, especially when the landscape contains numerous local minima. The acceptance probability P for transitioning from the current solution i to the new candidate solution j with objective function values $f(x_j)$ and $f(x_i)$ is determined as follows:

$$P = -\sum_{\substack{i=1\\j=1}}^{n}\left(\frac{f(x_i) - f(x_j)}{T}\right) \tag{18.28}$$

T represents the present temperature. The new solution is always accepted if $f(x_i)$ is less than $f(x_j)$. If the value of $f(x_i)$ is greater than or equal to the value of $f(x_j)$, the algorithm is allowed to escape local minima with a probability denoted as P.

In order to reduce the likelihood of settling for subpar solutions, the temperature T is lowered incrementally. A common cooling schedule is

$$T_{k+1} = \alpha T_k \tag{18.29}$$

where T_k is the temperature at iteration k, α (with $0 < \alpha < 10$) is the cooling rate, and T_{k+1} is the temperature for the next iteration. This cooling rate controls how quickly the temperature decreases. Simulated annealing is useful for finding good approximations to the global optimum in large, complex search spaces, especially when the problem landscape contains multiple local minima.

18.4.5.4 Genetic Algorithms

Genetic algorithms, also known as GAs, are a type of optimization strategy that originates from the principles that govern the process of evolution. Through processes that are analogous to mutation, natural selection, and crossover (recombination), they experiment with a number of different potential solutions, gradually improving them over the course of time. At the outset, a diverse range of potential solutions are generated. Each solution, referred to as an 'individual,' is then evaluated based on a fitness function, which assesses how well it addresses the optimization problem. The most successful individuals are chosen for reproduction, which involves the crossing of their characteristics in order to produce new generations. Random variations are also introduced by mutation, which allows for the exploration of previously uncharted regions of the solution space. The end result of this iterative process is a solution that is either optimal or almost optimal. This process continues with each new generation producing solutions that are marginally better than the problem that was solved in the previous generation [15–17].

The initial population P will consist of N individuals, and each individual x_i will represent a potential solution. Let us begin by initializing the population. Evaluating each individual with the help of a fitness function $f(x)$ provides a quantitative measure of how effective the solution is, and is given as Fitness$(x) = f(x)$. This function guides the selection process. Select individuals for reproduction based on their fitness. A commonly used approach is roulette wheel selection, where the probability of i is proportional to the achieved fitness, denoted as P_i:

$$P_i = \frac{f(x_i)}{\sum_{j=1}^{N} f(x_j)} \tag{18.30}$$

Create offspring by combining features from selected parents. For example, using single-point crossover, two parents x_1 and x_2 produce offspring x_1' and x_2':

$$x_1' = x_1^{[0:p]} \cup x_2^{[p+1:n]} \tag{18.31}$$

$$x_2' = x_2^{[0:p]} \cup x_1^{[p+1:n]} \tag{18.32}$$

where $x_1^{[0:p]}$ and $x_2^{[0:p]}$ are segments of the parent solutions. Introduce random changes to offspring to maintain diversity. For a binary representation, mutation might flip a bit x. GAs effectively search complex solution spaces by mimicking natural evolutionary processes, providing robust solutions to various optimization problems.

18.4.5.5 Hyperparameter Tuning

Hyperparameter tuning involves optimizing the configuration of a machine learning model to maximize its performance on a given task. In contrast, hyperparameters are pre-established and have influence over variables such as the learning rate, the number of layers in a neural network, and the level of regularization in a model. On the other hand, model parameters are obtained gradually. Cross-validation is a standard method for assessing the best combination of hyperparameter values during the process of fine-tuning a model. This is done to determine the optimal performance. An essential consideration in developing superior machine learning models is the optimization of hyperparameters to improve the model's robustness and precision.

18.4.5.6 Regularization Techniques

In neural networks, regularization techniques are absolutely necessary in order to avoid overfitting phenomena. It is possible to accomplish this by controlling the complexity of the model and making certain that it is able to successfully apply to data that is unfamiliar to it. One of the regularization methods that is frequently utilized is called $L1$, and it involves incorporating the absolute values of weights into the loss function in order to encourage sparsity in the model. On the other hand, $L2$ makes use of the squared values in order to discourage the use of large weights and to encourage the use of smaller weights that are more similarly distributed:

$$\mathrm{LOSS}_{L1} = \mathrm{LOSS}_{\mathrm{original}} + \lambda \sum i|w_i| \tag{18.33}$$

$$\mathrm{LOSS}_{L2} = \mathrm{LOSS}_{\mathrm{original}} + \lambda \sum i\, w_i^2 \tag{18.34}$$

During the training process, a technique known as dropout is utilized, in which a specific percentage of neurons are deactivated in a random manner. The ability to generalize is improved as a result of this phenomenon, which helps to prevent an excessive dependence on particular neurons [18–20]. Weight noise adds noise to the weights during training, preventing the model from overfitting. Norm constraints limit the magnitude of weight vectors to prevent them from growing too large. Label smoothing softens target labels, reducing the model's confidence and improving generalization. Finally, ensemble methods are used to further improve performance. These methods can be employed either separately or together to improve the resilience and adaptability of the model.

18.4.6 Analysis of Various Optimization Algorithms

When it comes to the training of deep learning models, optimization algorithms are absolutely necessary because their primary objective is to reduce the loss function and improve the accuracy of the model. The various algorithms used in deep learning are given in Table 18.1.

Table 18.1: Analysis of various optimization algorithms.

Optimization algorithm	Convergence speed	Memory usage	Hyperparameters	Stability	Use cases
Stochastic gradient descent (SGD)	Slow	Low	Learning rate	Moderate	Simple models, large-scale problem
Momentum	Faster than SGD	Low	Learning rate, momentum	Stable	Deep neural networks, where convergence speed matters
Nesterov accelerated gradient (NAG)	Faster than momentum	Low	Learning rate, momentum	More stable	Deep neural networks, harder-to-train models
Adagrad	Moderate	High	Learning rate	Less stable	Sparse data, text, or image data
RMSProp	Faster than Adagrad	Moderate	Learning rate, decay rate	Stable	Recurrent neural networks (RNNs)

Table 18.1 (continued)

Optimization algorithm	Convergence speed	Memory usage	Hyperparameters	Stability	Use cases
Adam (adaptive moment estimation)	Fast	Moderate	Learning rate, Beta1, Beta2	Very stable	General-purpose, complex models like CNNs and RNNs
AdaMax	Fast	Moderate	Learning rate, Beta1, Beta2	Very stable	General-purpose, complex models
Nadam	Faster than Adam	Moderate	Learning rate, Beta1, Beta2	Very stable	General-purpose, harder-to-train models
Adadelta	Moderate	Moderate	Learning rate, decay rate	Stable	General-purpose, real-time models
AMSGrad	Fast	Moderate	Learning rate, Beta1, Beta2	Very stable	General-purpose, long-term dependencies in data

From Table 18.2, it is clear that the adaptive learning rates and momentum terms present in Adam, Nadam, and AMSGrad are responsible for the acceleration of their convergence. It is not uncommon for the SGD algorithm to exhibit the worst performance when no optimizations or improvements are applied to the algorithm. Due to the fact that Adagrad must remember the squared gradients for each parameter, it consumes more memory than is strictly required. In contrast, SGD and momentum consume a negligible amount of memory while operating. As a result of the increased number of hyperparameters required by algorithms such as Adam, Nadam, and RMSProp, these algorithms are more difficult to fine-tune, but they also provide increased flexibility and adaptability. When it comes to training deep learning models, it is common knowledge that Adam, AMSGrad, and Nadam are stable. Adagrad, on the other hand, may exhibit slightly less stability as a result of its gradually decreasing learning rate. In the realm of deep learning models, Adam and Nadam are both general-purpose optimizers that have the potential to be utilized in a wide variety of applications. RMSProp and Adagrad are particularly effective in situations involving recurrent neural networks (RNNs) and sparse data. Table 18.2 lists accuracy values for various optimization algorithms used in deep learning across different tasks and models.

Considering the statistical parameters provided in Table 18.1, it is clear that Adam generally provides higher accuracy across different tasks compared to standard SGD, due to its adaptive learning rate and momentum. SGD with momentum and nesterov accelerated gradient are effective for image classification tasks, especially with large datasets like ImageNet. AdaGrad and RMSProp are useful for tasks with sparse data or nonstationary objectives, respectively, although they might not outperform Adam in general. Hessian-free optimization and L-BFGS are second-order methods that can

Table 18.2: Optimization algorithms versus accuracy.

Optimization algorithm	Inference	Datasets used	Model	Accuracy (%)
Stochastic gradient descent (SGD)	Standard SGD with momentum is typical for image classification	CIFAR-10	Convolutional neural network (CNN)	85.0
Adam	It often achieves faster convergence with better accuracy than vanilla SGD	CIFAR-10	Convolutional neural network (CNN)	87.5
RMSProp	Suitable for nonstationary objectives, it performs well with smaller learning rates	MNIST	Multilayer perceptron (MLP)	97.5
AdaGrad	It adjusts the learning rate adaptively, useful in sparse data settings, with slightly lower accuracy than Adam	IMDB (sentiment analysis)	Recurrent neural network (RNN)	86.5
Nesterov accelerated gradient (NAG)	It enhances SGD with momentum by anticipating future gradients, used in large-scale image tasks	ImageNet	ResNet-50	76.5
SGD with learning rate decay	Learning rate decay improves stability and convergence in sequence models like LSTM	Penn treebank (PTB)	LSTM	82.3
AdaDelta	It is an improvement over AdaGrad, particularly for datasets where learning rate decay is crucial	Fashion-MNIST	Convolutional neural network (CNN)	90.0
SGD with warm restarts (SGDR)	The cosine annealing schedule with restarts helps escape local minima for more complex datasets	CIFAR-100	WideResNet	75.0
L-BFGS	Second-order optimization is typically used for small-scale problems and provides high accuracy for convex tasks	Custom synthetic data	Logistic regression	92.0
Hessian-free optimization	Efficiently handles curvature information, achieving higher accuracy in deep networks compared to first-order methods	MNIST	Deep neural network (DNN)	98.2
Hyperband (for hyperparameter tuning)	It combines random search with early-stopping, optimizing hyperparameters can lead to improved accuracy	CIFAR-10	Convolutional neural network (CNN)	89.0

achieve high accuracy but are computationally intensive, making them less common for large-scale deep learning tasks. The graphical representation is shown in Figure 18.1.

Deep neural networks are trained by applying optimization techniques to modify weights and biases, with the goal of minimizing the loss function. Efficient optimiza-

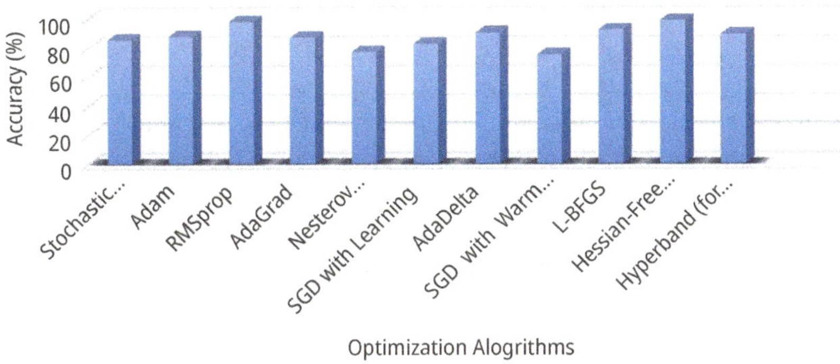

Figure 18.1: Accuracy values of different optimization algorithms.

tion ensures that the model is able to acquire knowledge from the data and demonstrates strong generalization capabilities when confronted with data that is new and unfamiliar to it. In practical applications, deep learning models frequently deal with large datasets or datasets that are extensive. The optimization process requires optimizers to navigate parameter spaces that contain a large number of dimensions in an efficient manner and to effectively allocate computational resources. Techniques such as mini-batch SGD and distributed training are utilized quite frequently. It is guaranteed that the model will arrive at a solution that is favorable and will remain stable throughout the entirety of the training procedure if effective optimization is performed. Convergence issues can be effectively addressed, and the occurrence of exploding or vanishing gradients can be mitigated through the utilization of strategies such as learning rate scheduling and gradient clipping. Several modern deep learning frameworks, including TensorFlow, PyTorch, and Keras, come with optimizers and tools that are already installed. These optimizers and tools make it possible to implement a variety of optimization strategies in an effective manner. Through the use of these frameworks, the application of complex optimization strategies to deep learning models is simplified.

18.5 Conclusion

In conclusion, optimization in deep learning is a critical component for developing effective and efficient models, driving advancements across numerous applications. Theoretical foundations, such as gradient-based methods and their variants, provide robust frameworks for training neural networks by minimizing complex loss functions. Techniques like SGD, momentum, and adaptive learning rates have revolutionized model optimization, addressing challenges like vanishing gradients and slow convergence. Additionally, gradient-free approaches, such as Bayesian optimization, and heuristic methods, like simulated annealing, offer alternative strategies for tuning hyperparameters and optimizing in scenarios where gradients are not available or are impractical to compute. Deep learning models are able to achieve exceptional performance, flexibility, and scalability across a variety of domains, such as computer vision, natural language processing, and reinforcement learning, when these optimization techniques are incorporated into the models. As the field of deep learning continues to develop, ongoing research and development in optimization methods will continue to play an essential role in pushing the boundaries of what can be accomplished by deep learning.

References

[1] Liu, L., Ouyang, W., Wang, X., Fieguth, P., Chen, J., Liu, X., & Pietikäinen, M. (2020). Deep learning for generic object detection: A survey. *International Journal of Computer Vision*, 128, 261–318.

[2] Sun, R. (2019). Optimization for deep learning: Theory and algorithms. arXiv Preprint arXiv:1912.08957.

[3] Ribeiro, A. H., Tiels, K., Aguirre, L. A., & Schön, T. (2020). Beyond exploding and vanishing gradients: Analyzing RNN training using attractors and smoothness. In International conference on artificial intelligence and statistics (pp. 2370–2380). PMLR.

[4] Andrychowicz, M., Denil, M., Gomez, S., Hoffman, M. W., Pfau, D., Schaul, T., Shillingford, B., & De Freitas, N. (2016). Learning to learn by gradient descent by gradient descent. Advances in Neural Information Processing Systems, 29.

[5] Saleh, H. M., Marouane, H., & Fakhfakh, A. (2024). Stochastic gradient descent intrusions detection for wireless sensor network attack detection system using machine learning. *IEEE Access*.

[6] Dhiman, H., Vyas, N., Pagrotra, A., & Dutt, V. (2023). Real-time traffic analysis and driver safety enhancement via IoT and computer vision with high-resolution project lamps for intelligent road illumination [Internet]. In 1st International Conference on Circuits, Power and Intelligent Systems (CCPIS). IEEE, 2023. Available from. http://dx.doi.org/10.1109/CCPIS59145.2023.10291872.

[7] Azari, V., Vazquez, O., Mackay, E., Sorbie, K., & Jordan, M. (2022). Gradient descent algorithm to optimize the offshore scale squeeze treatments. *Journal of Petroleum Science and Engineering, 208*, 109469.

[8] Franchini, G., Ruggiero, V., & Zanni, L. (2020). Steplength and mini-batch size selection in stochastic gradient methods. In International Conference on Machine Learning, Optimization, and Data Science (pp. 259–263). Springer International Publishing, Cham.

[9] Kingma, D. P., & Ba, J. (2014). Adam: A method for stochastic optimization. arXiv Preprint arXiv:1412.6980.

[10] Saab, S. Jr, Saab, K., Phoha, S., Zhu, M., & Ray, A. (2022). A multivariate adaptive gradient algorithm with reduced tuning efforts. *Neural Networks, 152*, 499–509.

[11] Dongpo, X., Zhang, S., Zhang, H., & Mandic, D. P. (2021). Convergence of the RMSProp deep learning method with penalty for nonconvex optimization. *Neural Networks, 139*, 17–23.

[12] Shi, N., & Li, D. (2021). RMSprop converges with proper hyperparameter. *International Conference on Learning Representation*.

[13] Bogolubsky, L., Dvurechenskii, P., Gasnikov, A., Gusev, G., Nesterov, Y., Raigorodskii, A. M., Tikhonov, A., & Zhukovskii, M. (2016). Learning supervised pagerank with gradient-based and gradient-free optimization methods. *Advances in Neural Information Processing Systems, 29*.

[14] Frazier, P. I. (2018). Bayesian optimization. In Recent advances in optimization and modeling of contemporary problems (pp. 255–278). Informs.

[15] Cavallaro, C., Cutello, V., Pavone, M., & Zito, F. (2024). Machine learning and genetic algorithms: A case study on image reconstruction. *Knowledge-Based Systems*, *284*, 111194.

[16] Alhijawi, B., & Awajan, A. (2024). Genetic algorithms: Theory, genetic operators, solutions, and applications. *Evolutionary Intelligence*, *17*(3), 1245–1256.

[17] Cavallaro, C., Cutello, V., Pavone, M., & Zito, F. (2024). Machine learning and genetic algorithms: A case study on image reconstruction. *Knowledge-Based Systems*, *284*, 111194.

[18] Apolinar-Fernández, A., Blázquez-Carmona, P., Ruiz-Mateos Brea, R., Barrasa-Fano, J., Van Oosterwyck, H., Reina-Romo, E., & Sanz-Herrera, J. A. (2024). Regularization techniques and inverse approaches in 3D traction force microscopy. *International Journal of Mechanical Sciences*, 109592.

[19] Sampaio, R. A., Dias Garcia, J., Poggi, M., & Vidal, T. (2024). Regularization and optimization in model-based clustering. *Pattern Recognition*, *150*, 110310.

[20] Hou, Z., Ling, Y., Liang, Y., Bohao, X., & Lei, Y. (2024). Integrating L1 and weighted L2 regularization for moving force identification from combined response measurements. *Measurement*, *228*, 114337.

Dankan Gowda V., Anand Polamarasetti, Pullela S. V. V. S. R. Kumar, Aparna Atul Junnarkar, and Bhaskar Roy

19 Principles and Applications of Bayesian Optimization in AI

Abstract: This chapter focuses on understanding the Bayesian optimization (BO) method and its use in the context of artificial intelligence (AI) with the aim of intelligently solving expensive problems. Thus, BO is now an important tool for improving different AI applications, including hyperparameter optimization in machine learning models, neural architecture search, and reinforcement learning. By using probabilistic models like Gaussian processes, BO is efficient when it comes to exploitation and exploration, therefore making it possible to arrive at an optimal solution within a few evaluations. This chapter further elaborates on other mathematical concepts in BO, such as surrogate modeling, acquisition functions, and the exploration–exploitation dilemma. It also provides examples of real problems that exemplify how BO can be helpful in increasing accuracy, speed, and reducing computational time when used in applications such as autonomous vehicles, healthcare, and robotics. Some of these challenges include scalability issues and computational burdens, as well as some future research and development areas. This brief survey, therefore, underscores the application of BO in the likely radical revolution of AI and how this innovation is critical for the progressive development of intelligent systems.

Keywords: Bayesian optimization, artificial intelligence, surrogate models, acquisition functions, hyperparameter tuning, neural architecture search, reinforcement learning, optimization techniques, machine learning, AI performance, real-time applications

19.1 Introduction

19.1.1 Overview of Bayesian Optimization (BO)

Bayesian optimization (BO) is a sequential design strategy to optimize the expensive black-box functions. Unlike other optimization methods that may take a number of function evaluations, BO constructively models the objective function through the process of search space exploration. Normally, the use of Gaussian processes (GP) as a surrogate model to represent the unknown function is preferred since it not only provides the prediction but also the uncertainty related to the prediction [1]. This probabilistic approach allows BO to explore and exploit new points effectively, which makes the algorithm useful in any scenario where the evaluation of the objective function is either complex or time-consuming. In artificial intelligence (AI), these evaluations can consist

https://doi.org/10.1515/9783111436180-019

of the training of large sophisticated machine learning models, setting hyperparameters or perhaps setting up reinforcement learning policies [2]. BO's major benefit is its sample efficiency, that is, achieving near optimality with as few evaluations as possible, which helps to save significant computational costs.

19.1.2 Importance of Optimization in AI

Different stages of AI are influenced by optimization, especially when developing or implementing machine learning models. Now, at the center of nearly every AI system, there is the idea of optimizing some kind of objective function, which can often mean accuracy, loss, and so on [3]. For instance, in supervised learning, the optimization part entails the finding of new parameters for a model that will lower the loss function, thus making the model more capable of predicting or classifying new data. Optimization is also central to hyperparameter tuning, which is another important aspect that cannot afford to be without optimization [4]. The efficacy of pursuing machine learning depends on factors such as learning rates, regularization conditions, and the type of network. Several older approaches such as grid search or even random search are not efficient and are rather time-consuming as the number of hyperparameters increases [5]. Unlike grid search, BO is better because it can determine which couple of hyperparameters to assess next, thereby improving the rate of finding the best configurations.

Optimization is another field that is relevant in areas such as neural network architecture search and is based on the search for the optimal architecture among a huge number of possible designs [6]. This task is by its nature nontrivial, especially due to the fact that it operates within a high-dimensional space. BO, being capable of efficiently optimizing large and complex search spaces, has proved beneficial in finding architectures that produce high performance [7]. Likewise, in reinforcement learning, for which the aim is to enhance the policy for the maximization of accrued reward, BO is useful for picking policy parameters or designing reward functions for a better learning process [8]. These different forms of optimization highlight how versatile BO is in tackling challenges that are diverse in nature, making it extremely crucial in bolstering development in AI.

Figure 19.1 presents the flow of BO, also known as the sequential design, which outlines the iterative optimization process of an expensive-to-optimize objective function. Shown in this figure is that the optimization first starts with the choice of a few initial sample points on which the GP surrogate model of the objective function is constructed [9]. The mean of the actual surrogate model is the equivalent of what it currently knows, while the confidence interval is the degree of uncertainty it holds. Another function that is employed in the selection process is an acquisition function, say, expected improvement (EI), with which one can select the next sample point with respect to exploration and exploitation [10]. This point is then assessed against the

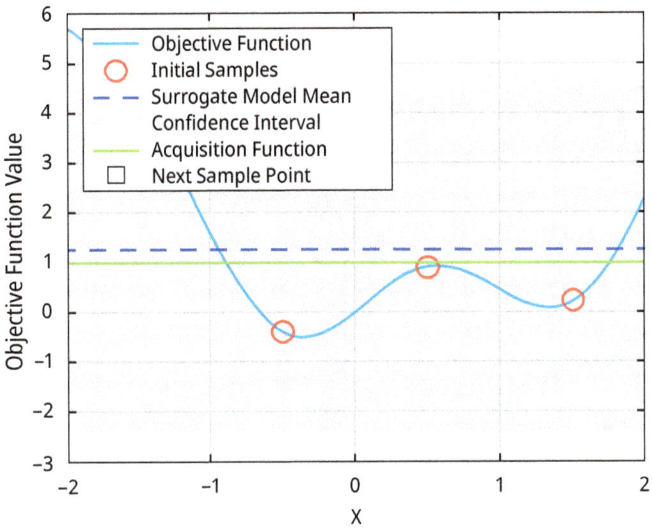

Figure 19.1: Overview of Bayesian optimization workflow.

actual objective function, and the outcome is then fed back into the GP model. This means the process continues in this manner with many iterations until a certain criterion that can be set by the programmer is met. In this way, the proficiency of BO in finding the best solution in the quickest way is evident.

The optimization accomplishment with regard to function value and time taken is presented in Figure 19.2, which consists of BO, grid search, and random search [11]. The plot on the right depicts the objective function measure against the number of function evaluations, which provides information on the convergence rate of each method to the optimal solution. From the results of Figure 19.2, it can be observed that BO achieves higher values near the true optimum, as opposed to algorithms like grid search and random search that require a relatively higher number of evaluations to reach a similar level of true optimum [12]. This figure also illustrates how the use of BO, in comparison with the other methods, is computationally less time-consuming and provides better results in terms of optimization speed and quality of the acquired solutions. This states the fact that if function evaluations are costly or require much time, then this is the most appropriate approach to such problems.

19.1.3 Purpose and Scope of the Chapter

The main goal of this chapter is twofold: first, to provide the reader with the necessary understanding of BO; second, to present three instances of usage of BO in AI. Thus, when working through the theoretical elements like surrogate modeling with

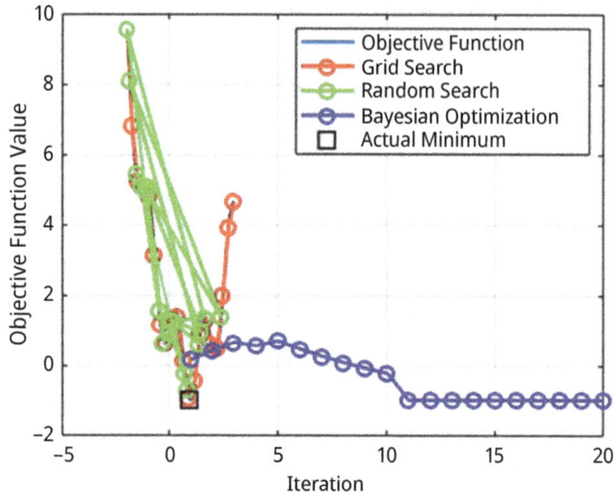

Figure 19.2: Comparison of optimization techniques.

GPs or the definition of the acquisition functions, this chapter is supposed to give a clear background on how BO works [13]. BO also goes through real-life examples where BO can be used to improve machine learning models, adjust their parameters, perform neural architecture search, and come up with better ways of learning in reinforcement learning [14]. These applications will be discussed in detail with the help of examples in order to demonstrate how BO can be used in different AI scenarios [15]. In addition, the chapter will also elaborate on the limitations as well as future prospects of BO while highlighting this research. Leaving the chapter, readers will be familiar with how BO can be applied to enhance AI systems, which will be resourceful for persons intending to develop the field [16]. This extensive review serves as the platform to propose the utilization of BO for the resolution of challenging AI problems in practice that have not seen a concrete application so far.

19.2 Principles of Bayesian Optimization

19.2.1 Mathematical foundation

BO relies on the construction and repeated improvement of a probabilistic model of the objective function; hence, it can be very useful for expensive or noisy functions. It is also important that some concepts of BO are based on two key mathematical tools: GPs and acquisition functions [17]. The information, which is the goal function, can be modeled with the help of GPs because they are flexible and able to estimate uncertainty [18]. A GP is a stochastic model for functions that captures the full distribution over functions characterized by mean and covariance functions, giving the measure of similarity between points in the input space.

Mathematically, a GP can be represented as follows:

$$f(x) \sim \mathrm{GP}(m(x), k(x, x'))$$

where $m(x)$ is the mean function, which is often assumed to be zero for simplicity, and $k(x, x')$ is the covariance function, which determines the shape and smoothness of the function being modeled [19]. With prior knowledge and by updating this model with the observed data, GPs regulate the objective function prediction at any point along with the certainty of it.

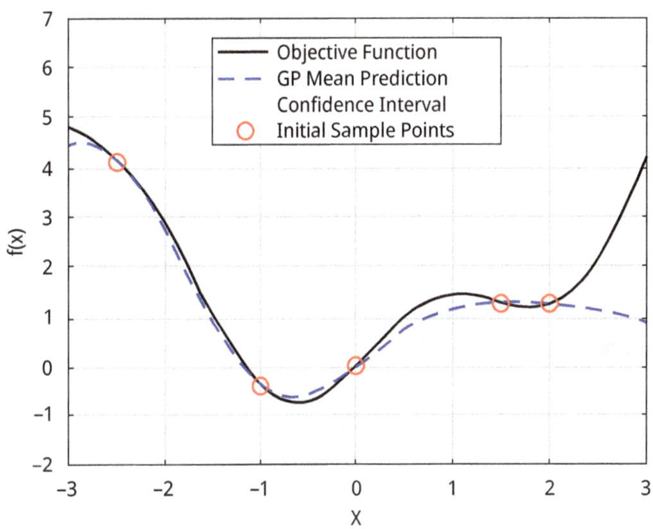

Figure 19.3: An example of a Gaussian process surrogate model.

Figure 19.3 illustrates the mapping of the GP model on BO as a surrogate model. The figure combines the actual objective function, the mean prediction of the GP, and the intervals of uncertainty [20]. It also demonstrates the method by which the GP model employs initial sample points to return uncertainty estimates of the objective function [21]. The confidence intervals shed light on which areas are characterized by a high level of uncertainty and are therefore used to select new sample points [22]. It can be seen from this figure that the surrogate model fits the objective function for the approximations and uses uncertainty as an important factor in determining the optimization process, which plays an important role in exploring the search space in BO.

Figure 19.4 shows the acquisition function based on the EI criterion. The figure adds the acquisition function to the GP surrogate model, mean prediction, and standard deviation. It illustrates how the acquisition function is used to decide where to sample next by incorporating the mean and variance [23]. The EI function recommends points with the largest potential to be better than the best-known value while controlling the exploration versus exploitation trade-off. As a result, this figure is important to illustrate how BO makes efficient steps to select new samples to update the objective function [24].

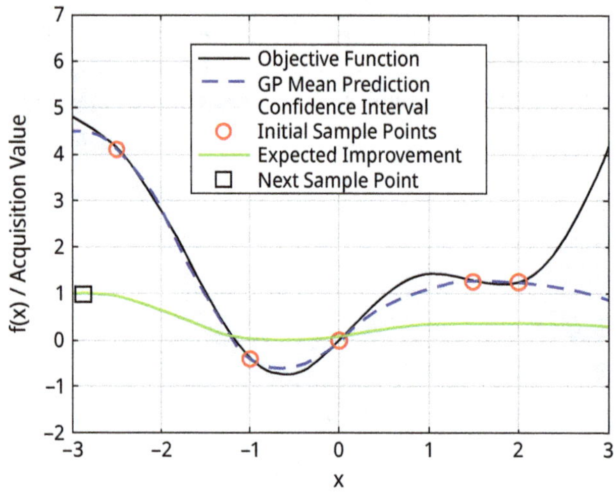

Figure 19.4: Acquisition function visualization.

Acquisition functions play a key role in BO, which involves determining how the exploration–exploitation trade-off is resolved [25]. These functions decide the next point to sample by estimating the improvement potential of the new sample solution over the best value found so far, taking into account the mean estimate and the variance of the surrogate model. Common acquisition functions include EI, probability of improvement (PI), and upper confidence bound (UCB). EI, for example, is defined as follows:

$$\mathrm{EI}(x) = E[\max(0, f(x) - f(x^{+}))]$$

where $f(x^{+})$ is the best-observed function value [26]. This work ensures that the areas with high potential improvement are selected, which means that there is always a way of symbolizing exploration of the unknown regions while still focusing on the promising regions.

19.2.2 Components of Bayesian Optimization

Surrogate Model: In the general context of the BO process, the surrogate model actually captures the function of interest. GP is used more often because it possesses a high degree of flexibility to approximate any arbitrary nonlinear function. GPs offer mean prediction of the given function along with an indication of the variability or uncertainty in that prediction, which is quantified by variance calculated using a probabilistic approach [27]. This double output enables BO to explore regions that the model assigns big standard deviations and exploit regions that the model assigns big means. In some cases, other models such as random forests or Bayesian neural networks can be used as surrogates depending on the nature of the problem or its dimensionality [28]. The choice of surrogate model plays a very crucial role in determining the efficiency and effectiveness of the optimization process that is undertaken in the BO [29].

Acquisition Function: Acquisition functions help in directing the optimization by indicating which points need to be evaluated next [30]. They are expected to work within the trade-off of exploration and exploitation to ensure that the search space is correctly explored as well as refined in promising areas [31]. The EI function, which most often serves as the acquisition function, is defined as the expected rise over the current best objective value, including both expected mean and variance [32]. PI centers on the likelihood that a point will enhance the current best solution by focusing on the most promising areas [33]. The UCB function combines the center of mass mean with the variability of the point's variance by fixing a level of confidence and directs the search toward the space with the highest upper bound of the estimated value of the objective function [34]. All these functions are different in nature, and their use may depend on the nature of the optimization problem and whether more exploration or more exploitation is required.

19.2.3 Algorithmic steps

Algorithms involved in the BO process are described as follows: Firstly, one has to choose an approximate model, which is usually a GP, for the objective function. Several starting points are taken in order to overcome this and optimize this model – a small number of samples serve as a starting point for training [35]. The next step is to choose an acquisition function, whereby the choice will define the next point to be sampled. A specific acquisition function is then used to maximize this quantity to find a point that has the most potential for improvement based on the predicted value and the level of uncertainty [36].

Once the point is selected, the objective function is computed, and the calculated data is used to update the surrogate model and enhance the model's accuracy as well as prediction uncertainties. This means that the process of choosing a point with acquisition function to get its value in the true objective function and then updating the surrogate model is repeated [37]. The iterations proceed until the convergence, that is, when a predefined number of iterations are reached, the amount of time is elapsed, or the improvement in the results is less than a defined amount. Over these iterations, the BO appropriately seeks the parameter space, and this is particularly useful where the function assessments are strenuous. This systematic approach enables BO to provide approximate solutions in relatively fewer numbers of attempts, thus making it an efficient approach toward optimization for AI and machine learning.

19.3 Applications of Bayesian Optimization in AI

19.3.1 Hyperparameter Tuning in Machine Learning

Hyperparameter selection is one of the most crucial steps when it comes to machine learning model implementation because the selection of many parameters directly influences the model outcomes. Some of the regular methods used in hyperparameter tuning include the grid search and the random search methods, and these may prove to be time-consuming and costly, where large model situations are invoked. When compared to this, BO is a more efficient approach since it takes an organized approach to evaluate the variation in the hyperparameters and uses a model to estimate its performance. This also helps to explore the best regions of the hyperparameter space while helping the optimization process to converge over time, hence leading to better models being found. For example, consider a machine learning model with hyperparameters θ that need to be optimized to minimize a loss function $L(\theta)$. BO constructs a probabilistic model of $L(\theta)$, using a GP to estimate the EI over the best-known loss. Thus, through the constant sampling and updating of the surrogate model used in BO, the best hyperparameter values or settings that promote a low loss function can be identified easily, enabling improvement of the general performance of the model.

Figure 19.5 illustrates the experimental study of hyperparameter tuning for the convolutional neural network (CNN) in an autonomous driving problem using BO, random search, and grid search. The plot demonstrates the object detection model accuracy of successive iterations and how different methods increase accuracy while searching the hyperparameter space. Hence, BO proves to be much faster and provides a much higher level of increased accuracy in comparison to both random search and grid search, which underscores the usefulness of the algorithm in the choice of efficient hyperparameters. This figure demonstrates the use of BO on how it can be integrated to solve real-world problems, especially where the performance of models is key.

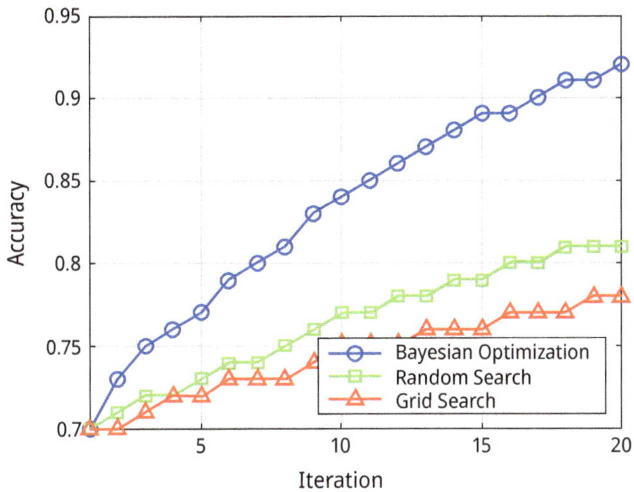

Figure 19.5: Hyperparameter tuning results.

Figure 19.6 illustrates how the search space for the neural architecture is explored using BO. Here, the figure represents different neural network architectures comprising layers and units per layer, as well as the performance of different architectures. Some of the sampled points by BO are depicted to show how the method reaches high-performing architectures. The figure also captures what defines BO, namely the trade-off between exploration and exploitation, as it navigates through the architectural space for the best-performing architectures in terms of accuracy with respect to the validation set and computational complexity.

19.3.2 Neural Architecture Search (NAS)

Neural architecture search (NAS) is the procedure of determining likely structures of neural networks that would provide optimal performance based on defined goals. NAS can be a very large space, especially for more complicated structures; it may involve factors such as the number of layers, type of activation functions, or connectivity patterns. BO is critical to NAS because it serves as a systematic way to navigate the defined search space. Applying a response surface in parallel to reflect the performance of various architectures, BO leads the search toward the best-performing architectures. This approach is also effective in ensuring that, on the one hand, the organization is trying out new, risky, and rather unproven architectures, on the other hand, it is capitalizing on proven, promising architectures. For instance, parameters that constitute a given neural network's architecture could include discrete and continuous constructs, while the technique of using BO involves the ability to maximizing

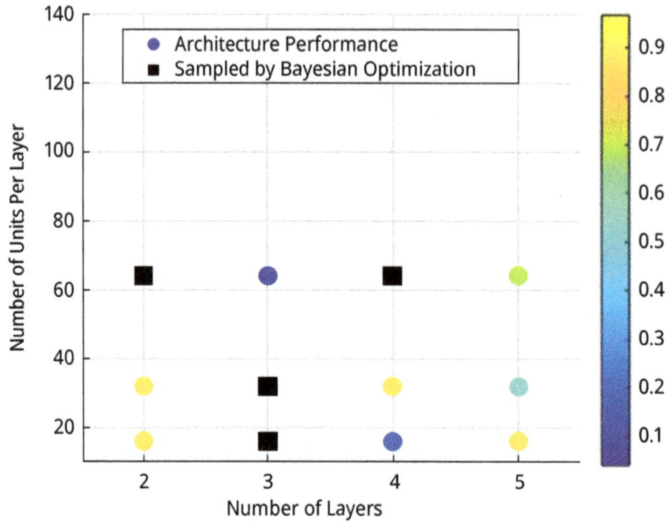

Figure 19.6: Neural architecture search space exploration.

an acquisition function that predicts the degree of improvement in performance. This approach to NAS enables one to strike a balance in arriving at the best-performing architecture faster and more economically.

19.3.3 Reinforcement learning

Reinforcement learning's objective is to find a policy that yields the highest cumulative reward in an environment. It can also be observed that there is a close relationship between parameters such as learning rates, discount factors, exploration strategies, and the performance of reinforcement learning algorithms. Thus, with the help of BO, one can improve these parameters and the process of learning, leading to the acquisition of a better policy. In essence, the reward function is modeled as a black-box function whose properties are uncertain, and since BO relies on this surrogate model, the latter's uncertainty guides the search process of the former. This is very useful, especially in reinforcement learning, as the reward signal can be virtually small and noisy. For instance, the optimization problem can be formulated in the context of the expected cumulative reward, $R(\theta)$, which is defined as the goal with regard to the parameter set θ. Through successive selection and evaluation of these parameters according to an acquisition function, BO ensures the discovery of settings optimal for policy learning, thereby enhancing the efficiency and performance of reinforcement learning techniques.

19.3.4 Automated Machine Learning (AutoML)

Automated machine learning (AutoML) is therefore the procedure of automating various aspects such as model selection, feature engineering, and hyperparameter tuning so that machine learning can be done by anyone. As mentioned, BO is one of the crucial parts of AutoML since it allows for a systematic search of the space of possible models and hyperparameters. In its deterministic nature, it assists the AutoML system in finding models and hyperparameters that are most likely to give good performance on a given dataset through the construction of a probabilistic model of the performance of different configurations. It also eliminates the time and human effort that would be needed to perform experiments manually while at the same time enhancing the quality of the generated models. For instance, in AutoML, BO can be applied for determining suitable algorithms, their hyperparameters, and preprocessing steps, thereby eliminating machine learning as a process. As a result of integrating BO into AutoML frameworks, it becomes easy to search through the large space of all model configurations, resulting in the creation of high-performing and resilient models with little to no human interference. As a result of utilizing BO, AutoML systems can be provided with the capability of reaching the level of performance, which is a state of the art in various tasks and domains, thereby proving the efficiency of BO in AI tasks.

19.4 Case Studies with Real-Time Applications

19.4.1 Case Study 1: Optimizing Hyperparameters for Deep Learning Models in Autonomous Driving

In studies of autonomous driving, object detection is one of the critical tasks, especially using CNNs. These models are used for real-time recognition and grouping of objects on the road, including pedestrians, vehicles, and traffic signs; therefore, it is extremely crucial that this model is very accurate. The goal of this case study is to enhance the detection models for objects in self-driving cars by increasing the efficiency of CNNs with adaptively set hyperparameters. BO is thus a commercialized technique that helps in hyperparameter tuning to find the best solution that optimizes the model while at the same time minimizing the cost of computations.

In this case, a BO approach is used to optimize parameters such as learning rate, batch size, the number of layers, and the filter sizes of CNNs. The GP helps in modeling the dependency of the detection accuracy upon the hyperparameter settings that BO selects during the process of navigating through the hyperparameter space. This is achieved through an acquisition function such as EI, which directs the selection of hyperparameters that are likely to produce higher accuracy. It tries out multiple configurations in order to identify those that are best suited and updates the surrogate model with this information. This approach clearly yields higher precision and recall rates than the models tuned using grid or random search methods. Furthermore, the results concern saving the processing time related to trained models, which is a key parameter when it comes to real-time applications such as self-driving cars, where quick decisions have to be made.

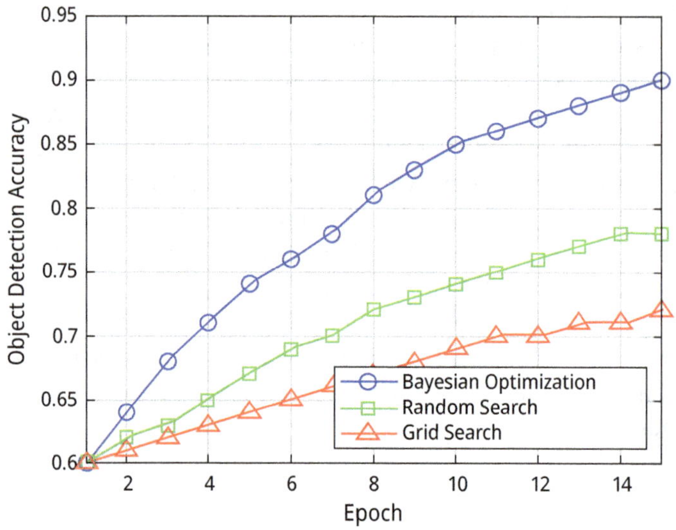

Figure 19.7: Object detection accuracy improvement in autonomous driving.

The plots represented in Figure 19.7 indicate how epoch-by-epoch object detection accuracy is enhanced for a CNN used in autonomous driving. The plot then analyzes the relative usefulness of BO compared to random search and grid search in fine-tuning the model's parameters. The outcomes emphasize how this method of BO improves accuracy with more speed and efficiency compared to conventional optimization of hyperparameters. As shown in this figure, BO is effective in improving the quality of models in sensitive areas like autonomous driving, supporting the notion that the method has many real-world applications.

In a reinforcement learning scenario, Figure 19.8 depicts the cumulative reward observed over episodes while using BO, random search, and grid search. Hence, the figure depicts how tuning of parameters using BO leads to faster learning and better cumulative rewards. As the new efficient method of optimization, BO enables the reinforcement learning policy to make better decisions, thus enhancing robotic systems' decision-making abilities. This figure shows how superior BO is in use cases that need fast learning or dynamic adjustment.

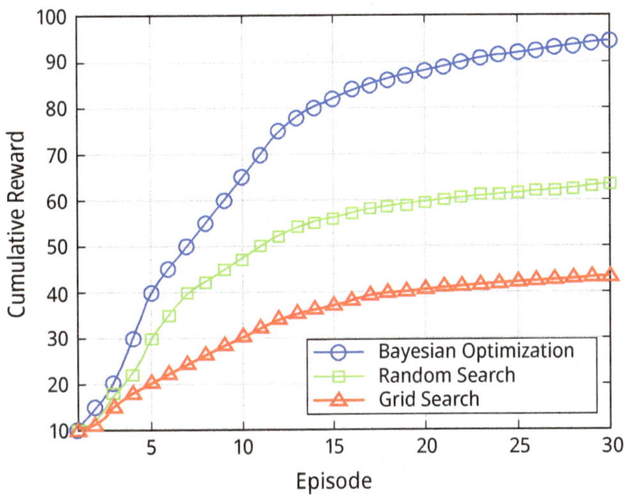

Figure 19.8: Reinforcement learning parameter tuning results.

19.4.2 Case Study 2: Neural Architecture Search for Image Classification in Healthcare

In the diagnosis of diseases and medical conditions, medical image classification involving MRI scans and X-rays, among others, requires accurate and efficient methods. The purpose of this case is to find the specific topologies of neural networks where the classification of medical images will give high accuracy and where BO works well. NAS is a task of searching for the network architecture from the huge space of possible architectures, where one has to decide about the number of layers, types of connections, and forms of activation functions. BO helps in this search by modeling the probability of architectures' performance and concentrating on compact architectures.

In this case study, BO is used for the automation of the search for corresponding neural network architectures for medical image classification. The process starts with the generation of the initial condition that defines the search space: the number of layers, the size of the filters used in the convolutional layers, the type of pooling operations, and so on. There is a probability distribution of the architecture's accuracy provided by using a GP model, given only a few evaluations. It is applied to focus the search toward architectures where the acquisition estimate is high, indicating efficient pruning of the search space. The findings show that for medical image classification problems, BO is able to find architectures with superior accuracy compared to previously reported architectures at a considerably lower computational expense. These optimized architectures not only improve diagnostic efficacy but also reduce the training time of the model, making them very useful in clinical settings.

19.4.3 Case Study 3: Tuning Reinforcement Learning Parameters for Real-Time Decision-Making in Robotics

For instance, in industrial applications like assembling, welding, or the lifting of materials, robotic arms prove very useful since real-time decisions need to be made rapidly. Specifically, the aim of this case study is to improve decision-making scenarios for robotic arms in manufacturing applications using principles of reinforcement learning with an appropriate choice of parameters. Reinforcement learning deals with an agent's ability to make decisions over time while trying to maximize cumulative reward, and the efficiency of these algorithms depends on factors including learning rate, discount factor, and the exploration-exploitation policy.

The use of such parameters needs to be tuned appropriately, and this is where BO offers a reliable means for fine-tuning reinforcement learning algorithms, so as to optimize the speed at which the robot can learn and/or make accurate decisions. Optimization precedes the parameter space identification that includes significant parameters influencing the policy learning of the robot. A meta-model, usually a GP, is built to evaluate the expected reward from different settings of the parameters. The acquisition function controls the exploration process, targeting configurations of the parameters with the highest potential for increasing the robot's performance. Using the surrogate model to always update new data of evaluations, BO optimally narrows down parameter values, thereby improving learning efficiency.

The findings presented in this case study show that optimistic features enhance the effectiveness of the robot's learning considerably, thus decreasing the time necessary to train it for the next tasks. Also, management achieves prognostic and prescriptive accuracy, owing to optimal parameters enhancing the efficiency of business processes and decreasing the rates of error in manufacturing processes. These results highlight the use of the BO technique in current decision-making systems due to the realistic adjustment of parameters used in reinforcement learning.

19.5 Results and Discussion

19.5.1 Summary of Findings

The case studies apply BO to machine learning and reinforcement learning problems and show dramatic enhancements in metrics such as accuracy and efficiency, as well as in decreasing computational costs. Focusing on hyperparameter tuning for deep learning models in autonomous driving, the BO resulted in significant increases in object detection accuracy, increasing the precision and recall ratio. The optimized models, as expected, showed higher accuracy in a shorter processing time, making them suitable for applications in self-driving cars. Applying BO in the search for neural architectures for image classification in healthcare, the model achieved high recognition accuracy in tasks related to medical image analysis, including MRI scans and X-rays. This approach drastically cut down the time consumption compared to other exhaustive search methods, hence enabling the implementation of complicated models in clinical practices. For example, when used in adjusting the parameters of reinforcement learning for robotic arms used in manufacturing, the stability and precision of the robots' learning process and decision-making were enhanced due to improvements in speed and accuracy, hence increasing the manufacturing output and decreasing error margins. These results illustrate the ability of BO to specify vast potential enhancements in numerous related AI applications, making it a useful tool for researchers and investigators.

Figure 19.9 presents an aggregate of optimization from various case studies pertaining to different fields such as object detection, NAS, and reinforcement learning. Here, the figure is drawn in bar charts to visually represent the relative values of different KPIs, including accuracy, efficiency, and computational cost of BO, random search, and grid search. This graph undoubtedly proves and shows how effectively BO helps to achieve more accuracy and efficiency at a lower computational cost, which make us sure that how effectively BO can be used in any field of AI. It also shows the general versatility of BO for the optimization of expensive and nonlinear functions.

Figure 19.10 represents the comparison of the heaviness of computations and the quality of optimization between various optimization methods. With the help of a scatter plot, it is evident how BO is able to provide an optimal solution with the least amount of computational work. Nonetheless, random search and grid search are found to have either higher costs or produce lower quality solutions. For instance, the above figure clearly conveys the promise that BO brings in leveraging the results while incurring little computational overhead to optimize the execution of most AI models and systems.

19.5.2 Comparison with Traditional Methods

In comparison to the conventional approaches such as grid search and random search, BO has certain benefits connected to the efficiency of the search and the quality of the obtained results. Earlier approaches merely entail the systematic or random search of the hyperspace, which is inherently slow and resource-intensive, particularly when dealing with spaces of relatively high dimensionality. Some algorithms such as grid search demand that the model tests all possible values of all hyperparameters; hence, with an increase in the number of hyperparameters, the number of times the model is tested increases exponentially. Random search, while more efficient, is also a time-consuming process since the user needs to consider numerous evaluations to define the finest setting. On the other hand, BO deliberately searches

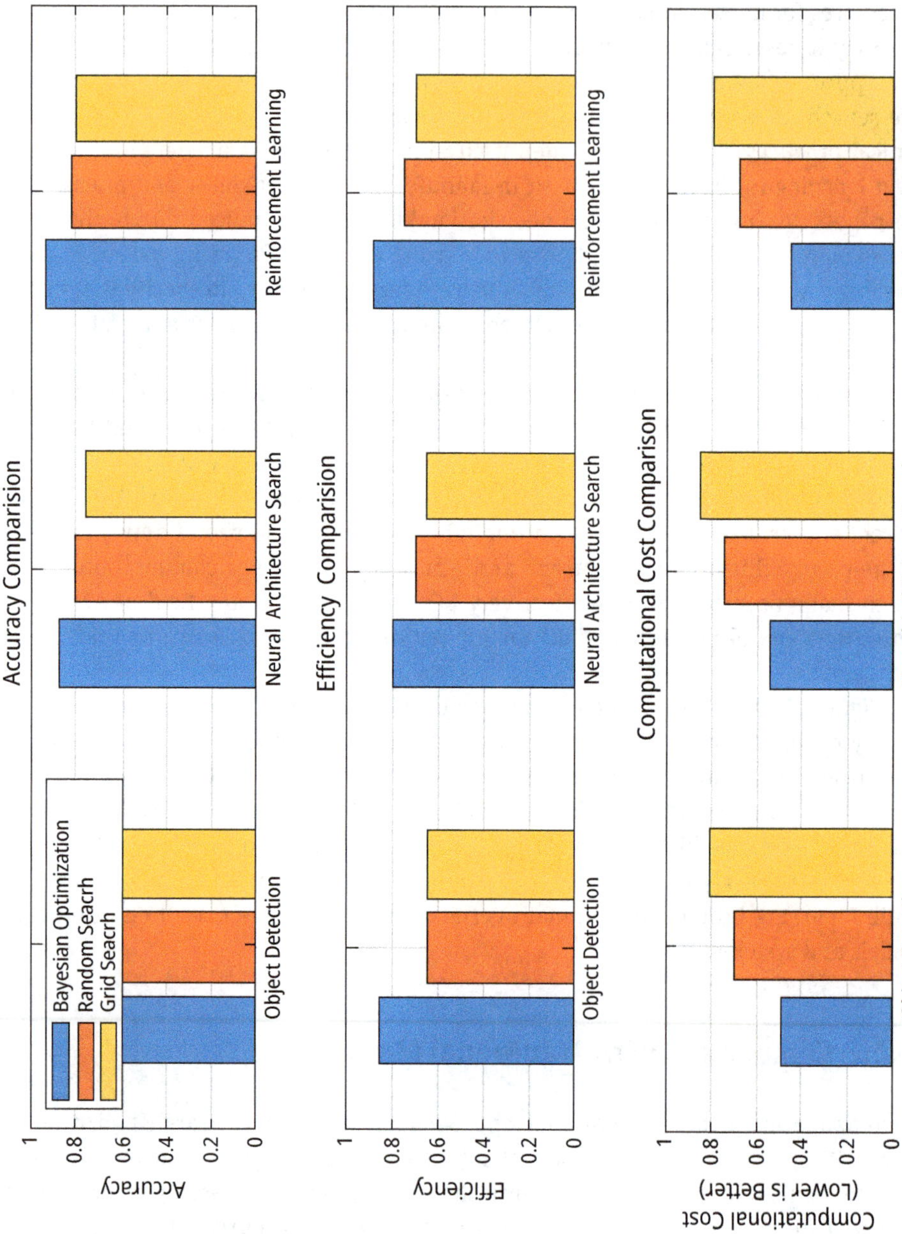

Figure 19.9: Summary of optimization results across case studies.

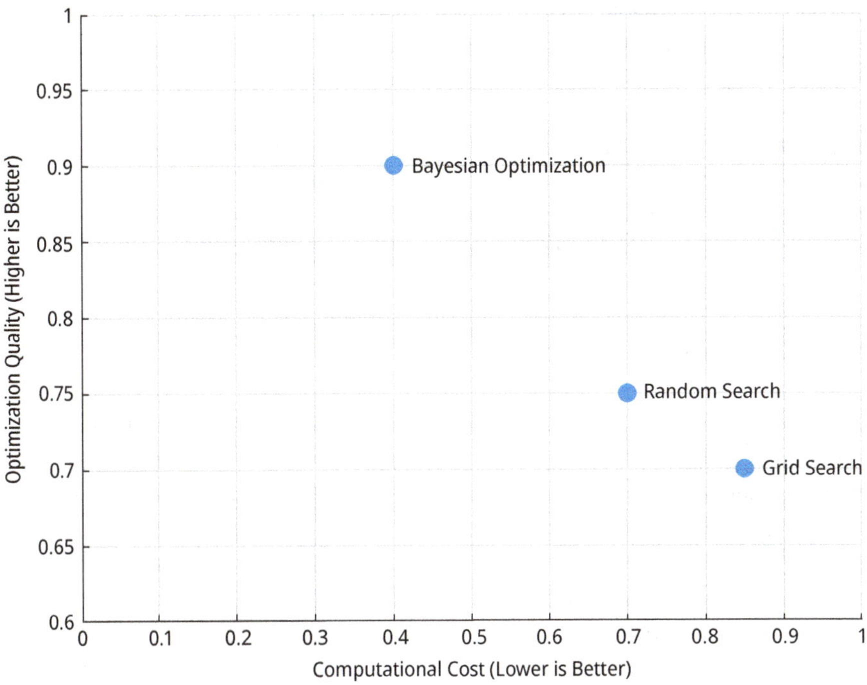

Figure 19.10: Computational cost versus optimization quality.

hypothesized regions of the parameter space by building a stochastic model of the objective function and concentrating on the areas with the highest profitability. With this targeted approach, new promising configurations with a significantly lesser number of evaluations can be discovered, making it less computationally intensive and faster. In addition, because BO includes probability estimates of uncertainty and optimizes the exploration–exploitation ratio, the solutions yield more enhanced performance measures, evident in the experiments conducted in the case studies.

19.5.3 Challenges and Limitations

However, like any optimization technique, it is not without problems when applied to BO. As is the case with many "big data" tools, the two main difficulties concern scalability: the dimensionality of the data and the models used. Although the use of such structures, such as GPs or others, may add flexibility in updating the surrogate model for the optimization task, the computational cost of these processes increases noticeably with the space dimensionality. This limitation comes at the cost of time required and may affect the efficiency of BO, especially when dealing with many hyperparameters. Furthermore, the overall success of the BO technique is highly sensitive to the quality of the surrogate model used in the process. These issues raise the question of whether and when surrogates are good enough, and in case the surrogate model does not accurately approximate the underlying function, highly suboptimal solutions could be accepted. Another difficulty is the computational cost used for the evaluation of the acquisition function, which aims to maximize in each step to determine which point has to be evaluated next. This can be time-consuming, especially when decisions have to be made within a short time frame. Solving these issues is crucial to the enlargement of the scope of BO application in various tasks.

19.5.4 Future Directions

Potential areas for future research in BO are better techniques for surrogate modeling, better acquisition functions, and use of this optimization in new environments. Enhancing surrogate models to improve their performance in high-dimensional spaces is another area of focus with methods such as sparse GP-based methods and deep kernel learning, as some of the approaches to improve the same. These techniques aim to make the advantages of GPs credible and reliable without compromising the computing time. As an ongoing promising approach, the design of acquisition functions with a better balance between exploration and exploitation is another important development direction. There are, for example, multifidelity optimization methods, which allow one to work with information of different levels of detail at once and, thus, make the process more efficient. Furthermore, the application of BO with other optimizing algorithms, for example, evolutionary ones, would be another enhancement factor. It is also helpful to search for new applications of BO in domains such as NLP or energy and financial modeling to better define its possible applications. Thus, by addressing the current difficulties and exploring new prospects, BO should be able to develop further as an effective technique for the optimization of large partially unknown and/or computationally costly functions in AI and other fields.

19.6 Conclusion

BO has become one of the most important optimization techniques in AI because of its several benefits over classical techniques. This chapter has discussed the basic ideas of the proposed approach, its use with examples, and the way it positively impacts AI systems' performance. Fundamentally, BO is good for settings where it takes a lot of resources to evaluate the performance of a given set of parameters because it uses statistical models such as GPs to make predictions. This characteristic to reach a balance in exploration and exploitation by implementing acquisition functions such as EI and PI enables BO to perform well in instances with large search spaces and thus obtain the solution in significantly fewer iterations. The real-world applications of BO used in this chapter from hyperparameter tuning in deep learning models for autonomous driving to NAS in healthcare, and reinforcement learning in robotics are evident from the following case studies. These applications have demonstrated gains in value measures, for example, in accuracy, learning rate, and compute, implying that the application of BO has the potential and capability to enhance the development of numerous advanced AI areas.

It is therefore imperative to note that the use of BO in AI goes beyond enhancing performance to encourage innovation and the development of new technologies. The time needed for optimizing large numbers of model parameters, which have highly nonlinear relationships between these parameters, is significantly reduced by the use of BO, thus opening up the opportunity for the development and application of more sophisticated AI systems in a range of industries. For instance, the enhancement of neural network architectures for medical image classification in the healthcare sector can greatly contribute to the precision of diagnosis, thereby enhancing the fulfillment of patients' needs appropriately. Thus, in applications such as autonomous driving, the deep learning hyperparameters that have been optimized can provide increased reliability and safety of self-driving vehicles, contributing to the penetration of autonomous technologies. In robotics, optimization of reinforcement learning enhances decision-making insights to bring the integration of intelligent robots in manufacturing, logistics, and service industries to higher levels of productivity.

Apart from particular examples, BO makes a contribution to the general field of AI by helping the development of generally superior and more versatile models. This is especially valuable because it has the potential to successfully solve high-dimensional optimization problems and release new possibilities for research, so scientists and engineers can work with models that were earlier considered to be too costly in terms of computing. Therefore, the appearance of this kind of algorithm is a breakthrough that gives a possibility to investigate different models that were earlier considered to be too complex for high-dimensional optimization. As this capability also propels an exponential advancement in AI innovations and discoveries, it also helps to identify new solutions to old challenges. Moreover, these ideas of BO can be combined with other optimization algorithms to make use of each algorithm's properties while at the same time using the strong aspects of the other algorithms, thereby expanding the envelope for AI.

The use of BO also affects the remaining aspect of AI capability, namely equality of access. This makes BO accessible to individuals who are not experts in machine learning and AI; it expands its use in many areas of application such that everyone can benefit from machine learning and AI. Such democratization can be beneficial and encourages the general development of further inclusion of AI into as many organizations as possible. The place of BO in the growth of the field of AI cannot be understated, as it is set to define the next steps of the AI evolution as well as incremental advances in the field.

In conclusion, one must acknowledge that BO is a powerful and valuable tool in an AI engineer's arsenal, as it provides a structured way of optimizing complex and expensive-to-evaluate functions. Applied in diverse contexts of AI, it proves its ability to boost model accuracy, optimize operations, and lower computational expenses. By overcoming the issue of scalability and further improving the model, as well as investigating new fields of application, BO will remain one of the fundamental tools, optimizing and contributing to the further development of AI and the creation of new, smart solutions. These reflections from this chapter emphasize the relevance of BO in the continuous endeavor of developing AI to solve the world's challenges.

References

[1] Ahmad, T., & Khan, M. I. (2023). A machine learning-based approach for predicting renewable energy production. In 2023 International Conference on Smart Grid and Clean Energy Technologies (ICSGCET) (pp. 150–156).

[2] Ali, S., & Ahmed, N. (2023). Blockchain technology for secure IoT applications in smart cities. In 2023 6th International Conference on Computer and Communication Engineering (ICCCE) (pp. 342–348).

[3] Brown, L. J., & Smith, R. K. (2023). Enhancing cloud computing security using deep learning techniques. In 2023 IEEE Symposium on Computers and Communications (ISCC) (pp. 784–790).

[4] Chen, Y., & Zhang, Q. (2023). AI-driven predictive maintenance for manufacturing industries. In 2023 IEEE International Conference on Industrial Technology (ICIT) (pp. 211–218).

[5] Manivasagam, G., & Prasad, K. D. V. (2024). Novel approaches to biometric security using enhanced session keys and elliptic curve cryptography. *Journal of Discrete Mathematical Sciences and Cryptography*, *27*(2–A), 477–488.

[6] Das, S., & Roy, A. (2023). An adaptive control system for autonomous vehicles using reinforcement learning. In 2023 9th International Conference on Advances in Computing, Communications, and Informatics (ICACCI) (pp. 305–311).

[7] Kumar, R. Senthil, & Prasad, K. D. V. (2024). Securing digital authentication with cryptographic innovations in intrinsic verification systems. *Journal of Discrete Mathematical Sciences and Cryptography*, *27*(2-A), 317–328.

[8] Punugoti, R., Vyas, N., Siddiqui, A. T., & Basit, A. (2023). The Convergence of Cutting-Edge Technologies: Leveraging AI and Edge Computing to Transform the Internet of Medical Things (IoMT). In 2023 4th International Conference on Electronics and Sustainable Communication Systems (ICESC). 2023 4th International Conference on Electronics and Sustainable Communication Systems (ICESC). IEEE. https://doi.org/10.1109/icesc57686.2023.10193047

[9] Sarma, Parismita, & Rahman, Mirzanur (2024). Mathematical analysis of wavelet-based multi-image compression in medical diagnostics. *Journal of Discrete Mathematical Sciences and Cryptography*, *27*(2–B), 675–687.

[10] Prasad, K. D. V. (2024). Cryptographic image-based data security strategies in wireless sensor networks. *Journal of Discrete Mathematical Sciences and Cryptography*, *27*(2–A), 293–304.

[11] Ding, H., & Li, X. (2023). A comprehensive review of deep learning applications in computer vision. In 2023 IEEE International Conference on Artificial Intelligence and Machine Learning (AIML) (pp. 100–106).

[12] Gao, L., & Chen, Z. (2023). IoT-based healthcare monitoring system using machine learning. In 2023 IEEE International Conference on Bioinformatics and Biomedicine (BIBM) (pp. 213–219).

[13] Gupta, R., & Patel, M. (2023). Securing IoT devices with blockchain: A new perspective. In 2023 International Conference on Internet of Things and Smart Systems (IoTSS) (pp. 450–456).

[14] Santosh Kumar, & Gupta, Anand Kumar (2023). A novel approach of unsupervised feature selection using iterative shrinking and expansion algorithm. *Journal of Interdisciplinary Mathematics*, *26*(3), 519–530.

[15] Hassan, M., & Ali, H. (2023). A smart agriculture framework using IoT and machine learning. In 2023 5th International Conference on Computing, Communication, and Security (ICCCS) (pp. 380–386).

[16] Hu, W., & Zhang, P. (2023). Big data analytics for improving supply chain management efficiency. In 2023 IEEE International Conference on Big Data (Big Data) (pp. 520–526).

[17] Jackson, R., & Davis, C. (2023). A deep learning-based approach for automated driving assistance systems. In 2023 8th International Conference on Intelligent Systems and Control (ISCO) (pp. 290–296).

[18] Jiang, H., & Wang, X. (2023). Enhancing network security using AI-based anomaly detection techniques. In 2023 IEEE International Conference on Communications (ICC) (pp. 600–606).

[19] Naziya, & Chinamuttevi, Veera Sivakumar (2023). A novel RF-SMOTE model to enhance the definite apprehensions for IoT security attacks. *Journal of Discrete Mathematical Sciences and Cryptography*, *26*(3), 861–873.

[20] John, R., & Kumar, S. (2023). Machine learning approaches for fraud detection in financial transactions. In 2023 International Conference on Information Systems and Computer Networks (ISCON) (pp. 320–326).

[21] Kim, J., & Lee, S. (2023). A novel approach to wireless sensor networks for smart home applications. In 2023 IEEE International Conference on Sensors (ICSENS) (pp. 400–406).

[22] Shekhar, R., & Chaturvedi, Abhay (2023). Securing networked image transmission using public-key cryptography and identity authentication. *Journal of Discrete Mathematical Sciences and Cryptography*, *26*(3), 779–791.

[23] Lee, H., & Park, J. (2023). Cloud-based AI solutions for intelligent traffic management. In 2023 IEEE Conference on Intelligent Transportation Systems (ITSC) (pp. 230–236).

[24] Li, Y., & Liu, X. (2023). A smart grid energy management system using IoT and machine learning. In 2023 International Conference on Smart Grid and Renewable Energy (SGRE) (pp. 540–546).

[25] Lin, D., & Wang, J. (2023). AI-powered predictive analytics for healthcare decision support systems. In 2023 7th International Conference on Health Informatics and Medical Systems (HIMS) (pp. 150–156).

[26] Liu, M., & Zhao, Y. (2023). An intelligent surveillance system using deep learning for anomaly detection. In 2023 IEEE International Conference on Image Processing (ICIP) (pp. 780–786).

[27] Miller, K., & White, A. (2023). AI-based solutions for optimizing manufacturing processes. In 2023 International Conference on Industrial Electronics and Applications (ICIEA) (pp. 310–316).

[28] Mohammed, A., & Sadiq, W. (2023). A scalable IoT architecture for smart city applications. In 2023 6th International Conference on Sustainable Energy and Environment (ICSEE) (pp. 410–416).

[29] Nguyen, T., & Tran, L. (2023). Deep reinforcement learning for robotics control applications. In 2023 IEEE International Conference on Robotics and Automation (ICRA) (pp. 690–696).

[30] Patel, D., & Mehta, V. (2023). A novel approach to cloud security using AI-based encryption techniques. In 2023 IEEE International Conference on Cloud Computing Technology and Science (CloudCom) (pp. 320–326).

[31] Rahman, S., & Islam, M. (2023). A comprehensive framework for smart farming using IoT. In 2023 International Conference on Agriculture and Bioengineering (ICAB) (pp. 290–296).

[32] Roberts, L., & Jones, M. (2023). Enhancing cybersecurity in smart grids using machine learning. In 2023 IEEE International Conference on Smart Grid Technologies (SGT) (pp. 420–426).

[33] Shah, P., & Desai, R. (2023). A deep learning-based predictive model for disease diagnosis. In 2023 IEEE International Conference on Bioinformatics and Bioengineering (BIBE) (pp. 330–336).

[34] Smith, J., & Johnson, T. (2023). AI-driven optimization for renewable energy systems. In 2023 International Conference on Sustainable Energy and Development (ICSED) (pp. 380–386).

[35] Tan, C., & Huang, L. (2023). IoT-enabled smart grid solutions for efficient energy management. In 2023 IEEE International Conference on Power and Energy Systems (ICPES) (pp. 510–516).

[36] Wang, L., & Wu, J. (2023). A new deep learning model for natural language processing applications. In 2023 International Conference on Computational Linguistics (COLING) (pp. 450–456).

[37] Zhang, T., & Chen, M. (2023). AI and machine learning for intelligent building management systems. In 2023 International Conference on Smart Building and Construction (ICSBC) (pp. 300–306).

Index

https://doi.org/10.1515/9783111436180-020

Mathematical Methods in the Digital Age

Volume 1
Satya Prakash Yadav, Abhishek Kumar, Ashutosh Kumar Dubey, Victor Hugo C. de Albuquerque, Adrian Rodriguez-Aguiñaga, Angeles Quezada
Medical Image Processing - Multimodal Fusion Techniques, 2024
ISBN 978-3-11-139549-4, e-ISBN (PDF) 978-3-11-143597-8, e-ISBN (EPUB) 978-3-11-143599-2

www.degruyter.com

www.ingramcontent.com/pod-product-compliance
Lightning Source LLC
Jackson TN
JSHW052046260325
81480JS00003B/8
* 9 7 8 3 1 1 1 4 3 6 0 5 0 *